SELECTED CHAPTERS FROM HUGHES-HALLETT: CALCULUS: SINGLE AND MULTIVARIABLE, 4TH EDITION

A Wiley Canada Custom Publication
for

Queen's University

Math 122 Multivariate Calculus Section

JOHN WILEY & SONS CANADA, LTD.

Copyright © 2009 by John Wiley & Sons Canada, Ltd

Hughes-Hallett, Calculus: Single and Multivariable (ISBN: 9780471472452)
Copyright © 2005 by John Wiley & Sons, Inc. All rights reserved. No part of this work covered by the copyright herein may be reproduced, transmitted, or used in any form or by any means—graphic, electronic or mechanical—without the prior written permission of the publisher. Any request for photocopying, recording, taping, or inclusion in information storage and retrieval systems of any part of this book shall be directed to The Canadian Copyright Licensing Agency (Access Copyright). For an Access Copyright licence, visit www.accesscopyright.ca or call toll-free, 1-800-893-5777. Care has been taken to trace ownership of copyright material contained in this text. The Publishers will gladly receive any information that will enable them to rectify any erroneous reference or credit line in subsequent editions.

Cover Photo: Pete Turner/The Image Bank/Getty Images

Marketing Manager: Anne-Marie Seymour
Custom Coordinator: Rachel Coffey

Printed and bound in the United States of America
10 9 8 7 6 5 4 3 2 1

John Wiley & Sons Canada, Ltd
6045 Freemont Blvd.
Mississauga, Ontario
L5R 4J3
Visit our website at: www.wiley.ca

Chapter Twelve
FUNCTIONS OF SEVERAL VARIABLES

Many quantities depend on more than one variable: the amount of food grown depends on the amount of rain and the amount of fertilizer used; the rate of a chemical reaction depends on the temperature and the pressure of the environment in which it proceeds; the strength of the gravitational attraction between two bodies depends on their masses and their distance apart; and the rate of fallout from a volcanic explosion depends on the distance from the volcano and the time since the explosion. Each example involves a function of two or more variables. In this chapter, we will see many different ways of looking at functions of several variables.

Chapter Twelve FUNCTIONS OF SEVERAL VARIABLES

12.1 FUNCTIONS OF TWO VARIABLES

Function Notation

Suppose you want to calculate your monthly payment on a five-year car loan; this depends on both the amount of money you borrow and the interest rate. These quantities can vary separately: the loan amount can change while the interest rate remains the same, or the interest rate can change while the loan amount remains the same. To calculate your monthly payment you need to know both. If the monthly payment is m, the loan amount is L, and the interest rate is $r\%$, then we express the fact that m is a function of L and r by writing:

$$m = f(L, r).$$

This is just like the function notation of one-variable calculus. The variable m is called the *dependent variable*, and the variables L and r are called the *independent variables*. The letter f stands for the *function* or rule that gives the value of m corresponding to given values of L and r.

A function of two variables can be represented graphically, numerically by a table of values, or algebraically by a formula. In this section, we give examples of each.

Graphical Example: A Weather Map

Figure 12.1 shows a weather map from a newspaper. What information does it convey? It displays the predicted high temperature, T, in degrees Fahrenheit (°F), throughout the US on that day. The curves on the map, called *isotherms*, separate the country into zones, according to whether T is in the 60s, 70s, 80s, 90s, or 100s. (*Iso* means same and *therm* means heat.) Notice that the isotherm separating the 80s and 90s zones connects all the points where the temperature is exactly 90°F.

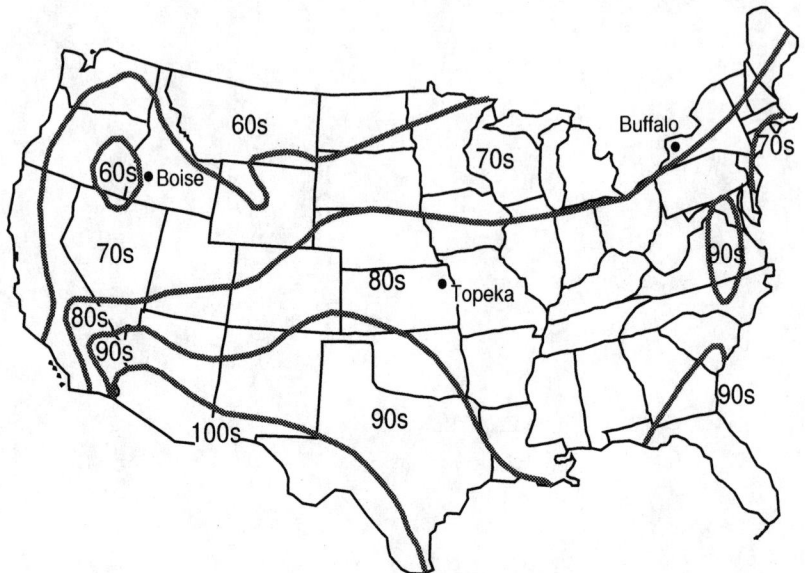

Figure 12.1: Weather map showing predicted high temperatures, T, on a summer day

Example 1 Estimate the predicted value of T in Boise, Idaho; Topeka, Kansas; and Buffalo, New York.

Solution   Boise and Buffalo are in the 70s region, and Topeka is in the 80s region. Thus, the predicted temperature in Boise and Buffalo is between 70 and 80 while the predicted temperature in Topeka is between 80 and 90. In fact, we can say more. Although both Boise and Buffalo are in the 70s, Boise is quite close to the $T = 70$ isotherm, whereas Buffalo is quite close to the $T = 80$ isotherm. So we estimate the temperature to be in the low 70s in Boise and in the high 70s in Buffalo. Topeka is about halfway between the $T = 80$ isotherm and the $T = 90$ isotherm. Thus, we guess the temperature in Topeka to be in the mid 80s. In fact, the actual high temperatures for that day were 71°F for Boise, 79°F for Buffalo, and 86°F for Topeka.

The predicted high temperature, T, illustrated by the weather map is a function of (that is, depends on) two variables, often longitude and latitude, or miles east-west and miles north-south of a fixed point, say, Topeka. The weather map in Figure 12.1 is called a *contour map* or *contour diagram* of that function. Section 12.2 shows another way of visualizing functions of two variables using surfaces; Section 12.3 looks at contour maps in detail.

Numerical Example: Beef Consumption

Suppose you are a beef producer and you want to know how much beef people will buy. This depends on how much money people have and on the price of beef. The consumption of beef, C (in pounds per week per household) is a function of household income, I (in thousands of dollars per year), and the price of beef, p (in dollars per pound). In function notation, we write:

$$C = f(I, p).$$

Table 12.1 contains values of this function. Values of p are shown across the top, values of I are down the left side, and corresponding values of $f(I, p)$ are given in the table.[1] For example, to find the value of $f(40, 3.50)$, we look in the row corresponding to $I = 40$ under $p = 3.50$, where we find the number 4.05. Thus,

$$f(40, 3.50) = 4.05.$$

This means that, on average, if a household's income is $40,000 a year and the price of beef is $3.50/lb, the family will buy 4.05 lbs of beef per week.

Table 12.1 *Quantity of beef bought (pounds/household/week)*

		\multicolumn{4}{c}{Price of beef, ($/lb)}
		3.00	3.50	4.00	4.50
Household	20	2.65	2.59	2.51	2.43
income	40	4.14	4.05	3.94	3.88
per year,	60	5.11	5.00	4.97	4.84
I	80	5.35	5.29	5.19	5.07
(1000)	100	5.79	5.77	5.60	5.53

Notice how this differs from the table of values of a one-variable function, where one row or one column is enough to list the values of the function. Here many rows and columns are needed because the function has a value for every *pair* of values of the independent variables.

Algebraic Examples: Formulas

In both the weather map and beef consumption examples, there is no formula for the underlying function. That is usually the case for functions representing real-life data. On the other hand, for many idealized models in physics, engineering, or economics, there are exact formulas.

---

[1] Adapted from Richard G. Lipsey, *An Introduction to Positive Economics*, 3rd ed., (London: Weidenfeld and Nicolson, 1971).

Example 2 Give a formula for the function $M = f(B, t)$ where M is the amount of money in a bank account t years after an initial investment of B dollars, if interest is accrued at a rate of 5% per year compounded annually.

Solution Annual compounding means that M increases by a factor of 1.05 every year, so
$$M = f(B, t) = B(1.05)^t.$$

Example 3 A cylinder with closed ends has radius r and height h. If its volume is V and its surface area is A, find formulas for the functions $V = f(r, h)$ and $A = g(r, h)$.

Solution Since the area of the circular base is πr^2, we have
$$V = f(r, h) = \text{Area of base} \cdot \text{Height} = \pi r^2 h.$$

The surface area of the side is the circumference of the bottom, $2\pi r$, times the height h, giving $2\pi rh$. Thus,
$$A = g(r, h) = 2 \cdot \text{Area of base} + \text{Area of side} = 2\pi r^2 + 2\pi rh.$$

A Tour of 3-Space

In Section 12.2 we see how to visualize a function of two variables as a surface in space. Now we see how to locate points in three-dimensional space (3-space).

Imagine three coordinate axes meeting at the *origin*: a vertical axis, and two horizontal axes at right angles to each other. (See Figure 12.2.) Think of the xy-plane as being horizontal, while the z-axis extends vertically above and below the plane. The labels x, y, and z show which part of each axis is positive; the other side is negative. We generally use *right-handed axes* in which looking down the positive z-axis gives the usual view of the xy-plane. We specify a point in 3-space by giving its coordinates (x, y, z) with respect to these axes. Think of the coordinates as instructions telling you how to get to the point; start at the origin, go x units along the x-axis, then y units in the direction parallel to the y-axis and finally z units in the direction parallel to the z-axis. The coordinates can be positive, zero or negative; a zero coordinate means "don't move in this direction," and a negative coordinate means "go in the negative direction parallel to this axis." For example, the origin has coordinates $(0, 0, 0)$, since we get there from the origin by doing nothing at all.

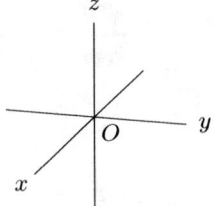
Figure 12.2: Coordinate axes in three-dimensional space

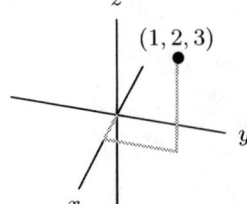
Figure 12.3: The point $(1, 2, 3)$ in 3-space

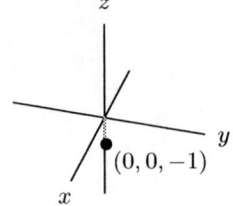
Figure 12.4: The point $(0, 0, -1)$ in 3-space

Example 4 Describe the position of the points with coordinates $(1, 2, 3)$ and $(0, 0, -1)$.

Solution We get to the point $(1, 2, 3)$ by starting at the origin, going 1 unit along the x-axis, 2 units in the direction parallel to the y-axis, and 3 units up in the direction parallel to the z-axis. (See Figure 12.3.)

To get to $(0, 0, -1)$, we don't move at all in the x and y directions, but move 1 unit in the negative z direction. So the point is on the negative z-axis. (See Figure 12.4.) You can check that the position of the point is independent of the order of the x, y, and z displacements.

Example 5 You start at the origin, go along the y-axis a distance of 2 units in the positive direction, and then move vertically upward a distance of 1 unit. What are the coordinates of your final position?

Solution   You started at the point $(0, 0, 0)$. When you went along the y-axis, your y-coordinate increased to 2. Moving vertically increased your z-coordinate to 1; your x-coordinate did not change because you did not move in the x direction. So your final coordinates are $(0, 2, 1)$. (See Figure 12.5.)

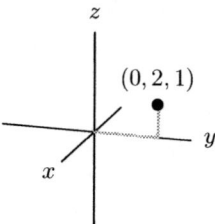

Figure 12.5: The point $(0, 2, 1)$ is reached by moving 2 along the y-axis and 1 upward

It is often helpful to picture a three dimensional coordinate system in terms of a room. The origin is a corner at floor level where two walls meet the floor. The z-axis is the vertical intersection of the two walls; the x- and y-axes are the intersections of each wall with the floor. Points with negative coordinates lie behind a wall in the next room or below the floor.

Graphing Equations in 3 Space

We can graph an equation involving the variables x, y, and z in 3-space; such a graph is a picture of all points (x, y, z) that satisfy the equation.

Example 6 What do the graphs of the equations $z = 0$, $z = 3$, and $z = -1$ look like?

Solution   To graph $z = 0$, we visualize the set of points whose z-coordinate is zero. If the z-coordinate is 0, then we must be at the same vertical level as the origin, that is, we are in the horizontal plane containing the origin. So the graph of $z = 0$ is the middle plane in Figure 12.6. The graph of $z = 3$ is a plane parallel to the graph of $z = 0$, but three units above it. The graph of $z = -1$ is a plane parallel to the graph of $z = 0$, but one unit below it.

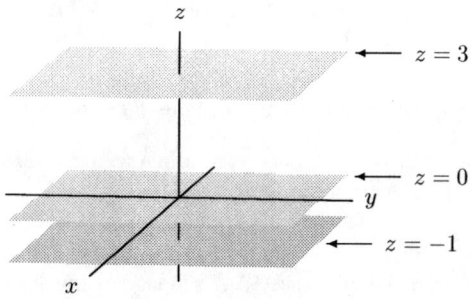

Figure 12.6: The planes $z = -1$, $z = 0$, and $z = 3$

The plane $z = 0$ contains the x- and y-coordinate axes, and is called the xy-plane. There are two other coordinate planes. The yz-plane contains both the y- and the z-axes, and the xz-plane contains the x- and z-axes. (See Figure 12.7.)

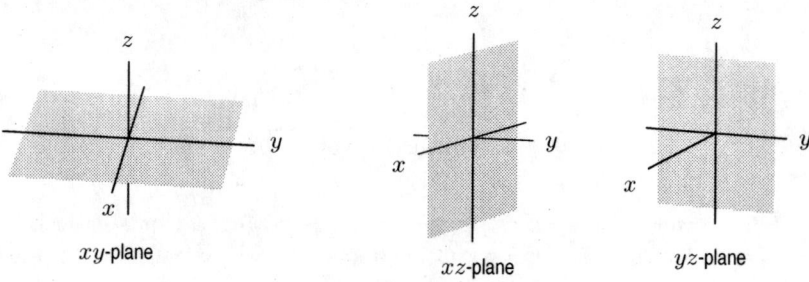

Figure 12.7: The three coordinate planes

608   Chapter Twelve   FUNCTIONS OF SEVERAL VARIABLES

Example 7 Which of the points $A = (1, -1, 0)$, $B = (0, 3, 4)$, $C = (2, 2, 1)$, and $D = (0, -4, 0)$ lies closest to the xz-plane? Which point lies on the y-axis?

Solution The magnitude of the y-coordinate gives the distance to the xz-plane. The point A lies closest to that plane, because it has the smallest y-coordinate in magnitude. To get to a point on the y-axis, we move along the y-axis, but we don't move at all in the x or z directions. Thus, a point on the y-axis has both its x- and z-coordinates equal to zero. The only point of the four that satisfies this is D. (See Figure 12.8.)

In general, if a point has one of its coordinates equal to zero, it lies in one of the coordinate planes. If a point has two of its coordinates equal to zero, it lies on one of the coordinate axes.

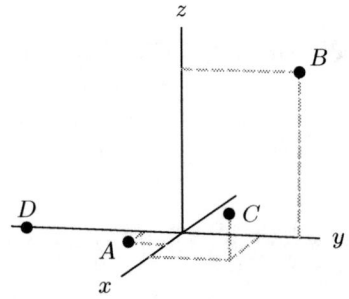

Figure 12.8: Which point lies closest to the xz-plane? Which point lies on the y-axis?

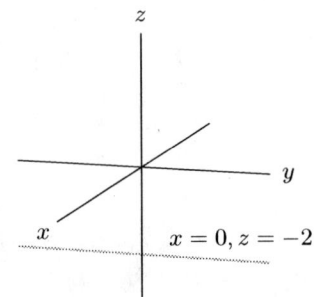

Figure 12.9: The line $x = 0$, $z = -2$

Example 8 You are 2 units below the xy-plane and in the yz-plane. What are your coordinates?

Solution Since you are 2 units below the xy-plane, your z-coordinate is -2. Since you are in the yz-plane, your x-coordinate is 0; your y-coordinate can be anything. Thus, you are at the point $(0, y, -2)$. The set of all such points forms a line parallel to the y-axis, 2 units below the xy-plane, and in the yz-plane. (See Figure 12.9.)

Example 9 You are standing at the point $(4, 5, 2)$, looking at the point $(0.5, 0, 3)$. Are you looking up or down?

Solution The point you are standing at has z-coordinate 2, whereas the point you are looking at has z-coordinate 3; hence you are looking up.

Example 10 Imagine that the yz-plane in Figure 12.7 is a page of this book. Describe the region behind the page algebraically.

Solution The positive part of the x-axis pokes out of the page; moving in the positive x direction brings you out in front of the page. The region behind the page corresponds to negative values of x, so it is the set of all points in 3-space satisfying the inequality $x < 0$.

Distance Between Two Points

In 2-space, the formula for the distance between two points (x, y) and (a, b) is given by

$$\text{Distance} = \sqrt{(x-a)^2 + (y-b)^2}.$$

The distance between two points (x, y, z) and (a, b, c) in 3-space is represented by PG in Figure 12.10. The side PE is parallel to the x-axis, EF is parallel to the y-axis, and FG is parallel to the z-axis.

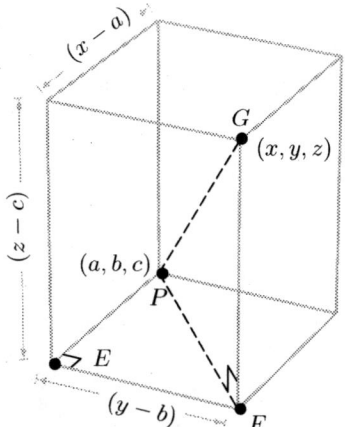

Figure 12.10: The diagonal PG gives the distance between the points (x, y, z) and (a, b, c)

Using Pythagoras' theorem twice gives

$$(PG)^2 = (PF)^2 + (FG)^2 = (PE)^2 + (EF)^2 + (FG)^2 = (x-a)^2 + (y-b)^2 + (z-c)^2.$$

Thus, a formula for the distance between the points (x, y, z) and (a, b, c) in 3-space is

$$\text{Distance} = \sqrt{(x-a)^2 + (y-b)^2 + (z-c)^2}.$$

Example 11 Find the distance between $(1, 2, 1)$ and $(-3, 1, 2)$.

Solution Distance $= \sqrt{(-3-1)^2 + (1-2)^2 + (2-1)^2} = \sqrt{18} = 4.24$.

Example 12 Find an expression for the distance from the origin to the point (x, y, z).

Solution The origin has coordinates $(0, 0, 0)$, so the distance from the origin to (x, y, z) is given by

$$\text{Distance} = \sqrt{(x-0)^2 + (y-0)^2 + (z-0)^2} = \sqrt{x^2 + y^2 + z^2}.$$

Example 13 Find an equation for a sphere of radius 1 with center at the origin.

Solution The sphere consists of all points (x, y, z) whose distance from the origin is 1, that is, which satisfy the equation

$$\sqrt{x^2 + y^2 + z^2} = 1.$$

This is an equation for the sphere. If we square both sides we get the equation in the form

$$x^2 + y^2 + z^2 = 1.$$

Note that this equation represents the *surface* of the sphere. The solid ball enclosed by the sphere is represented by the inequality $x^2 + y^2 + z^2 \leq 1$.

Chapter Twelve FUNCTIONS OF SEVERAL VARIABLES

Exercises and Problems for Section 12.1

Exercises

1. Which of the points $A = (1.3, -2.7, 0)$, $B = (0.9, 0, 3.2)$, $C = (2.5, 0.1, -0.3)$ is closest to the yz-plane? Which one lies on the xz-plane? Which one is farthest from the xy-plane?

2. Which of the points $A = (23, 92, 48)$, $B = (-60, 0, 0)$, $C = (60, 1, -92)$ is closest to the yz-plane? Which lies on the xz-plane? Which is farthest from the xy-plane?

3. You are at the point $(-1, -3, -3)$, standing upright and facing the yz-plane. You walk 2 units forward, turn left, and walk for another 2 units. What is your final position? From the point of view of an observer looking at the coordinate system in Figure 12.2 on page 606, are you in front of or behind the yz-plane? To the left or to the right of the xz-plane? Above or below the xy-plane?

4. You are at the point $(3, 1, 1)$, standing upright and facing the yz-plane. You walk 2 units forward, turn left, and walk another 2 units. What is your final position? From the point of view of an observer looking at the coordinate system in Figure 12.2 on page 606, are you in front of or behind the yz-plane? To the left or to the right of the xz-plane? Above or below the xy-plane?

5. Which of the points $P = (1, 2, 1)$ and $Q = (2, 0, 0)$ is closest to the origin?

6. Which two of the three points $P_1 = (1, 2, 3)$, $P_2 = (3, 2, 1)$ and $P_3 = (1, 1, 0)$ are closest to each other?

7. Find the equation of the sphere of radius 5 centered at the origin.

8. Find the equation of the sphere of radius 5 centered at $(1, 2, 3)$.

Sketch graphs of the equations in Exercises 9–11 in 3-space.

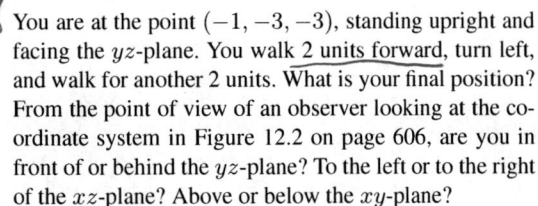

9. $x = -3$
10. $y = 1$
11. $z = 2$ and $y = 4$

Exercises 12–14 refer to the map in Figure 12.1 on page 604.

12. Give the range of daily high temperatures for
    (a) Pennsylvania (b) North Dakota
    (c) California.

13. Sketch a possible graph of the predicted high temperature T on a line north-south through Topeka.

14. Sketch possible graphs of the predicted high temperature on a north-south line and an east-west line through Boise.

For Exercises 15–17, refer to Table 12.1 on page 605, where p is the price of beef and I is annual household income.

15. Make a table showing the amount of money, M, that the average household spends on beef (in dollars per household per week) as a function of the price of beef and household income.

16. Give tables for beef consumption as a function of p, with I fixed at $I = 20$ and $I = 100$. Give tables for beef consumption as a function of I, with p fixed at $p = 3.00$ and $p = 4.00$. Comment on what you see in the tables.

17. How does beef consumption vary as a function of household income if the price of beef is held constant?

Problems

18. The gravitational force, F newtons, exerted on an object by the earth depends on its mass, m kilograms, and its distance, r meters, from the center of the earth, so $F = f(m, r)$. Interpret the following statement in terms of gravitation: $f(100, 7000000) \approx 820$.

19. The balance, B, in dollars, in a bank account depends on the amount deposited, A dollars, the annual interest rate, $r\%$, and the time, t, in months since the deposit, so $B = f(A, r, t)$.
    (a) Is f an increasing or decreasing function of A? Of r? Of t?
    (b) Interpret the statement $f(1250, 1, 25) \approx 1276$. Give units.

20. The monthly payments, P dollars, on a mortgage in which A dollars were borrowed at an annual interest rate of $r\%$ for t years is given by $P = f(A, r, t)$. Is f an increasing or decreasing function of A? Of r? Of t?

21. Consider the acceleration due to gravity, g, at a height h above the surface of a planet of mass m.
    (a) If m is held constant, is g an increasing or decreasing function of h? Why?
    (b) If h is held constant, is g an increasing or decreasing function of m? Why?

22. A car rental company charges $40 a day and 15 cents a mile for its cars.
    (a) Write a formula for the cost, C, of renting a car as a function, f, of the number of days, d, and the number of miles driven, m.
    (b) If $C = f(d, m)$, find $f(5, 300)$ and interpret it.

23. The temperature adjusted for wind-chill is a temperature which tells you how cold it feels, as a result of the combination of wind and temperature.[2] See Table 12.2.
    (a) If the temperature is 0°F and the wind speed is 15 mph, how cold does it feel?

---

[2] Data from www.nws.noaa.gov/om/windchill, accessed on May 30, 2004.

(b) If the temperature is 35°F, what wind speed makes it feel like 24°F?
(c) If the temperature is 25°F, what wind speed makes it feel like 12°F?
(d) If the wind is blowing at 20 mph, what temperature feels like 0°F?

Table 12.2 *Temperature adjusted for wind-chill (°F) as a function of wind speed and temperature*

	35	30	25	20	15	10	5	0
5	31	25	19	13	7	1	-5	-11
10	27	21	15	9	3	-4	-10	-16
15	25	19	13	6	0	-7	-13	-19
20	24	17	11	4	-2	-9	-15	-22
25	23	16	9	3	-4	-11	-17	-24

Temperature (°F) across top; Wind Speed (mph) down side.

24. Using Table 12.2, make tables of the temperature adjusted for wind-chill as a function of wind speed for temperatures of 20°F and 0°F.

25. Find a possible formula for the function $f(x, y)$ whose values are in Table 12.3.

Table 12.3

	-2	-1	0	1	2
-2	4	1	0	1	4
-1	4	1	0	1	4
0	4	1	0	1	4
1	4	1	0	1	4
2	4	1	0	1	4

26. A cube is located such that its top four corners have the coordinates $(-1, -2, 2)$, $(-1, 3, 2)$, $(4, -2, 2)$ and $(4, 3, 2)$. Give the coordinates of the center of the cube.

27. Describe the set of points whose distance from the x-axis is 2.

28. Describe the set of points whose distance from the x-axis equals the distance from the yz-plane.

29. Find a formula for the shortest distance between a point (a, b, c) and the y-axis.

30. Find the equations of planes that just touch the sphere $(x - 2)^2 + (y - 3)^2 + (z - 3)^2 = 16$ and are parallel to

(a) The xy-plane (b) The yz-plane
(c) The xz-plane

31. Find an equation of the largest sphere contained in the cube determined by the planes $x = 2$, $x = 6$; $y = 5$, $y = 9$; and $z = -1$, $z = 3$.

32. Let $f(x, y, z) = x^2 + y^2 + z^2 - 2x + 3y + z$.

(a) Find a point (a, b, c) so that f can be expressed in terms of distance from (a, b, c). [Hint: Complete the square.]
(b) Express f in terms of the distance d from the point you found in part (a).

33. Which of the following functions can be expressed in terms of the distance from the y-axis?

(a) $f(x, y, z) = 10e^y$
(b) $g(x, y, z) = x^2 z^2$
(c) $h(x, y, z) = 1/\sqrt{x^2 + b^2 + z^2}$, for constant b.

12.2 GRAPHS OF FUNCTIONS OF TWO VARIABLES

The weather map on page 604 is one way of visualizing a function of two variables. In this section we see how to visualize a function of two variables in another way, using a surface in 3-space.

Visualizing a Function of Two Variables Using a Graph

For a function of one variable, $y = f(x)$, the graph of f is the set of all points (x, y) in 2-space such that $y = f(x)$. In general, these points lie on a curve in the plane. When a computer or calculator graphs f, it approximates by plotting points in the xy-plane and joining consecutive points by line segments. The more points, the better the approximation.

Now consider a function of two variables.

> The **graph** of a function of two variables, f, is the set of all points (x, y, z) such that $z = f(x, y)$. In general, the graph of a function of two variables is a surface in 3-space.

Plotting the Graph of the Function $f(x, y) = x^2 + y^2$

To sketch the graph of f we connect points as for a function of one variable. We first make a table of values of f, such as in Table 12.4.

612 Chapter Twelve FUNCTIONS OF SEVERAL VARIABLES

Table 12.4 *Table of values of* $f(x,y) = x^2 + y^2$

	y						
x		−3	−2	−1	0	1	2	3
	−3	18	13	10	9	10	13	18
	−2	13	8	5	4	5	8	13
	−1	10	5	2	1	2	5	10
	0	9	4	1	0	1	4	9
	1	10	5	2	1	2	5	10
	2	13	8	5	4	5	8	13
	3	18	13	10	9	10	13	18

Now we plot points. For example, we plot $(1,2,5)$ because $f(1,2) = 5$ and we plot $(0,2,4)$ because $f(0,2) = 4$. Then, we connect the points corresponding to the rows and columns in the table. The result is called a *wire-frame* picture of the graph. Filling in between the wires gives a surface. That is the way a computer drew the graphs in Figure 12.11 and 12.12. As more points are plotted, we get the surface in Figure 12.13, called a *paraboloid*.

You should check to see if the sketches make sense. Notice that the graph goes through the origin since $(x,y,z) = (0,0,0)$ satisfies $z = x^2 + y^2$. Observe that if x is held fixed and y is allowed to vary, the graph dips down and then goes back up, just like the entries in the rows of Table 12.4. Similarly, if y is held fixed and x is allowed to vary, the graph dips down and then goes back up, just like the columns of Table 12.4.

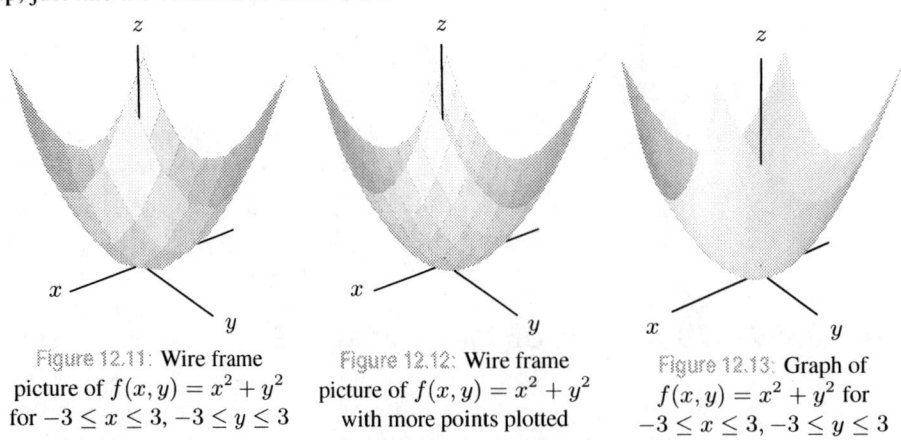

Figure 12.11: Wire frame picture of $f(x,y) = x^2 + y^2$ for $-3 \le x \le 3, -3 \le y \le 3$

Figure 12.12: Wire frame picture of $f(x,y) = x^2 + y^2$ with more points plotted

Figure 12.13: Graph of $f(x,y) = x^2 + y^2$ for $-3 \le x \le 3, -3 \le y \le 3$

New Graphs from Old

We can use the graph of a function to visualize the graphs of related functions.

Example 1 Let $f(x,y) = x^2 + y^2$. Describe in words the graphs of the following functions:
(a) $g(x,y) = x^2 + y^2 + 3$,   (b) $h(x,y) = 5 - x^2 - y^2$,   (c) $k(x,y) = x^2 + (y-1)^2$.

Solution We know from Figure 12.13 that the graph of f is a paraboloid, or bowl with its vertex at the origin. From this we can work out what the graphs of g, h, and k will look like.
(a) The function $g(x,y) = x^2 + y^2 + 3 = f(x,y) + 3$, so the graph of g is the graph of f, but raised by 3 units. See Figure 12.14.
(b) Since $-x^2 - y^2$ is the negative of $x^2 + y^2$, the graph of $-x^2 - y^2$ is an upside down paraboloid. Thus, the graph of $h(x,y) = 5 - x^2 - y^2 = 5 - f(x,y)$ looks like an upside down paraboloid with vertex at $(0,0,5)$, as in Figure 12.15.
(c) The graph of $k(x,y) = x^2 + (y-1)^2 = f(x, y-1)$ is a paraboloid with vertex at $x = 0, y = 1$, since that is where $k(x,y) = 0$, as in Figure 12.16.

12.2 GRAPHS OF FUNCTIONS OF TWO VARIABLES

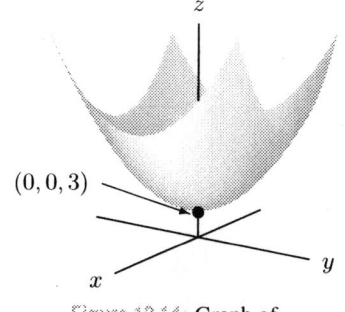

Figure 12.14: Graph of $g(x,y) = x^2 + y^2 + 3$

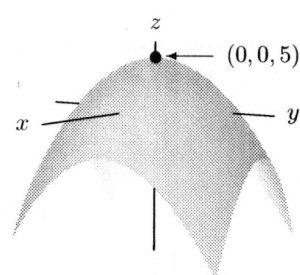

Figure 12.15: Graph of $h(x,y) = 5 - x^2 - y^2$

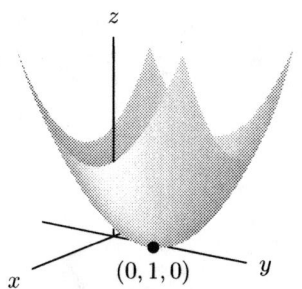

Figure 12.16: Graph of $k(x,y) = x^2 + (y-1)^2$

Example 2 Describe the graph of $G(x,y) = e^{-(x^2+y^2)}$. What symmetry does it have?

Solution Since the exponential function is always positive, the graph lies entirely above the xy-plane. From the graph of $x^2 + y^2$ we see that $x^2 + y^2$ is zero at the origin and gets larger as we move farther from the origin in any direction. Thus, $e^{-(x^2+y^2)}$ is 1 at the origin, and gets smaller as we move away from the origin in any direction. It can't go below the xy-plane; instead it flattens out, getting closer and closer to the plane. We say the surface is *asymptotic* to the xy-plane. (See Figure 12.17.)

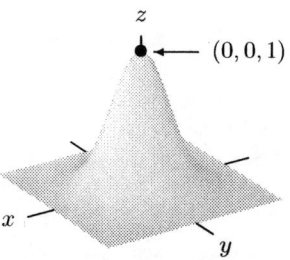

Figure 12.17: Graph of $G(x,y) = e^{-(x^2+y^2)}$

Now consider a point (x,y) on the circle $x^2 + y^2 = r^2$. Since

$$G(x,y) = e^{-(x^2+y^2)} = e^{-r^2},$$

the value of the function G is the same at all points on this circle. Thus, we say the graph of G has *circular symmetry*.

Cross Sections and the Graph of a Function

We have seen that a good way to analyze a function of two variables is to let one variable vary at a time while the other is kept fixed.

> For a function $f(x,y)$, the function we get by holding x fixed and letting y vary is called a **cross-section** of f with x fixed. The graph of the cross-section of $f(x,y)$ with $x = c$ is the curve, or cross-section, we get by intersecting the graph of f with the plane $x = c$. We define a cross-section of f with y fixed similarly.

For example, the cross-section of $f(x, y) = x^2 + y^2$ with $x = 2$ is $f(2, y) = 4 + y^2$. The graph of this cross-section is the curve we get by intersecting the graph of f with the plane perpendicular to the x-axis at $x = 2$. (See Figure 12.18.)

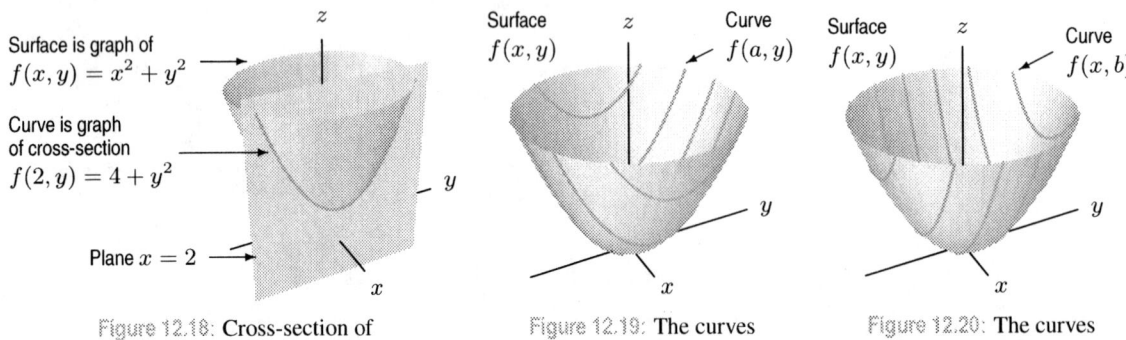

Figure 12.18: Cross-section of the surface $z = f(x, y)$ by the plane $x = 2$

Figure 12.19: The curves $z = f(a, y)$ with a constant: cross-sections with x fixed

Figure 12.20: The curves $z = f(x, b)$ with b constant: cross-sections with y fixed

Figure 12.19 shows graphs of other cross-sections of f with x fixed; Figure 12.20 shows graphs of cross-sections with y fixed.

Example 3 Describe the cross-sections of the function $g(x, y) = x^2 - y^2$ with y fixed and then with x fixed. Use these cross-sections to describe the shape of the graph of g.

Solution The cross-sections with y fixed at $y = b$ are given by

$$z = g(x, b) = x^2 - b^2.$$

Thus, each cross-section with y fixed gives a parabola opening upward, with minimum $z = -b^2$. The cross-sections with x fixed are of the form

$$z = g(a, y) = a^2 - y^2,$$

which are parabolas opening downward with a maximum of $z = a^2$. (See Figures 12.21 and 12.22.) The graph of g is shown in Figure 12.23. Notice the upward opening parabolas in the x-direction and the downward opening parabolas in the y-direction. We say that the surface is *saddle-shaped*.

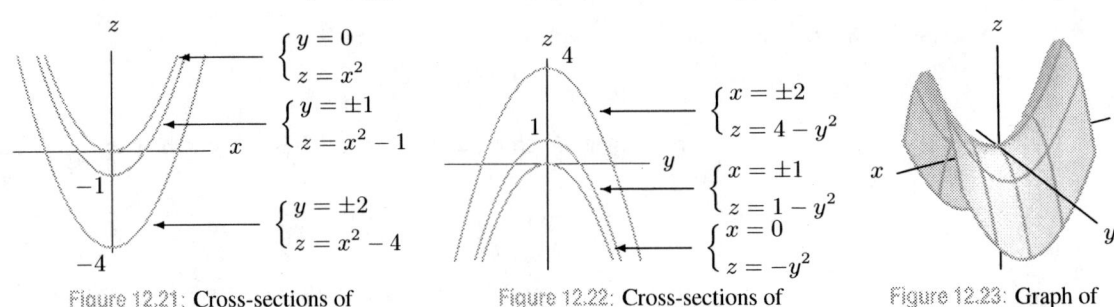

Figure 12.21: Cross-sections of $g(x, y) = x^2 - y^2$ with y fixed

Figure 12.22: Cross-sections of $g(x, y) = x^2 - y^2$ with x fixed

Figure 12.23: Graph of $g(x, y) = x^2 - y^2$ showing cross sections

Linear Functions

Linear functions are central to single variable calculus; they are equally important in multivariable calculus. You may be able to guess the shape of the graph of a linear function of two variables. (It's a plane.) Let's look at an example.

Example 4 Describe the graph of $f(x, y) = 1 + x - y$.

Solution The plane $x = a$ is vertical and parallel to the yz-plane. Thus, the cross-section with $x = a$ is the line $z = 1 + a - y$ which slopes downward in the y-direction. Similarly, the plane $y = b$ is parallel to the xz-plane. Thus, the cross-section with $y = b$ is the line $z = 1 + x - b$ which slopes upward in the x-direction. Since all the cross-sections are lines, you might expect the graph to be a flat plane, sloping down in the y-direction and up in the x-direction. This is indeed the case. (See Figure 12.24.)

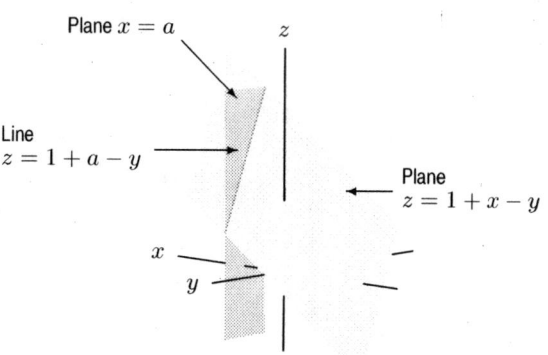

Figure 12.24: Graph of the plane $z = 1 + x - y$ showing cross-section with $x = a$

When One Variable is Missing: Cylinders

Suppose we graph an equation like $z = x^2$ which has one variable missing. What does the surface look like? Since y is missing from the equation, the cross-sections with y fixed are all the same parabola, $z = x^2$. Letting y vary up and down the y-axis, this parabola sweeps out the trough-shaped surface shown in Figure 12.25. The cross-sections with x fixed are horizontal lines obtained by cutting the surface by a plane perpendicular to the x-axis. This surface is called a *parabolic cylinder*, because it is formed from a parabola in the same way that an ordinary cylinder is formed from a circle; it has a parabolic cross-section instead of a circular one.

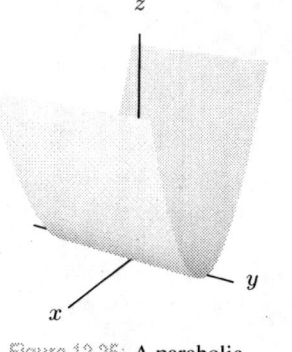

Figure 12.25: A parabolic cylinder $z = x^2$

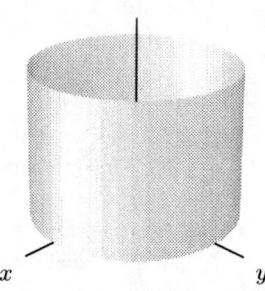

Figure 12.26: Circular cylinder $x^2 + y^2 = 1$

Example 5 Graph the equation $x^2 + y^2 = 1$ in 3-space.

Solution Although the equation $x^2 + y^2 = 1$ does not represent a function, the surface representing it can be graphed by the method used for $z = x^2$. The graph of $x^2 + y^2 = 1$ in the xy-plane is a circle. Since z does not appear in the equation, the intersection of the surface with any horizontal plane will be the same circle $x^2 + y^2 = 1$. Thus, the surface is the cylinder shown in Figure 12.26.

Chapter Twelve FUNCTIONS OF SEVERAL VARIABLES

Exercises and Problems for Section 12.2

Exercises

1. Without a calculator or computer, match the functions with their graphs in Figure 12.27.

   (a) $z = 2 + x^2 + y^2$ (b) $z = 2 - x^2 - y^2$
   (c) $z = 2(x^2 + y^2)$ (d) $z = 2 + 2x - y$
   (e) $z = 2$

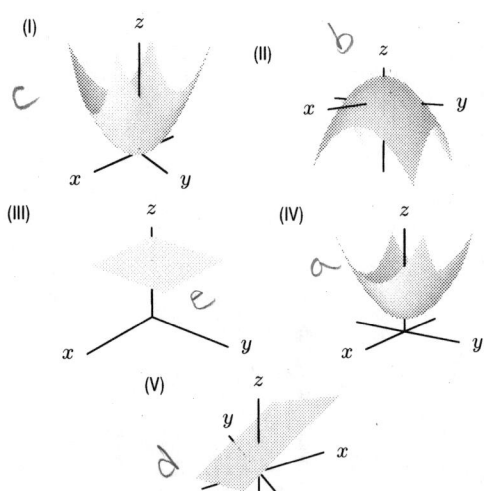

Figure 12.27

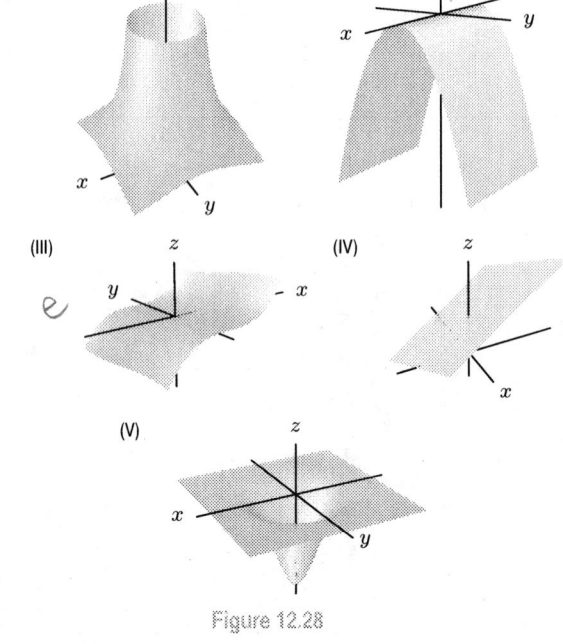

Figure 12.28

2. Without a calculator or computer, match the functions with their graphs in Figure 12.28.

   (a) $z = \dfrac{1}{x^2+y^2}$ (b) $z = -e^{-x^2-y^2}$
   (c) $z = x + 2y + 3$ (d) $z = -y^2$
   (e) $z = x^3 - \sin y$.

In Exercises 3–10, sketch a graph of the surface and briefly describe it in words.

3. $z = 3$ 4. $x^2 + y^2 + z^2 = 9$
5. $z = x^2 + y^2 + 4$ 6. $z = 5 - x^2 - y^2$
7. $z = y^2$ 8. $2x + 4y + 3z = 12$
9. $x^2 + y^2 = 4$ 10. $x^2 + z^2 = 4$

Problems

11. For each of the following functions, decide whether it could be a bowl, a plate, or neither. Consider a plate to be any fairly flat surface and a bowl to be anything that could hold water, assuming the positive z-axis is up.

    (a) $z = x^2 + y^2$ (b) $z = 1 - x^2 - y^2$
    (c) $x + y + z = 1$ (d) $z = -\sqrt{5 - x^2 - y^2}$
    (e) $z = 3$

12. For each function in Problem 11 sketch cross-sections.

13. Consider the function f given by $f(x, y) = y^3 + xy$. Draw graphs of cross-sections with:

    (a) x fixed at $x = -1$, $x = 0$, and $x = 1$.
    (b) y fixed at $y = -1$, $y = 0$, and $y = 1$.

14. Without a computer or calculator, match the equations (a)–(i) with the graphs (I)–(IX).

    (a) $z = xye^{-(x^2+y^2)}$ (b) $z = \cos(\sqrt{x^2+y^2})$
    (c) $z = \sin y$ (d) $z = -\dfrac{1}{x^2+y^2}$
    (e) $z = \cos^2 x \cos^2 y$ (f) $z = \dfrac{\sin(x^2+y^2)}{x^2+y^2}$
    (g) $z = \cos(xy)$ (h) $z = |x||y|$
    (i) $z = (2x^2+y^2)e^{1-x^2-y^2}$

12.2 GRAPHS OF FUNCTIONS OF TWO VARIABLES

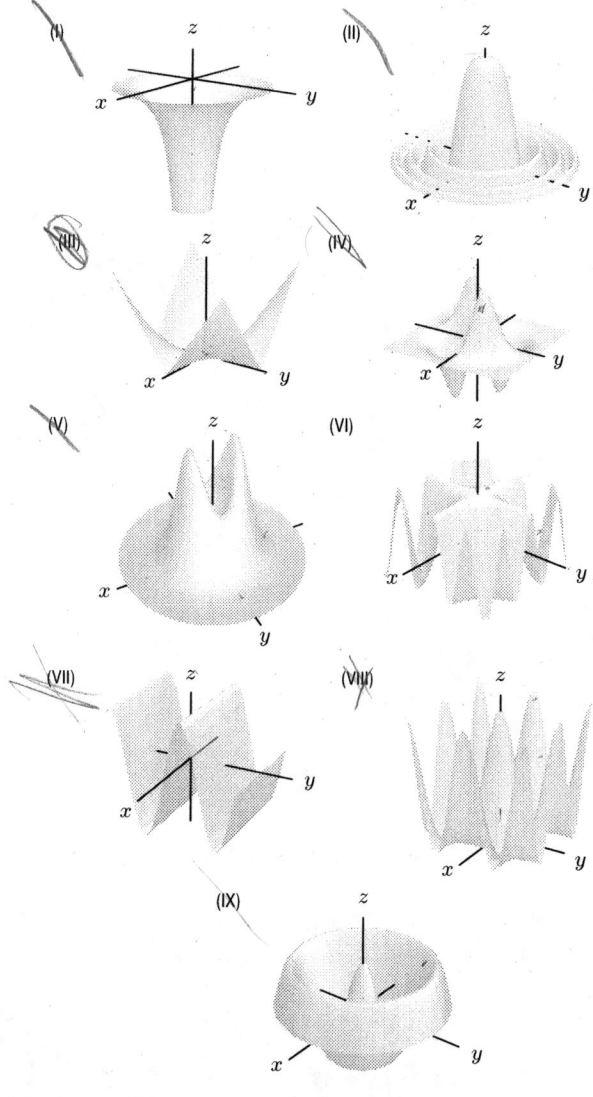

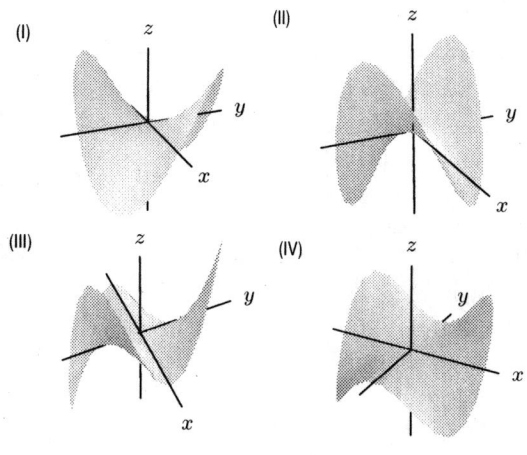

Figure 12.30

Problems 16–18 concern the concentration, C, in mg per liter, of a drug in the blood as a function of x, the amount, in mg, of the drug given and t, the time in hours since the injection. For $0 \leq x \leq 4$ and $t \geq 0$, we have $C = f(x,t) = te^{-t(5-x)}$.

16. Find $f(3,2)$. Give units and interpret in terms of drug concentration.

17. Graph the following two single variable functions and explain their significance in terms of drug concentration.

    (a) $f(4,t)$ (b) $f(x,1)$

18. Graph $f(a,t)$ for $a = 1, 2, 3, 4$ on the same axes. Describe how the graph changes as a increases and explain what this means in terms of drug concentration.

19. You like pizza and you like cola. Which of the graphs in Figure 12.31 represents your happiness as a function of how many pizzas and how much cola you have if

    (a) There is no such thing as too many pizzas and too much cola?
    (b) There is such a thing as too many pizzas or too much cola?
    (c) There is such a thing as too much cola but no such thing as too many pizzas?

15. Figure 12.29 contains graphs of the parabolas $z = f(x,b)$ for $b = -2, -1, 0, 1, 2$. Which of the graphs of $z = f(x,y)$ in Figure 12.30 best fits this information?

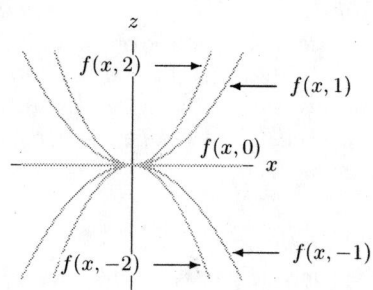

Figure 12.29

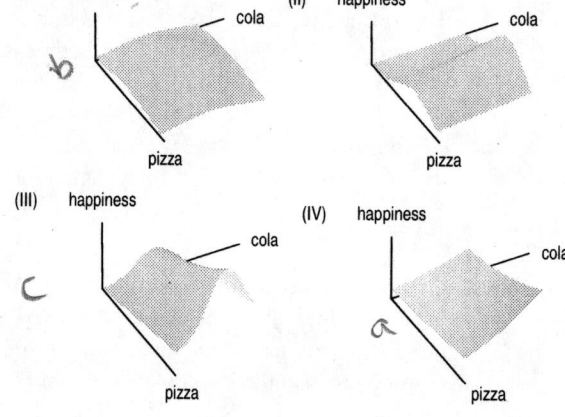

Figure 12.31

20. For each of the graphs I-IV in Problem 19 draw:
  (a) Two cross-sections with pizza fixed
  (b) Two cross-sections with cola fixed.

21. By setting one variable constant, find a plane that intersects the graph of $z = 4x^2 - y^2 + 1$ in a:
  (a) Parabola opening upward
  (b) Parabola opening downward
  (c) Pair of intersecting straight lines

22. By setting one variable constant, find a plane that intersects the graph of $z = (x^2 + 1)\sin y + xy^2$ in a:
  (a) Parabola
  (b) Straight line
  (c) Sine curve

23. A wave travels along a canal. Let x be the distance along the canal from the middle, t be the time, and z be the height of the water above the equilibrium level. The graph of z as a function of x and t is in Figure 12.32.
  (a) Draw the profile of the wave for $t = -1, 0, 1, 2$. (Put the x-axis to the right and the z-axis vertical.)
  (b) Is the wave traveling in the direction of increasing or decreasing x?
  (c) Sketch a surface representing a wave traveling in the opposite direction.

24. A swinging pendulum consists of a mass at the end of a string. At one moment the string makes an angle x with the vertical and the mass has speed y. At that time, the energy, E, of the pendulum is given by the expression[3]

$$E = 1 - \cos x + \frac{y^2}{2}.$$

  (a) Consider the surface representing the energy. Sketch a cross-section of the surface:
    (i) Perpendicular to the x-axis at $x = c$.
    (ii) Perpendicular to the y-axis at $y = c$.
  (b) For each of the graphs in Figures 12.33 and 12.34 use your answer to part (a) to decide which is the x-axis and which is the y-axis and to put reasonable units on each one.

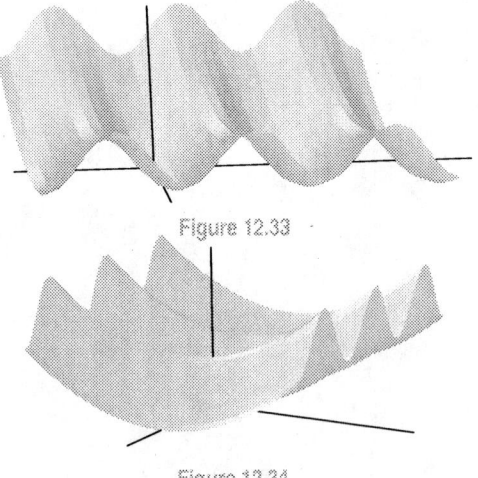

Figure 12.33

Figure 12.34

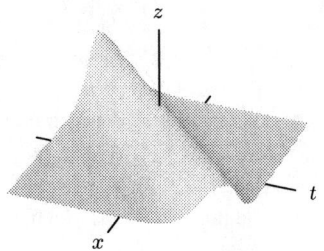

Figure 12.32

12.3 CONTOUR DIAGRAMS

The surface which represents a function of two variables often gives a good idea of the function's general behavior—for example, whether it is increasing or decreasing as one of the variables increases. However it is difficult to read numerical values off a surface and it can be hard to see all of the function's behavior from a surface. Thus, functions of two variables are often represented by contour diagrams like the weather map on page 604. Contour diagrams have the additional advantage that they can be extended to functions of three variables.

Topographical Maps

One of the most common examples of a contour diagram is a topographical map like that shown in Figure 12.35. It gives the elevation in the region and is a good way of getting an overall picture of the terrain: where the mountains are, where the flat areas are. Such topographical maps are frequently colored green at the lower elevations and brown, red, or white at the higher elevations.

[3] Adapted from *Calculus in Context*, by James Callahan, Kenneth Hoffman, (New York: W.H. Freeman, 1995).

12.3 CONTOUR DIAGRAMS

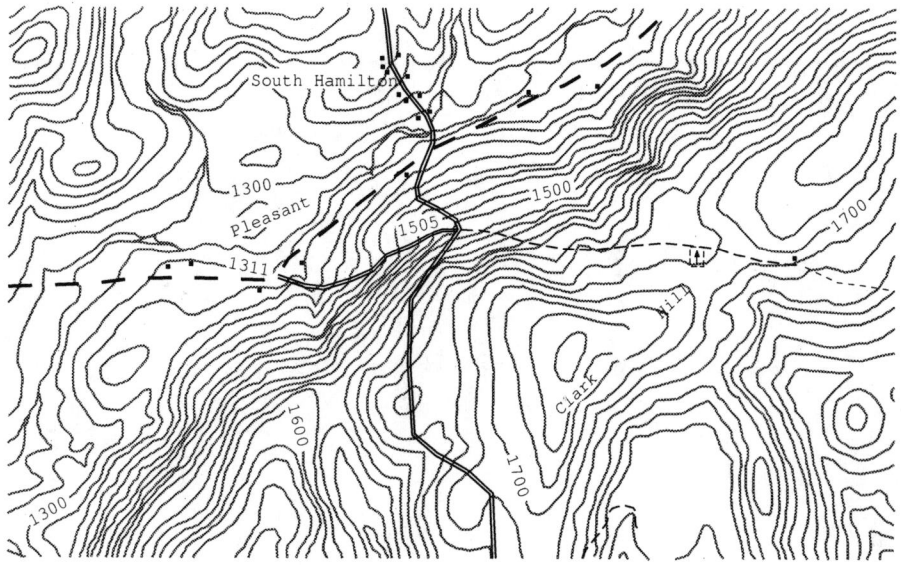

Figure 12.35: A topographical map showing the region around South Hamilton, NY

The curves on a topographical map that separate lower elevations from higher elevations are called *contour lines* because they outline the contour or shape of the land.[4] Because every point along the same contour has the same elevation, contour lines are also called *level curves* or *level sets*. The more closely spaced the contours, the steeper the terrain; the more widely spaced the contours, the flatter the terrain (provided, of course, that the elevation between contours varies by a constant amount). Certain features have distinctive characteristics. A mountain peak is typically surrounded by contour lines like those in Figure 12.36. A pass in a range of mountains may have contours that look like Figure 12.37. A long valley has parallel contour lines indicating the rising elevations on both sides of the valley (see Figure 12.38); a long ridge of mountains has the same type of contour lines, only the elevations decrease on both sides of the ridge. Notice that the elevation numbers on the contour lines are as important as the curves themselves. We usually draw contours for equally spaced values of z.

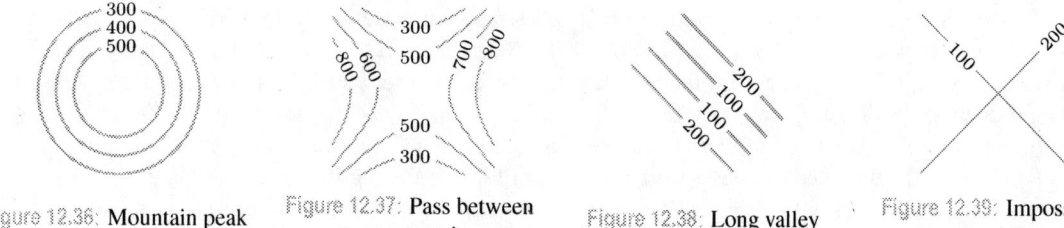

Figure 12.36: Mountain peak Figure 12.37: Pass between two mountains Figure 12.38: Long valley Figure 12.39: Impossible contour lines

Notice that two contours corresponding to different elevations cannot cross each other as shown in Figure 12.39. If they did, the point of intersection of the two curves would have two different elevations, which is impossible (assuming the terrain has no overhangs).

Corn Production

Contour maps can display information about a function of two variables without reference to a surface. Consider the effect of weather conditions on US corn production. Figure 12.40 gives corn production $C = f(R, T)$ as a function of the total rainfall, R, in inches, and average temperature, T, in degrees Fahrenheit, during the growing season.[5] At the present time, $R = 15$ inches and

---

[4] In fact they are usually not straight lines, but curves. They may also be in disconnected pieces.
[5] Adapted from S. Beaty and R. Healy, "The Future of American Agriculture," *Scientific American* 248, No.2, February 1983.

$T = 76°F$. Production is measured as a percentage of the present production; thus, the contour through $R = 15, T = 76$, has value 100, that is, $C = f(15, 76) = 100$.

Example 1 Use Figure 12.40 to estimate $f(18, 78)$ and $f(12, 76)$ and interpret in terms of corn production.

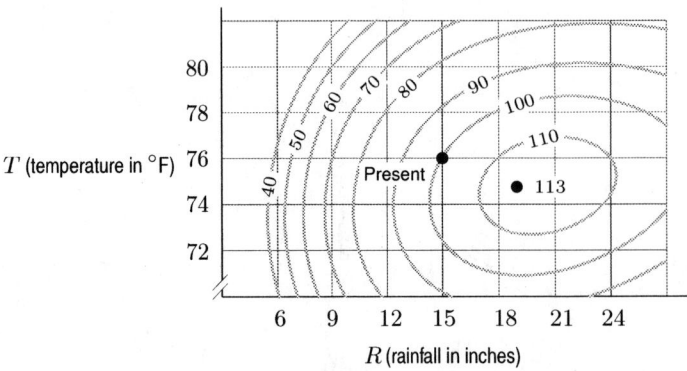

Figure 12.40: Corn production, C, as a function of rainfall and temperature

Solution The point with R-coordinate 18 and T-coordinate 78 is on the contour $C = 100$, so $f(18, 78) = 100$. This means that if the annual rainfall were 18 inches and the temperature were 78°F, the country would produce about the same amount of corn as at present, although it would be wetter and warmer than it is now.

The point with R-coordinate 12 and T-coordinate 76 is about halfway between the $C = 80$ and the $C = 90$ contours, so $f(12, 76) \approx 85$. This means that if the rainfall fell to 12 inches and the temperature stayed at 76°, then corn production would drop to about 85% of what it is now.

Example 2 Use Figure 12.40 to describe in words the cross-sections with T and R constant through the point representing present conditions. Give a common sense explanation of your answer.

Solution To see what happens to corn production if the temperature stays fixed at 76°F but the rainfall changes, look along the horizontal line $T = 76$. Starting from the present and moving left along the line $T = 76$, the values on the contours decrease. In other words, if there is a drought, corn production decreases. Conversely, as rainfall increases, that is, as we move from the present to the right along the line $T = 76$, corn production increases, reaching a maximum of more than 110% when $R = 21$, and then decreases (too much rainfall floods the fields).

If, instead, rainfall remains at the present value and temperature increases, we move up the vertical line $R = 15$. Under these circumstances corn production decreases; a 2° increase causes a 10% drop in production. This makes sense since hotter temperatures lead to greater evaporation and hence drier conditions, even with rainfall constant at 15 inches. Similarly, a decrease in temperature leads to a very slight increase in production, reaching a maximum of around 102% when $T = 74$, followed by a decrease (the corn won't grow if it is too cold).

Contour Diagrams and Graphs

Contour diagrams and graphs are two different ways of representing a function of two variables. How do we go from one to the other? In the case of the topographical map, the contour diagram was created by joining all the points at the same height on the surface and dropping the curve into the xy-plane.

How do we go the other way? Suppose we wanted to plot the surface representing the corn production function $C = f(R, T)$ given by the contour diagram in Figure 12.40. Along each contour the function has a constant value; if we take each contour and lift it above the plane to a height equal to this value, we get the surface in Figure 12.41.

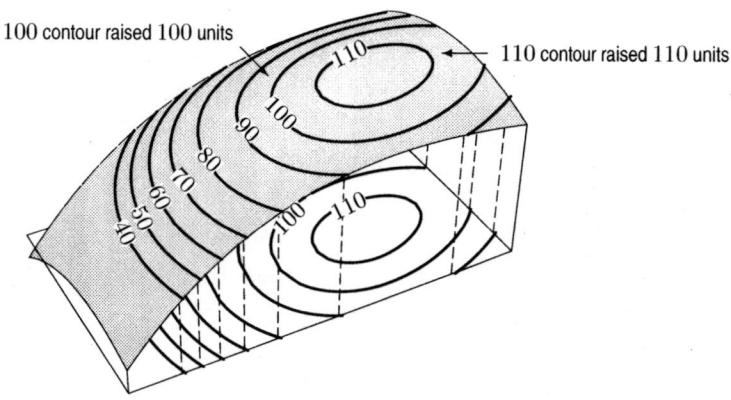

Figure 12.41: Getting the graph of the corn yield function from the contour diagram

Notice that the raised contours are the curves we get by slicing the surface horizontally. In general, we have the following result:

> Contour lines, or level curves, are obtained from a surface by slicing it with horizontal planes.

Finding Contours Algebraically

Algebraic equations for the contours of a function f are easy to find if we have a formula for $f(x, y)$. Suppose the surface has equation

$$z = f(x, y).$$

A contour is obtained by slicing the surface with a horizontal plane with equation $z = c$. Thus, the equation for the contour at height c is given by:

$$f(x, y) = c.$$

Example 3 Find equations for the contours of $f(x, y) = x^2 + y^2$ and draw a contour diagram for f. Relate the contour diagram to the graph of f.

Solution The contour at height c is given by

$$f(x, y) = x^2 + y^2 = c.$$

This is a contour only for $c \geq 0$. For $c > 0$ it is a circle of radius $\sqrt{c}$. For $c = 0$, it is a single point (the origin). Thus, the contours at an elevation of $c = 1, 2, 3, 4, \ldots$ are all circles centered at the origin of radius $1, \sqrt{2}, \sqrt{3}, 2, \ldots$. The contour diagram is shown in Figure 12.42. The bowl-shaped graph of f is shown in Figure 12.43. Notice that the graph of f gets steeper as we move further away from the origin. This is reflected in the fact that the contours become more closely packed as we move further from the origin; for example, the contours for $c = 6$ and $c = 8$ are closer together than the contours for $c = 2$ and $c = 4$.

622 Chapter Twelve FUNCTIONS OF SEVERAL VARIABLES

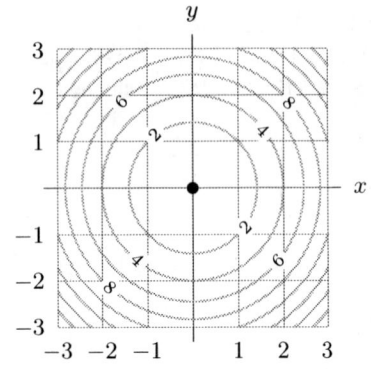

Figure 12.42: Contour diagram for $f(x,y) = x^2 + y^2$ (even values of c only)

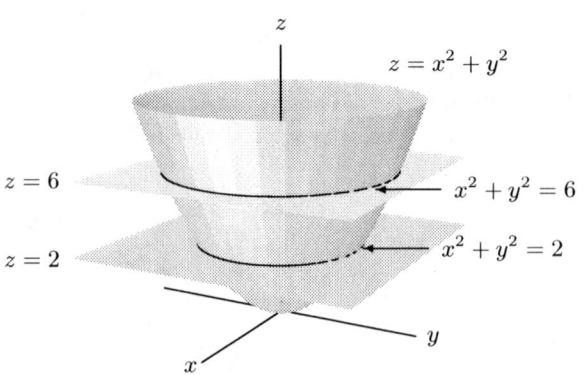

Figure 12.43: The graph of $f(x,y) = x^2 + y^2$

Example 4 Draw a contour diagram for $f(x,y) = \sqrt{x^2 + y^2}$ and relate it to the graph of f.

Solution The contour at level c is given by
$$f(x,y) = \sqrt{x^2 + y^2} = c.$$
For $c > 0$ this is a circle, just as in the previous example, but here the radius is c instead of $\sqrt{c}$. For $c = 0$, it is the origin. Thus, if the level c increases by 1, the radius of the contour increases by 1. This means the contours are equally spaced concentric circles (see Figure 12.44) which do not become more closely packed further from the origin. Thus, the graph of f has the same constant slope as we move away from the origin (see Figure 12.45), making it a cone rather than a bowl.

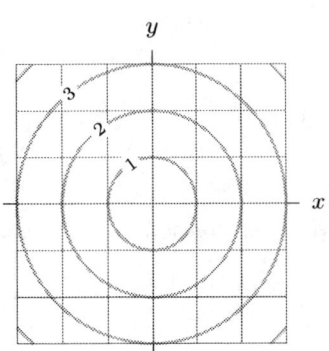

Figure 12.44: A contour diagram for $f(x,y) = \sqrt{x^2 + y^2}$

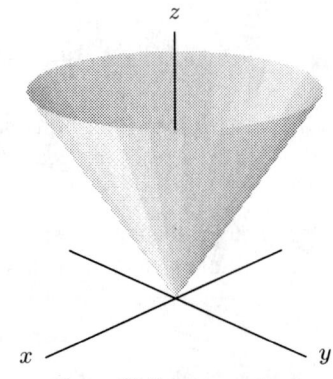

Figure 12.45: The graph of $f(x,y) = \sqrt{x^2 + y^2}$

In both of the previous examples the level curves are concentric circles because the surfaces have circular symmetry. Any function of two variables which depends only on the quantity (x^2+y^2) has such symmetry: for example, $G(x,y) = e^{-(x^2+y^2)}$ or $H(x,y) = \sin(\sqrt{x^2 + y^2})$.

Example 5 Draw a contour diagram for $f(x,y) = 2x + 3y + 1$.

Solution The contour at level c has equation $2x + 3y + 1 = c$. Rewriting this as $y = -(2/3)x + (c-1)/3$, we see that the contours are parallel lines with slope $-2/3$. The y-intercept for the contour at level c is $(c-1)/3$; each time c increases by 3, the y-intercept moves up by 1. The contour diagram is shown in Figure 12.46.

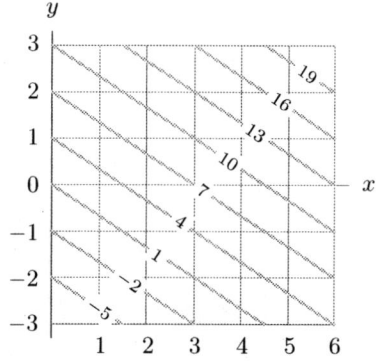

Figure 12.46: A contour diagram for $f(x,y) = 2x + 3y + 1$

Contour Diagrams and Tables

Sometimes we can get an idea of what the contour diagram of a function looks like from its table.

Example 6 Relate the values of $f(x, y) = x^2 - y^2$ in Table 12.5 to its contour diagram in Figure 12.47.

Table 12.5 *Table of values of $f(x,y) = x^2 - y^2$*

	3	0	−5	−8	−9	−8	−5	0
	2	5	0	−3	−4	−3	0	5
	1	8	3	0	−1	0	3	8
y	0	9	4	1	0	1	4	9
	−1	8	3	0	−1	0	3	8
	−2	5	0	−3	−4	−3	0	5
	−3	0	−5	−8	−9	−8	−5	0
		−3	−2	−1	0	1	2	3
					x			

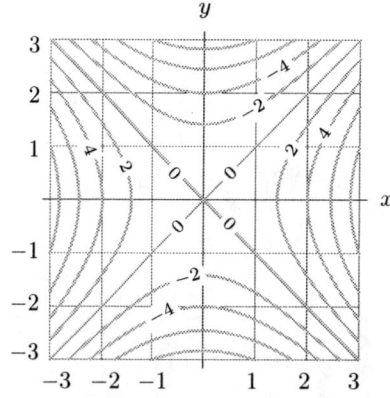

Figure 12.47: Contour map of $f(x,y) = x^2 - y^2$

Solution One striking feature of the values in Table 12.5 is the zeros along the diagonals. This occurs because $x^2 - y^2 = 0$ along the lines $y = x$ and $y = -x$. So the $z = 0$ contour consists of these two lines. In the triangular region of the table that lies to the right of both diagonals, the entries are positive. To the left of both diagonals, the entries are also positive. Thus, in the contour diagram, the positive contours lie in the triangular regions to the right and left of the lines $y = x$ and $y = -x$. Further, the table shows that the numbers on the left are the same as the numbers on the right; thus, each contour has two pieces, one on the left and one on the right. See Figure 12.47. As we move away from the origin along the x-axis, we cross contours corresponding to successively larger values. On the saddle-shaped graph of $f(x,y) = x^2 - y^2$ shown in Figure 12.48, this corresponds to climbing out of the saddle along one of the ridges. Similarly, the negative contours occur in pairs in the top and bottom triangular regions; the values get more and more negative as we go out along the y-axis. This corresponds to descending from the saddle along the valleys that are submerged below the xy-plane in Figure 12.48. Notice that we could also get the contour diagram by graphing the family of hyperbolas $x^2 - y^2 = 0, \pm 2, \pm 4, \ldots$.

624   Chapter Twelve FUNCTIONS OF SEVERAL VARIABLES

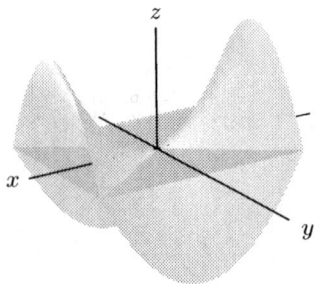

Figure 12.48: Graph of $f(x, y) = x^2 - y^2$ showing plane $z = 0$

Using Contour Diagrams: The Cobb Douglas Production Function

Suppose you decide to expand your small printing business. How should you expand? Should you start a night-shift and hire more workers? Should you buy more expensive but faster computers which will enable the current staff to keep up with the work? Or should you do some combination of the two?

Obviously, the way such a decision is made in practice involves many other considerations— such as whether you could get a suitably trained night shift, or whether there are any faster computers available. Nevertheless, you might model the quantity, P, of work produced by your business as a function of two variables: your total number, N, of workers, and the total value, V, of your equipment.

How would you expect such a production function to behave? In general, having more equipment and more workers enables you to produce more. However, increasing equipment without increasing the number of workers will increase production a bit, but not beyond a point. (If equipment is already lying idle, having more of it won't help.) Similarly, increasing the number of workers without increasing equipment will increase production, but not past the point where the equipment is fully utilized, as any new workers would have no equipment available to them.

Example 7   Explain why the contour diagram in Figure 12.49 does not model the behavior expected of the production function, whereas the contour diagram in Figure 12.50 does.

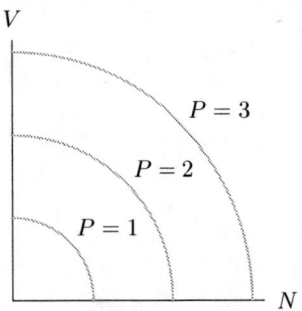
Figure 12.49: Incorrect contours for printing production

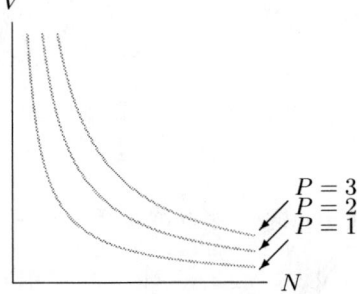
Figure 12.50: Correct contours for printing production

Solution   Look at Figure 12.49. Fixing V and letting N increase corresponds to moving to the right on the contour diagram. As you do so, you cross contours with larger and larger P values, meaning that production increases indefinitely. On the other hand, in Figure 12.50, as you move in the same direction you move nearly parallel to the contours, crossing them less and less frequently. Therefore, production increases more and more slowly as N increases with V fixed. Similarly, if you fix N and let V increase, the contour diagram in Figure 12.49 shows production increasing at a steady rate, whereas Figure 12.50 shows production increasing, but at a decreasing rate. Thus, Figure 12.50 fits the expected behavior of the production function best.

Formula for a Production Function

Production functions are often approximated by formulas of the form

$$P = f(N, V) = cN^\alpha V^\beta$$

where P is the quantity produced and c, α, and β are positive constants, $0 < \alpha < 1$ and $0 < \beta < 1$.

Example 8 Show that the contours of the function $P = cN^\alpha V^\beta$ have approximately the shape of the contours in Figure 12.50.

Solution The contours are the curves where P is equal to a constant value, say P_0, that is, where

$$cN^\alpha V^\beta = P_0.$$

Solving for V we get

$$V = \left(\frac{P_0}{c}\right)^{1/\beta} N^{-\alpha/\beta}.$$

Thus, V is a power function of N with a negative exponent, so its graph has the shape shown in Figure 12.50.

The Cobb-Douglas Production Model

In 1928, Cobb and Douglas used a similar function to model the production of the entire US economy in the first quarter of this century. Using government estimates of P, the total yearly production between 1899 and 1922, of K, the total capital investment over the same period, and of L, the total labor force, they found that P was well approximated by the *Cobb-Douglas production function*

$$P = 1.01 L^{0.75} K^{0.25}.$$

This function turned out to model the US economy surprisingly well, both for the period on which it was based, and for some time afterward.

Exercises and Problems for Section 12.3

Exercises

In Exercises 1–9, sketch a contour diagram for the function with at least four labeled contours. Describe in words the contours and how they are spaced.

1. $f(x, y) = x + y$
2. $f(x, y) = 3x + 3y$
3. $f(x, y) = x^2 + y^2$
4. $f(x, y) = -x^2 - y^2 + 1$
5. $f(x, y) = xy$
6. $f(x, y) = y - x^2$
7. $f(x, y) = x^2 + 2y^2$
8. $f(x, y) = \sqrt{x^2 + 2y^2}$
9. $f(x, y) = \cos\sqrt{x^2 + y^2}$

10. (a) For $z = f(x, y) = xy$, sketch and label the level curves $z = \pm 1$, $z = \pm 2$.
    (b) Sketch and label cross-sections of f with $x = \pm 1$, $x = \pm 2$.
    (c) The surface $z = xy$ is cut by a vertical plane containing the line $y = x$. Sketch the cross-section.

11. Total sales, Q, of a product is a function of its price and the amount spent on advertising. Figure 12.51 shows a contour diagram for total sales. Which axis corresponds to the price of the product and which to the amount spent on advertising? Explain.

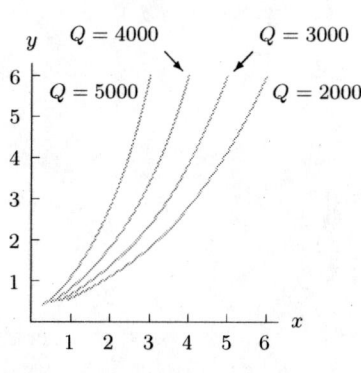

Figure 12.51

626   Chapter Twelve FUNCTIONS OF SEVERAL VARIABLES

For each of the surfaces in Exercises 12–15, sketch a possible contour diagram, marked with reasonable z-values. (Note: There are many possible answers.)

17. Match the surfaces (a)–(e) in Figure 12.53 with the contour diagrams (I)–(V) in Figure 12.54.

12.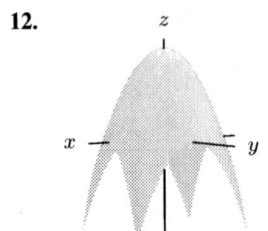

13.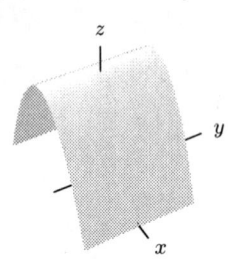

(a)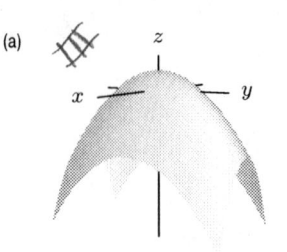

(b)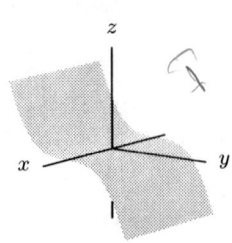

14.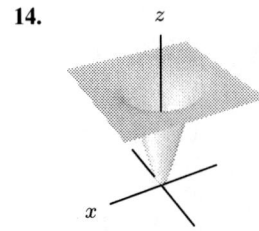

15.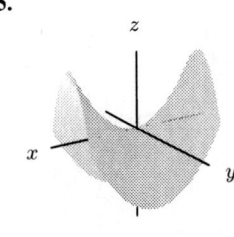

(c)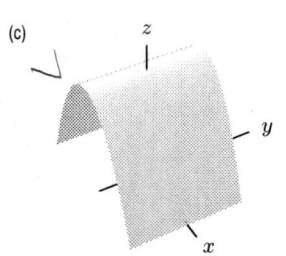

(d)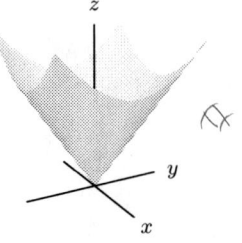

(e)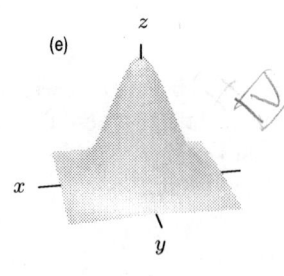

Figure 12.53

16. Match Tables 12.6–12.9 with the contour diagrams (I)–(IV) in Figure 12.52.

Table 12.6

$y \backslash x$	-1	0	1
-1	2	1	2
0	1	0	1
1	2	1	2

Table 12.7

$y \backslash x$	-1	0	1
-1	0	1	0
0	1	2	1
1	0	1	0

Table 12.8

$y \backslash x$	-1	0	1
-1	2	0	2
0	2	0	2
1	2	0	2

Table 12.9

$y \backslash x$	-1	0	1
-1	2	2	2
0	0	0	0
1	2	2	2

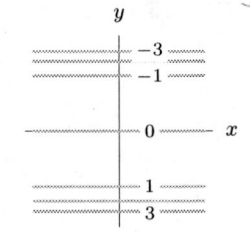

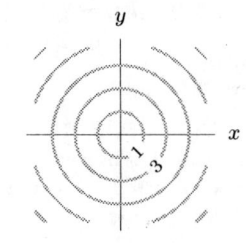

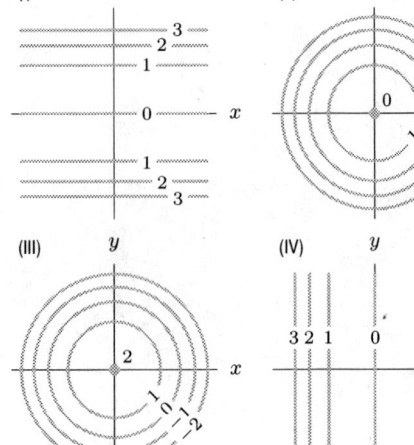

Figure 12.52

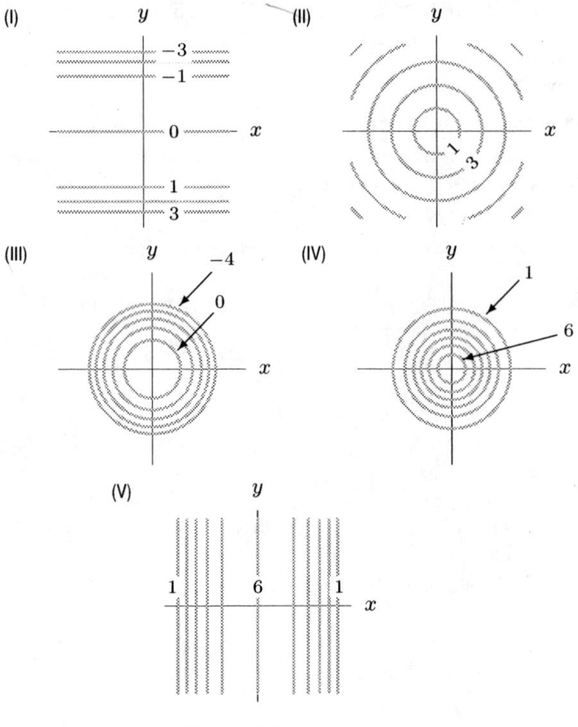

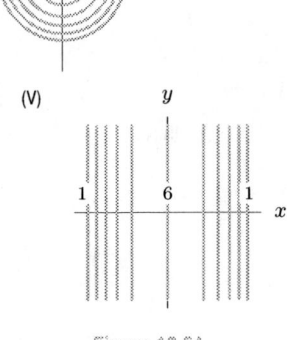

Figure 12.54

Problems

18. Match the contour diagrams (a)–(d) with the surfaces (I)–(IV). Give reasons for your choice.

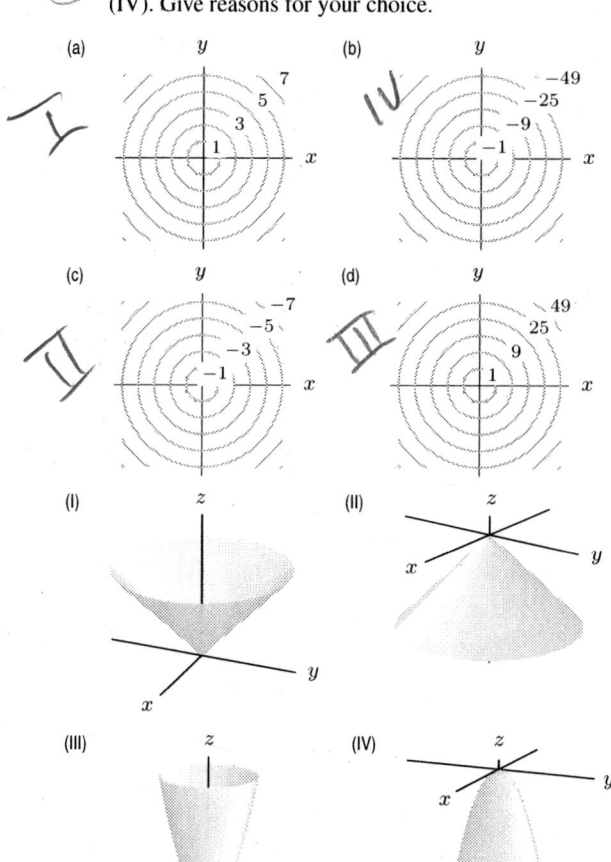

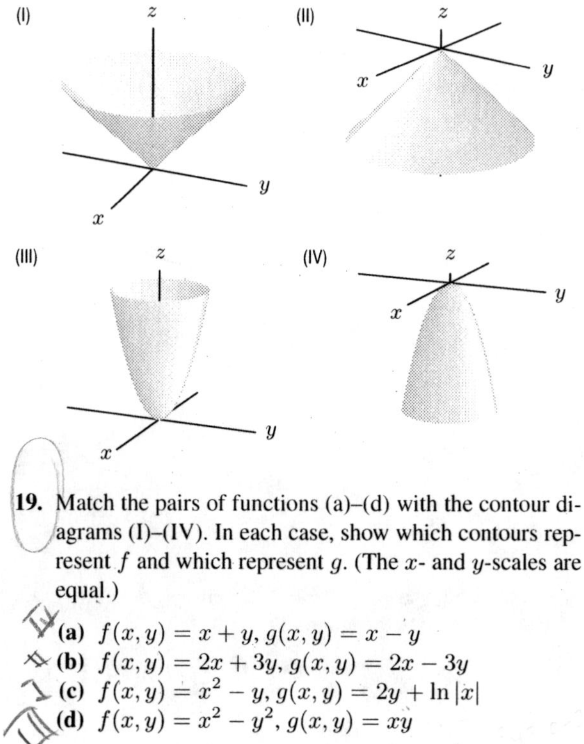

19. Match the pairs of functions (a)–(d) with the contour diagrams (I)–(IV). In each case, show which contours represent f and which represent g. (The x- and y-scales are equal.)

(a) $f(x,y) = x+y$, $g(x,y) = x-y$
(b) $f(x,y) = 2x+3y$, $g(x,y) = 2x-3y$
(c) $f(x,y) = x^2 - y$, $g(x,y) = 2y + \ln|x|$
(d) $f(x,y) = x^2 - y^2$, $g(x,y) = xy$

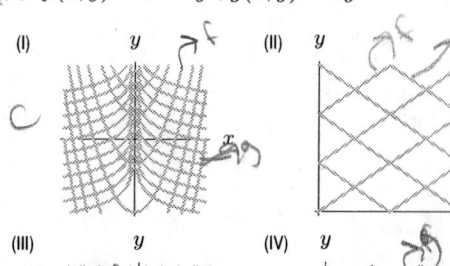
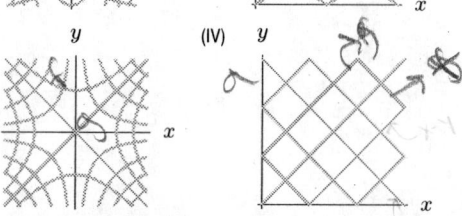

20. Figure 12.55 is a contour diagram of the monthly payment on a 5-year car loan as a function of the interest rate and the amount you borrow. The interest rate is 13% and you borrow $6000.

(a) What is your monthly payment?
(b) If interest rates drop to 11%, how much more can you borrow without increasing your monthly payment?
(c) Make a table of how much you can borrow, without increasing your monthly payment, as a function of the interest rate.

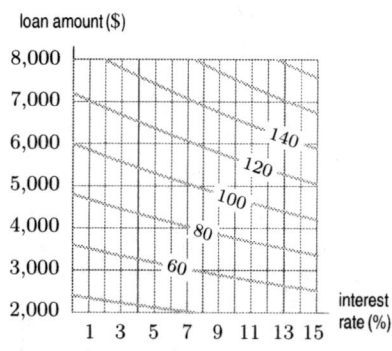

Figure 12.55

21. Figure 12.56 shows a contour map of a hill with two paths, A and B.

(a) On which path, A or B, will you have to climb more steeply?
(b) On which path, A or B, will you probably have a better view of the surrounding countryside? (Assuming trees do not block your view.)
(c) Alongside which path is there more likely to be a stream?

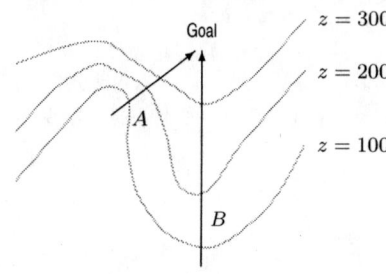

Figure 12.56

22. Match the functions (a)–(d) with the shapes of their level curves (I)–(IV). Sketch each contour diagram.

(a) $f(x,y) = x^2$
(b) $f(x,y) = x^2 + 2y^2$
(c) $f(x,y) = y - x^2$
(d) $f(x,y) = x^2 - y^2$

I. Lines
II. Parabolas
III. Hyperbolas
IV. Ellipses

Chapter Twelve FUNCTIONS OF SEVERAL VARIABLES

23. Figure 12.57 shows the density of the fox population P (in foxes per square kilometer) for southern England. Draw two different cross-sections along a north-south line and two different cross-sections along an east-west line of the population density P.

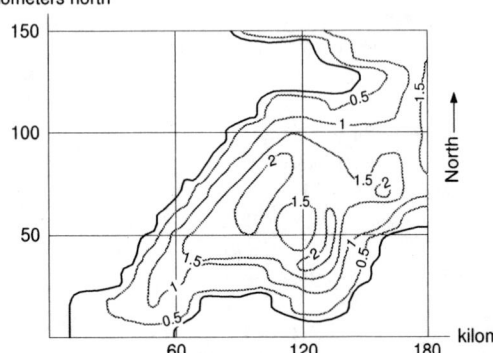

Figure 12.57

24. A manufacturer sells two goods, one at a price of \$3000 a unit and the other at a price of \$12,000 a unit. A quantity q_1 of the first good and q_2 of the second good are sold at a total cost of \$4000 to the manufacturer.

(a) Express the manufacturer's profit, π, as a function of q_1 and q_2.
(b) Sketch curves of constant profit in the q_1q_2-plane for $\pi = 10{,}000$, $\pi = 20{,}000$, and $\pi = 30{,}000$ and the break-even curve $\pi = 0$.

25. Consider the Cobb-Douglas production function $P = f(L, K) = 1.01L^{0.75}K^{0.25}$. What is the effect on production of doubling both labor and capital?

26. Match each Cobb-Douglas production function (a)–(c) with a graph in Figure 12.58 and a statement (D)–(G).

(a) $F(L, K) = L^{0.25}K^{0.25}$
(b) $F(L, K) = L^{0.5}K^{0.5}$
(c) $F(L, K) = L^{0.75}K^{0.75}$
(D) Tripling each input triples output.
(E) Quadrupling each input doubles output.
(G) Doubling each input almost triples output.

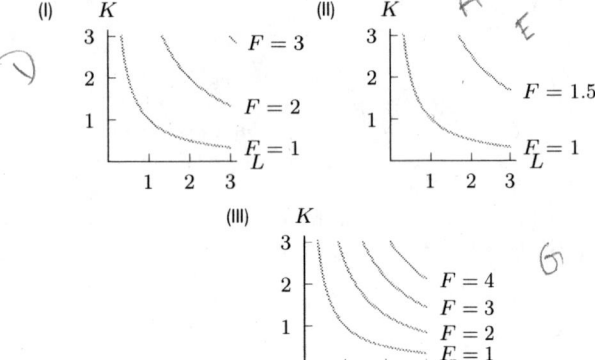

Figure 12.58

27. A general Cobb-Douglas production function has the form
$$P = cL^\alpha K^\beta.$$
What happens to production if labor and capital are both scaled up? For example, does production double if both labor and capital are doubled? Economists talk about

- *increasing returns to scale* if doubling L and K more than doubles P,
- *constant returns to scale* if doubling L and K exactly doubles P,
- *decreasing returns to scale* if doubling L and K less than doubles P.

What conditions on α and β lead to increasing, constant, or decreasing returns to scale?

28. (a) Sketch level curves of $f(x, y) = \sqrt{x^2 + y^2} + x$ for $f = 1, 2, 3$.
(b) For what values of c can level curves $f = c$ be drawn?

29. Figure 12.59 is the contour diagram of $f(x, y)$. Sketch the contour diagram of each of the following functions.

(a) $3f(x, y)$ (b) $f(x, y) - 10$
(c) $f(x - 2, y - 2)$ (d) $f(-x, y)$

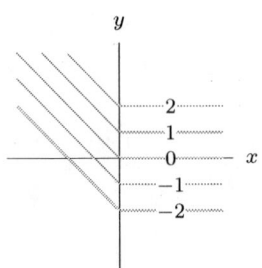

Figure 12.59

30. Figure 12.60 shows part of the contour diagram of $f(x, y)$. Complete the diagram for $x < 0$ if

(a) $f(-x, y) = f(x, y)$ (b) $f(-x, y) = -f(x, y)$

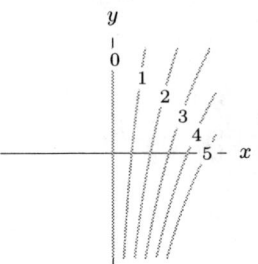

Figure 12.60

31. (a) Draw the contour diagram of $f(x, y) = g(y - x)$ if
    (i) $g(t) = t^2$ (ii) $g(t) = \sin t$
    (b) What can you say about the level curves of a function of the form $f(x, y) = g(y - x)$ where $g(t)$ is a one-variable function?

32. Use the factored form of $f(x, y) = x^2 - y^2 = (x - y)(x + y)$ to sketch the contour $f(x, y) = 0$ and to find the regions in the xy-plane where $f(x, y) > 0$ and the regions where $f(x, y) < 0$. Explain how this sketch shows that the graph of $f(x, y)$ is saddle-shaped at the origin.

33. Use Problem 32 to find a formula for a "monkey-saddle" surface $z = g(x, y)$ which has three regions with $g(x, y) > 0$ and three with $g(x, y) < 0$.

12.4 LINEAR FUNCTIONS

What is a Linear Function of Two Variables?

Linear functions played a central role in one-variable calculus because many one-variable functions have graphs that look like a line when we zoom in. In two-variable calculus, a *linear function* is one whose graph is a plane. In Chapter 14, we see that many two-variable functions have graphs which look like planes when we zoom in.

What Makes a Plane Flat?

What makes the graph of the function $z = f(x, y)$ a plane? Linear functions of *one* variable have straight line graphs because they have constant slope. On a plane, the situation is a bit more complicated. If we walk around on a tilted plane, the slope is not always the same: it depends on the direction in which we walk. However, at every point on the plane, the slope is the same as long as we choose the same direction. If we walk parallel to the x-axis, we always find ourselves walking up or down with the same slope; the same is true if we walk parallel to the y-axis. In other words, the slope ratios $\Delta z/\Delta x$ (with y fixed) and $\Delta z/\Delta y$ (with x fixed) are each constant.

Example 1 A plane cuts the z-axis at $z = 5$, has slope 2 in the x direction and slope -1 in the y direction. What is the equation of the plane?

Solution Finding the equation of the plane means constructing a formula for the z-coordinate of the point on the plane directly above the point (x, y) in the xy-plane. To get to that point start from the point above the origin, where $z = 5$. Then walk x units in the x direction. Since the slope in the x direction is 2, the height increases by $2x$. Then walk y units in the y direction; since the slope in the y direction is -1, the height decreases by y units. Since the height has changed by $2x - y$ units, the z-coordinate is $5 + 2x - y$. Thus, the equation for the plane is

$$z = 5 + 2x - y.$$

For any linear function, if we know its value at a point (x_0, y_0), its slope in the x direction, and its slope in the y direction, then we can write the equation of the function. This is just like the equation of a line in the one-variable case, except that there are two slopes instead of one.

If a **plane** has slope m in the x direction, slope n in the y direction, and passes through the point (x_0, y_0, z_0), then its equation is

$$z = z_0 + m(x - x_0) + n(y - y_0).$$

This plane is the graph of the **linear function**

$$f(x, y) = z_0 + m(x - x_0) + n(y - y_0).$$

If we write $c = z_0 - mx_0 - ny_0$, then we can write $f(x, y)$ in the equivalent form

$$f(x, y) = c + mx + ny.$$

Chapter Twelve FUNCTIONS OF SEVERAL VARIABLES

Just as in 2-space a line is determined by two points, so in 3-space a plane is determined by three points, provided they do not lie on a line.

Example 2 Find the equation of the plane passing through the points $(1, 0, 1)$, $(1, -1, 3)$, and $(3, 0, -1)$.

Solution The first two points have the same x-coordinate, so we use them to find the slope of the plane in the y-direction. As the y-coordinate changes from 0 to -1, the z-coordinate changes from 1 to 3, so the slope in the y-direction is $n = \Delta z/\Delta y = (3-1)/(-1-0) = -2$. The first and third points have the same y-coordinate, so we use them to find the slope in the x-direction; it is $m = \Delta z/\Delta x = (-1-1)/(3-1) = -1$. Because the plane passes through $(1, 0, 1)$, its equation is

$$z = 1 - (x-1) - 2(y-0) \quad \text{or} \quad z = 2 - x - 2y.$$

You should check that this equation is also satisfied by the points $(1, -1, 3)$ and $(3, 0, -1)$.

Example 2 was made easier by the fact that two of the points had the same x-coordinate and two had the same y-coordinate. An alternative method, which works for any three points, is to substitute the x, y, and z-values of each of the three points into the equation $z = c + mx + ny$. The resulting three equations in c, m, n are then solved simultaneously.

Linear Functions from a Numerical Point of View

To avoid flying planes with empty seats, airlines sell some tickets at full price and some at a discount. Table 12.10 shows an airline's revenue in dollars from tickets sold on a particular route, as a function of the number of full-price tickets sold, f, and the number of discount tickets sold, d.

Table 12.10 *Revenue from ticket sales (dollars)*

		\multicolumn{4}{c}{Full-price tickets (f)}
		100	200	300	400
	200	39,700	63,600	87,500	111,400
	400	55,500	79,400	103,300	127,200
Discount tickets (d)	600	71,300	95,200	119,100	143,000
	800	87,100	111,000	134,900	158,800
	1000	102,900	126,800	150,700	174,600

In every column, the revenue jumps by \$15,800 for each extra 200 discount tickets. Thus, each column is a linear function of the number of discount tickets sold. In addition, every column has the same slope, $15{,}800/200 = 79$ dollars/ticket. This is the price of a discount ticket. Similarly, each row is a linear function and all the rows have the same slope, 239, which is the price in dollars of a full-fare ticket. Thus, R is a linear function of f and d, given by:

$$R = 239f + 79d.$$

We have the following general result:

A **linear function** can be recognized from its table by the following features:
- Each row and each column is linear.
- All the rows have the same slope.
- All the columns have the same slope (although the slope of the rows and the slope of the columns are generally different).

12.4 LINEAR FUNCTIONS

Example 3 The table contains values of a linear function. Fill in the blank and give a formula for the function.

$x \backslash y$	1.5	2.0
2	0.5	1.5
3	−0.5	?

Solution In the first column the function decreases by 1 (from 0.5 to −0.5) as x goes from 2 to 3. Since the function is linear, it must decrease by the same amount in the second column. So the missing entry must be $1.5 - 1 = 0.5$. The slope of the function in the x-direction is -1. The slope in the y-direction is 2, since in each row the function increases by 1 when y increases by 0.5. From the table we get $f(2, 1.5) = 0.5$. Therefore, the formula is

$$f(x,y) = 0.5 - (x - 2) + 2(y - 1.5) = -0.5 - x + 2y.$$

What Does the Contour Diagram of a Linear Function Look Like?

The formula for the airline revenue function in Table 12.10 is $R = 239f + 79d$, where f is the number of full-fares and d is the number of discount fares sold.

Notice that the contours of this function in Figure 12.61 are parallel straight lines. What is the practical significance of the slope of these contour lines? Consider the contour $R = 100{,}000$; that means we are looking at combinations of ticket sales that yield \$100,000 in revenue. If we move down and to the right on the contour, the f-coordinate increases and the d-coordinate decreases, so we sell more full-fares and fewer discount fares. This is because to receive a fixed revenue of \$100,000, we must sell more full-fares if we sell fewer discount fares. The exact trade-off depends on the slope of the contour; the diagram shows that each contour has a slope of about -3. This means that for a fixed revenue, we must sell three discount fares to replace one full-fare. This can also be seen by comparing prices. Each full fare brings in \$239; to earn the same amount in discount fares we need to sell $239/79 \approx 3.03 \approx 3$ fares. Since the price ratio is independent of how many of each type of fare we sell, this slope remains constant over the whole contour map; thus, the contours are all parallel straight lines.

Notice also that the contours are evenly spaced. Thus, no matter which contour we are on, a fixed increase in one of the variables causes the same increase in the value of the function. In terms of revenue, no matter how many fares we have sold, an extra fare, whether full or discount, brings the same revenue as before.

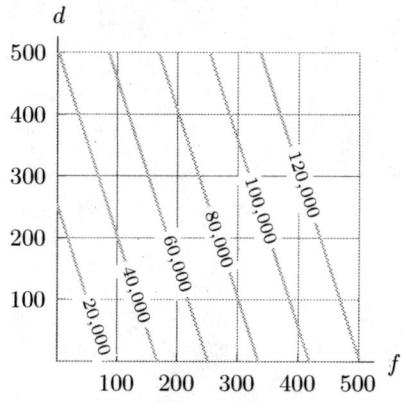

Figure 12.61: Revenue as a function of full and discount fares, $R = 239f + 79d$

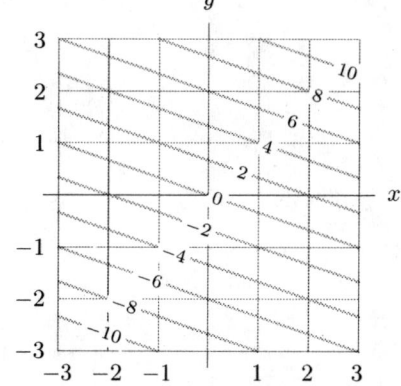

Figure 12.62: Contour map of linear function $f(x, y)$

Example 4 Find the equation of the linear function whose contour diagram is in Figure 12.62.

Solution Suppose we start at the origin on the $z = 0$ contour. Moving 2 units in the y direction takes us to the $z = 6$ contour; so the slope in the y direction is $\Delta z/\Delta y = 6/2 = 3$. Similarly, a move of 2 in the x-direction from the origin takes us to the $z = 2$ contour, so the slope in the x direction is $\Delta z/\Delta x = 2/2 = 1$. Since $f(0, 0) = 0$, we have $f(x, y) = x + 3y$.

Exercises and Problems for Section 12.4

Exercises

1. The charge, C, in dollars, to use an Internet service is a function of m, the number of months of use, and t, the total number of minutes on-line:

$$C = f(m, t) = 35 + 15m + 0.05t.$$

   (a) Is f a linear function?
   (b) Give units for the coefficients of m and t, and interpret them as charges.
   (c) Interpret the intercept 35 as a charge.
   (d) Find $f(3, 800)$ and interpret your answer.

2. Suppose that z is a linear function of x and y with slope 2 in the x direction and slope 3 in the y direction.

   (a) A change of 0.5 in x and -0.2 in y produces what change in z?
   (b) If $z = 2$ when $x = 5$ and $y = 7$, what is the value of z when $x = 4.9$ and $y = 7.2$?

3. (a) Find a formula for the linear function whose graph is a plane passing through point $(4, 3, -2)$ with slope 5 in the x-direction and slope -3 in the y-direction.
   (b) Sketch the contour diagram for this function.

Which of the tables of values in Exercises 4–7 could represent linear functions?

4.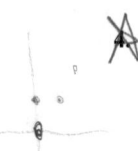

	y		
	0	1	2
x 0	10	13	16
1	6	9	12
2	2	5	8

5.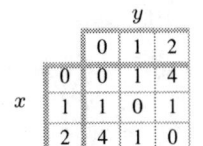

	y		
	0	1	2
x 0	0	1	4
1	1	0	1
2	4	1	0

6.

	y		
	0	1	2
x 0	0	5	10
1	2	7	12
2	4	9	14

7.

	y		
	0	1	2
x 0	5	7	9
1	6	9	12
2	7	11	15

8. Find the equation of the linear function $z = c + mx + ny$ whose graph contains the points $(0, 0, 0)$, $(0, 2, -1)$, and $(-3, 0, -4)$.

9. 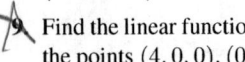 Find the linear function whose graph is the plane through the points $(4, 0, 0)$, $(0, 3, 0)$ and $(0, 0, 2)$.

10. Find an equation for the plane containing the line in the xy-plane where $y = 1$, and the line in the xz-plane where $z = 2$.

11. 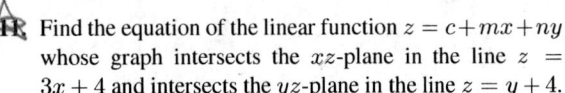 Find the equation of the linear function $z = c + mx + ny$ whose graph intersects the xz-plane in the line $z = 3x + 4$ and intersects the yz-plane in the line $z = y + 4$.

Which of the contour diagrams in Exercises 12–15 could represent linear functions?

12.

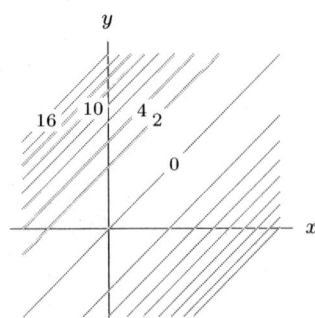

13.

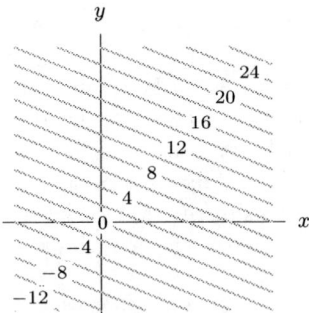

14.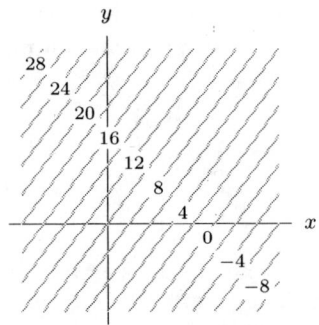

15.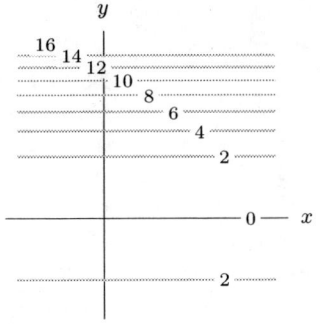

Problems

For Problems 16–17, find possible equations for linear functions with the given contour diagrams.

16.

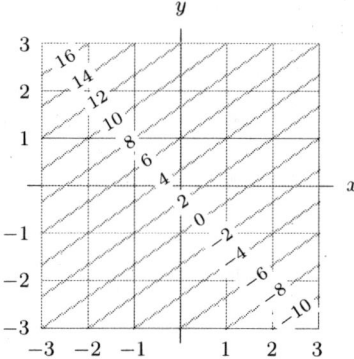

17.

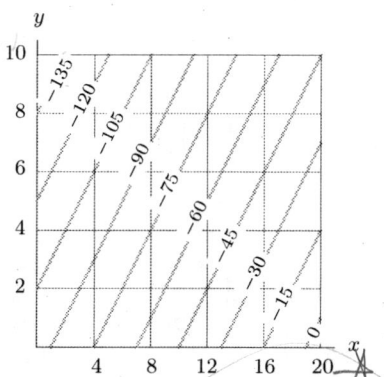

For Problems 18–19, find equations for linear functions with the given values.

18.

$x \backslash y$	−1	0	1	2
0	1.5	1	0.5	0
1	3.5	3	2.5	2
2	5.5	5	4.5	4
3	7.5	7	6.5	6

19.

$x \backslash y$	10	20	30	40
100	3	6	9	12
200	2	5	8	11
300	1	4	7	10
400	0	3	6	9

It is difficult to graph a linear function by hand. One method that works if the x, y, and z-intercepts are positive is to plot the intercepts and join them by a triangle as shown in Figure 12.63; this shows the part of the plane in the octant where $x \geq 0$, $y \geq 0$, $z \geq 0$. If the intercepts are not all positive, the same method works if the x, y, and z-axes are drawn from a different perspective. Use this method to graph the linear functions in Problems 20–23.

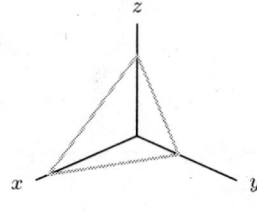

Figure 12.63

20. $z = 2 - x - 2y$

21. $z = 6 - 2x - 3y$

22. $z = 4 + x - 2y$

23. $z = 2 - 2x + y$

Problems 24–26 concern Table 12.11, which gives the number of calories burned per minute for someone roller-blading, as a function of the person's weight and speed.[6]

Table 12.11

	Calories burned per minute			
Weight	8 mph	9 mph	10 mph	11 mph
120 lbs	4.2	5.8	7.4	8.9
140 lbs	5.1	6.7	8.3	9.9
160 lbs	6.1	7.7	9.2	10.8
180 lbs	7.0	8.6	10.2	11.7
200 lbs	7.9	9.5	11.1	12.6

24. Does the data in Table 12.11 look approximately linear? Give a formula for B, the number of calories burned per minute in terms of the weight, w, and the speed, s. Does the formula make sense for all weights or speeds?

25. Who burns more total calories to go 10 miles: A 120 lb person going 10 mph or a 180 lb person going 8 mph? Which of these two people burns more calories per pound for the 10-mile trip?

26. Use Problem 24 to give a formula for P, the number of calories burned per pound, in terms of w and s, for a person weighing w lbs roller-blading 10 miles at s mph.

27. A manufacturer makes two products out of two raw materials. Let q_1, q_2 be the quantities sold of the two products, p_1, p_2 their prices, and m_1, m_2 the quantities purchased of the two raw materials. Which of the following functions do you expect to be linear, and why? In each case, assume that all variables except the ones mentioned are held fixed.

(a) Expenditure on raw materials as a function of m_1 and m_2.
(b) Revenue as a function of q_1 and q_2.
(c) Revenue as a function of p_1 and q_1.

28. Let f be the linear function $f(x, y) = c + mx + ny$, where c, m, n are constants and $n \neq 0$.

(a) Show that all the contours of f are lines of slope $-m/n$.
(b) For all x and y, show $f(x + n, y - m) = f(x, y)$.
(c) Explain the relation between parts (a) and (b).

[6]From the August 28, 1994, issue of *Parade Magazine*.

Problems 29–30 refer to the linear function $z = f(x,y)$ whose values are in Table 12.12.

Table 12.12

	y				
	4	6	8	10	12
5	3	6	9	12	15
10	7	10	13	16	19
x 15	11	14	17	20	23
20	15	18	21	24	27
25	19	22	25	28	31

29. Each column of Table 12.12 is linear with the same slope, $m = \Delta z/\Delta x = 4/5$. Each row is linear with the same slope, $n = \Delta z/\Delta y = 3/2$. We now investigate the slope obtained by moving through the table along lines that are neither rows nor columns.

(a) Move down the diagonal of the table from the upper left corner ($z = 3$) to the lower right corner ($z = 31$). What do you notice about the changes in z? Now move diagonally from $z = 6$ to $z = 27$. What do you notice about the changes in z now?

(b) Move in the table along a line right one step, up two steps from $z = 19$ to $z = 9$. Then move in the same direction from $z = 22$ to $z = 12$. What do you notice about the changes in z?

(c) Show that $\Delta z = m\Delta x + n\Delta y$. Use this to explain what you observed in parts (a) and (b).

30. If we hold y fixed, that is we keep $\Delta y = 0$, and step in the positive x-direction, we get the x-slope, m. If instead we keep $\Delta x = 0$ and step in the positive y-direction, we get the y-slope, n. Fix a step in which neither $\Delta x = 0$ nor $\Delta y = 0$. The slope in the $\Delta x, \Delta y$ direction is

$$\text{Slope} = \frac{\text{Rise}}{\text{Run}} = \frac{\Delta z}{\text{Length of step}}$$
$$= \frac{\Delta z}{\sqrt{(\Delta x)^2 + (\Delta y)^2}}.$$

(a) Compute the slopes for the linear function in Table 12.12 in the direction of $\Delta x = 5, \Delta y = 2$.

(b) Compute the slopes for the linear function in Table 12.12 in the direction of $\Delta x = -10, \Delta y = 2$.

12.5 FUNCTIONS OF THREE VARIABLES

In applications of calculus, functions of any number of variables can arise. The density of matter in the universe is a function of three variables, since it takes three numbers to specify a point in space. Models of the US economy often use functions of ten or more variables. We need to be able to apply calculus to functions of arbitrarily many variables.

One difficulty with functions of more than two variables is that it is hard to visualize them. The graph of a function of one variable is a curve in 2-space, the graph of a function of two variables is a surface in 3-space, so the graph of a function of three variables would be a solid in 4-space. Since we can't easily visualize 4-space, we won't use the graphs of functions of three variables. On the other hand, it is possible to draw contour diagrams for functions of three variables, only now the contours are surfaces in 3-space.

Representing a Function of Three Variables Using a Family of Level Surfaces

A function of two variables, $f(x,y)$, can be represented by a family of level curves of the form $f(x,y) = c$ for various values of the constant, c.

> A **level surface**, or **level set** of a function of three variables, $f(x,y,z)$, is a surface of the form $f(x,y,z) = c$, where c is a constant. The function f can be represented by the family of level surfaces obtained by allowing c to vary.

The value of the function, f, is constant on each level surface.

Example 1 The temperature, in °C, at a point (x,y,z) is given by $T = f(x,y,z) = x^2 + y^2 + z^2$. What do the level surfaces of the function f look like and what do they mean in terms of temperature?

Solution The level surface corresponding to $T = 100$ is the set of all points where the temperature is 100°C. That is, where $f(x,y,z) = 100$, so

$$x^2 + y^2 + z^2 = 100.$$

12.5 FUNCTIONS OF THREE VARIABLES

This is the equation of a sphere of radius 10, with center at the origin. Similarly, the level surface corresponding to $T = 200$ is the sphere with radius $\sqrt{200}$. The other level surfaces are concentric spheres. The temperature is constant on each sphere. We may view the temperature distribution as a set of nested spheres, like concentric layers of an onion, each one labeled with a different temperature, starting from low temperatures in the middle and getting hotter as we go out from the center. (See Figure 12.64.) The level surfaces become more closely spaced as we move farther from the origin because the temperature increases more rapidly the farther we get from the origin.

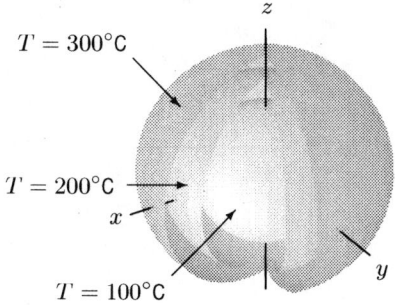

Figure 12.64: Level surfaces of $T = f(x, y, z) = x^2 + y^2 + z^2$, each one having a constant temperature

Example 2 What do the level surfaces of $f(x, y, z) = x^2 + y^2$ and $g(x, y, z) = z - y$ look like?

Solution The level surface of f corresponding to the constant c is the surface consisting of all points satisfying the equation
$$x^2 + y^2 = c.$$
Since there is no z-coordinate in the equation, z can take any value. For $c > 0$, this is a circular cylinder of radius $\sqrt{c}$ around the z-axis. The level surfaces are concentric cylinders; on the narrow ones near the z-axis, f has small values; on the wider ones, f has larger values. See Figure 12.65.

The level surface of g corresponding to the constant c is the surface
$$z - y = c.$$
This time there is no x variable, so this surface is the one we get by taking each point on the straight line $z - y = c$ in the yz-plane and letting x vary. We get a plane which cuts the yz-plane diagonally; the x-axis is parallel to this plane. See Figure 12.66.

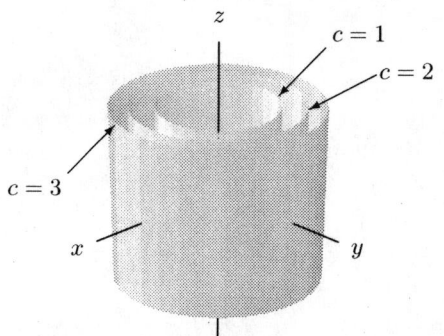
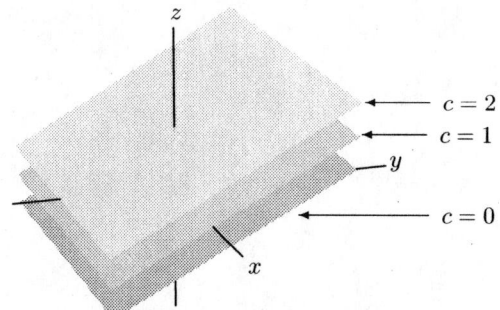

Figure 12.65: Level surfaces of $f(x, y, z) = x^2 + y^2$

Figure 12.66: Level surfaces of $g(x, y, z) = z - y$

Example 3 What do the level surfaces of $f(x, y, z) = x^2 + y^2 - z^2$ look like?

Solution In Section 12.3, we saw that the two-variable quadratic function $g(x, y) = x^2 - y^2$ has a saddle-shaped graph and three types of contours. The contour equation $x^2 - y^2 = c$ gives a hyperbola opening right-left when $c > 0$, a hyperbola opening up-down when $c < 0$, and a pair of intersecting lines when $c = 0$. Similarly, the three-variable quadratic function $f(x, y, z) = x^2 + y^2 - z^2$ has three types of level surfaces depending on the value of c in the equation $x^2 + y^2 - z^2 = c$.

Suppose that $c > 0$, say $c = 1$. Rewrite the equation as $x^2 + y^2 = z^2 + 1$ and think of what happens as we cut the surface perpendicular to the z-axis by holding z fixed. The result is a circle, $x^2 + y^2 =$ constant, of radius at least 1 (since the constant $z^2 + 1 \geq 1$). The circles get larger as z gets larger. If we take the $x = 0$ cross-section instead, we get the hyperbola $y^2 - z^2 = 1$. The result is shown in Figure 12.70, with $a = b = c = 1$.

Suppose instead $c < 0$, say $c = -1$. Then the horizontal cross-sections of $x^2 + y^2 = z^2 - 1$ are again circles except that the radii shrink to 0 at $z = \pm 1$ and between $z = -1$ and $z = 1$ there are no cross-sections at all. The result is shown in Figure 12.71 with $a = b = c = 1$.

When $c = 0$, we get the equation $x^2 + y^2 = z^2$. Again the horizontal cross-sections are circles, this time with the radius shrinking down to exactly 0 when $z = 0$. The resulting surface, shown in Figure 12.72 with $a = b = c = 1$, is the cone $z = \sqrt{x^2 + y^2}$ studied in Section 12.3, together with the lower cone $z = -\sqrt{x^2 + y^2}$.

A Catalog of Surfaces

For later reference, here is a small catalog of the surfaces we have encountered.

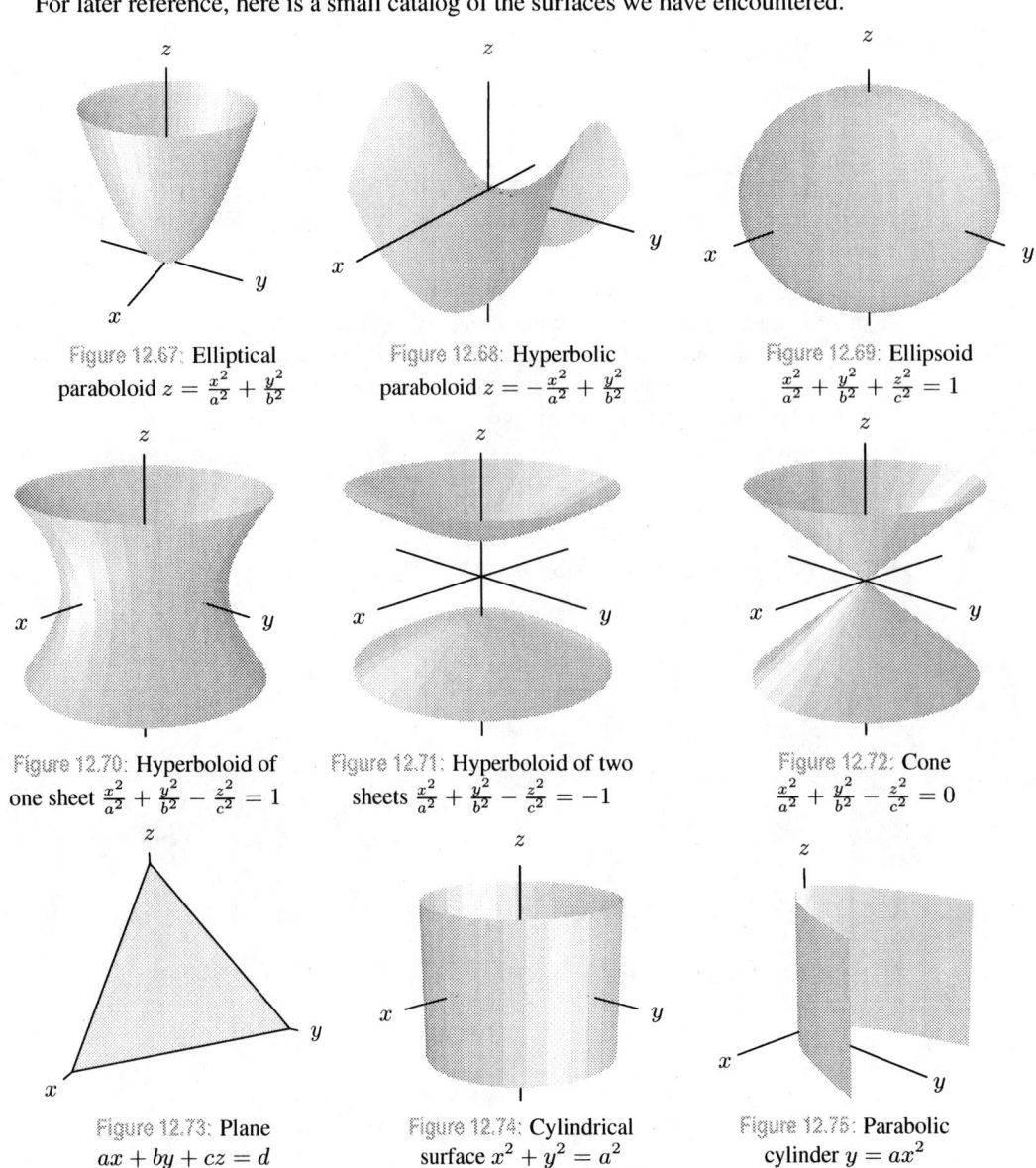

Figure 12.67: Elliptical paraboloid $z = \frac{x^2}{a^2} + \frac{y^2}{b^2}$

Figure 12.68: Hyperbolic paraboloid $z = -\frac{x^2}{a^2} + \frac{y^2}{b^2}$

Figure 12.69: Ellipsoid $\frac{x^2}{a^2} + \frac{y^2}{b^2} + \frac{z^2}{c^2} = 1$

Figure 12.70: Hyperboloid of one sheet $\frac{x^2}{a^2} + \frac{y^2}{b^2} - \frac{z^2}{c^2} = 1$

Figure 12.71: Hyperboloid of two sheets $\frac{x^2}{a^2} + \frac{y^2}{b^2} - \frac{z^2}{c^2} = -1$

Figure 12.72: Cone $\frac{x^2}{a^2} + \frac{y^2}{b^2} - \frac{z^2}{c^2} = 0$

Figure 12.73: Plane $ax + by + cz = d$

Figure 12.74: Cylindrical surface $x^2 + y^2 = a^2$

Figure 12.75: Parabolic cylinder $y = ax^2$

(These are viewed as equations in three variables x, y, and z)

How Surfaces Can Represent Functions of Two Variables and Functions of Three Variables

You may have noticed that we have used surfaces to represent functions in two different ways. First, we used a *single* surface to represent a two-variable function $f(x, y)$. Second, we used a *family* of level surfaces to represent a three-variable function $g(x, y, z)$. These level surfaces have equation $g(x, y, z) = c$.

What is the relation between these two uses of surfaces? For example, consider the function
$$f(x, y) = x^2 + y^2 + 3.$$

Define
$$g(x, y, z) = x^2 + y^2 + 3 - z$$

The points on the graph of f satisfy $z = x^2 + y^2 + 3$, so they also satisfy $x^2 + y^2 + 3 - z = 0$. Thus the graph of f is the same as the level surface
$$g(x, y, z) = x^2 + y^2 + 3 - z = 0.$$

In general, we have the following result:

> A single surface that is the graph of a two-variable function $f(x, y)$ can be thought of as one member of the family of level surfaces representing the three-variable function
> $$g(x, y, z) = f(x, y) - z.$$
> The graph of f is the level surface $g = 0$.

Conversely, a single level surface $g(x, y, z) = c$ can be regarded as the graph of a function $f(x, y)$ if it is possible to solve for z. Sometimes the level surface is pieced together from the graphs of two or more two-variable functions. For example, if $g(x, y, z) = x^2 + y^2 + z^2$, then one member of the family of level surfaces is the sphere
$$x^2 + y^2 + z^2 = 1.$$

This equation defines z implicitly as a function of x and y. Solving it gives two functions
$$z = \sqrt{1 - x^2 - y^2} \quad \text{and} \quad z = -\sqrt{1 - x^2 - y^2}.$$

The graph of the first function is the top half of the sphere and the graph of the second function is the bottom half.

Exercises and Problems for Section 12.5

Exercises

1. Match the following functions with the level surfaces in Figure 12.76.

   (a) $f(x, y, z) = y^2 + z^2$ (b) $h(x, y, z) = x^2 + z^2$.

2. Find a formula for a function $f(x, y, z)$ whose level surfaces look like those in Figure 12.77.

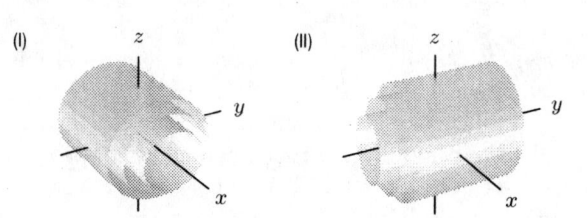

Figure 12.76

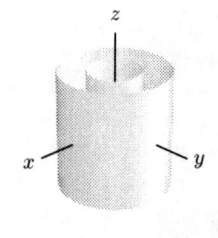

Figure 12.77

In Exercises 3–5, represent the surface whose equation is given as the graph of a two-variable function, $f(x, y)$, and as the level surface of a three-variable function, $g(x, y, z) = c$. There are many possible answers.

3. The plane $4x - y - 2z = 6$
4. The top half of the sphere $x^2 + y^2 + z^2 - 10 = 0$
5. The bottom half of the ellipsoid $x^2 + y^2 + z^2/2 = 1$

In Exercises 6–9, decide if the given level surface can be expressed as the graph of a function, $f(x, y)$.

6. $2x + 3y - 5z - 10 = 0$ **7.** $x^2 + y^2 + z^2 - 1 = 0$
8. $z - x^2 - 3y^2 = 0$ **9.** $z^2 = x^2 + 3y^2$

10. Find a formula for a function $f(x, y, z)$ whose level surface $f = 4$ is a sphere of radius 2, centered at the origin.

11. Find a formula for a function $f(x, y, z)$ whose level surfaces are spheres centered at the point (a, b, c).

12. Write the level surface $x + 2y + 3z = 5$ as the graph of a function $f(x, y)$.

13. Write the level surface $x^2 + y + \sqrt{z} = 1$ as the graph of a function $f(x, y)$.

14. Which of the graphs in the catalog of surfaces on page 636 is the graph of a function of x and y?

Use the catalog on page 636 to identify the surfaces in Exercises 15–18.

15. $-x^2 - y^2 + z^2 = 1$ **16.** $-x^2 + y^2 - z^2 = 0$
17. $x^2 + y^2 - z = 0$ **18.** $x^2 + z^2 = 1$

Problems

19. Find a function $f(x, y, z)$ whose level surface $f = 1$ is the graph of the function $g(x, y) = x + 2y$.

20. Find two functions $f(x, y)$ and $g(x, y)$ so that the graphs of both together form the ellipsoid $x^2 + y^2/4 + z^2/9 = 1$.

21. Describe, in words, the level surface $f(x, y, z) = x^2/4 + z^2 = 1$.

22. Describe, in words, the level surface $g(x, y, z) = x^2 + y^2/4 + z^2 = 1$. [Hint: Look at cross-sections with constant x, y and z values.]

23. Describe in words the level surfaces of the function $g(x, y, z) = x + y + z$.

24. Find a formula for a function $g(x, y, z)$ whose level surfaces are planes parallel to the plane $z = 2x + 3y - 5$.

25. Describe in words the level surfaces of $f(x, y, z) = \sin(x + y + z)$.

26. Describe in words the level surfaces of $g(x, y, z) = e^{-(x^2+y^2+z^2)}$.

27. The surface S is the graph of $f(x, y) = \sqrt{1 - x^2 - y^2}$.

(a) Explain why S is the upper hemisphere of radius 1, with equator in the xy-plane, centered at the origin.

(b) Find a level surface $g(x, y, z) = c$ representing S.

28. The surface S is the graph of $f(x, y) = \sqrt{1 - y^2}$.

(a) Explain why S is the upper half of a circular cylinder of radius 1, centered along the x-axis.
(b) Find a level surface $g(x, y, z) = c$ representing S.

29. A cone C, with height 1 and radius 1, has its base in the xz-plane and its vertex on the positive y-axis. Find a function $g(x, y, z)$ such that C is part of the level surface $g(x, y, z) = 0$. [Hint: The graph of $f(x, y) = \sqrt{x^2 + y^2}$ is a cone which opens up and has vertex at the origin.]

30. What do the level surfaces of $f(x, y, z) = x^2 - y^2 + z^2$ look like? [Hint: Use cross-sections with y constant instead of cross-sections with z constant.]

31. Describe the surface $x^2 + y^2 = (2 + \sin z)^2$. In general, if $f(z) \geq 0$ for all z, describe the surface $x^2 + y^2 = (f(z))^2$.

12.6 LIMITS AND CONTINUITY

The sheer vertical face of Half Dome, in Yosemite National Park in California, was caused by glacial activity during the Ice Age. (See Figure 12.78.) The height of the terrain rises abruptly by nearly 1000 feet as we scale the rock from the west, whereas it is possible to make a gradual climb to the top from the east.

If we consider the function h giving the height of the terrain above sea level in terms of longitude and latitude, then h has a *discontinuity* along the path at the base of the cliff of Half Dome. Looking at the contour map of the region in Figure 12.79, we see that in most places a small change in position results in a small change in height, except near the cliff. There, no matter how small a step we take, we get a large change in height. (You can see how crowded the contours get near the cliff; some end abruptly along the discontinuity.)

This geological feature illustrates the ideas of continuity and discontinuity. Roughly speaking, a function is said to be *continuous* at a point if its values at places near the point are close to the value at the point. If this is not the case, the function is said to be *discontinuous*.

12.6 LIMITS AND CONTINUITY

Figure 12.78: Half Dome in Yosemite National Park

Figure 12.79: A contour map of Half Dome

The property of continuity is one that, practically speaking, we usually assume of the functions we are studying. Informally, we expect (except under special circumstances) that values of a function do not change drastically when making small changes to the input variables. Whenever we model a one-variable function by an unbroken curve, we are making this assumption. Even when functions come to us as tables of data, we usually make the assumption that the missing function values between data points are close to the measured ones.

In this section we study limits and continuity a bit more formally in the context of functions of several variables. For simplicity we study these concepts for functions of two variables, but our discussion can be adapted to functions of three or more variables.

One can show that sums, products, and compositions of continuous functions are continuous, while the quotient of two continuous functions is continuous everywhere the denominator function is nonzero. Thus, each of the functions

$$\cos(x^2 y), \quad \ln(x^2 + y^2), \quad \frac{e^{x+y}}{x+y}, \quad \ln(\sin(x^2 + y^2))$$

is continuous at all points (x, y) where it is defined. As for functions of one variable, the graph of a continuous function over an unbroken domain is unbroken—that is, the surface has no holes or rips in it.

Example 1 From Figures 12.80–12.83, which of the following functions appear to be continuous at $(0, 0)$?

(a) $f(x, y) = \begin{cases} \dfrac{x^2 y}{x^2 + y^2}, & (x, y) \neq (0, 0), \\ 0, & (x, y) = (0, 0). \end{cases}$

(b) $g(x, y) = \begin{cases} \dfrac{x^2}{x^2 + y^2}, & (x, y) \neq (0, 0), \\ 0, & (x, y) = (0, 0). \end{cases}$

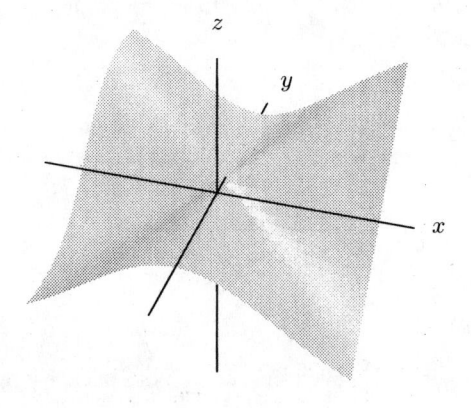

Figure 12.80: Graph of $z = x^2 y/(x^2 + y^2)$

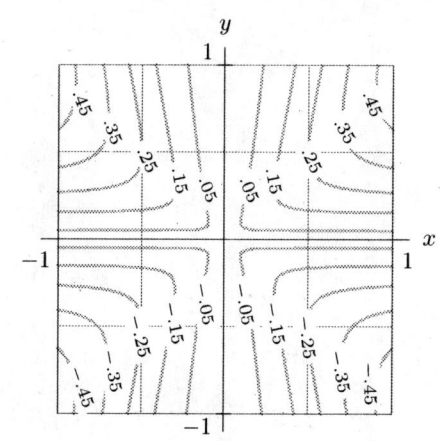

Figure 12.81: Contour diagram of $z = x^2 y/(x^2 + y^2)$

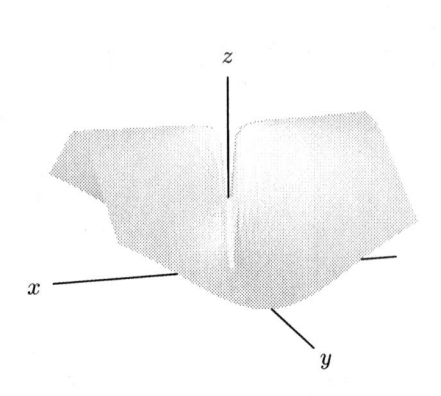

Figure 12.82: Graph of $z = x^2/(x^2 + y^2)$

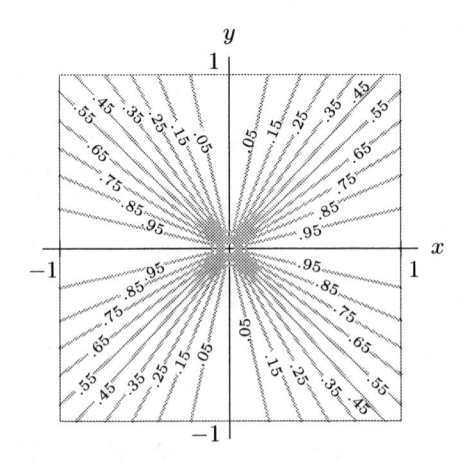

Figure 12.83: Contour diagram of $z = x^2/(x^2 + y^2)$

Solution (a) The graph and contour diagram of f in Figures 12.80 and 12.81 suggest that f is close to 0 when (x, y) is close to $(0, 0)$. That is, the figures suggest that f is continuous at the point $(0, 0)$; the graph appears to have no rips or holes there.

However, the figures cannot tell us for sure whether f is continuous. To be certain we must investigate the limit analytically, as is done in Example 2(a) on page 641.

(b) The graph of g and its contours near $(0, 0)$ in Figure 12.82 and 12.83 suggest that g behaves differently from f: The contours of g seem to "crash" at the origin and the graph rises rapidly from 0 to 1 near $(0, 0)$. Small changes in (x, y) near $(0, 0)$ can yield large changes in g, so we expect that g is not continuous at the point $(0, 0)$. Again, a more precise analysis is given in Example 2(b) on page 641.

The previous example suggests that continuity *at* a point depends on a function's behavior *near* the point. To study behavior near a point more carefully we need the idea of a limit of a function of two variables. Suppose that $f(x, y)$ is a function defined on a set in 2-space, not necessarily containing the point (a, b), but containing points (x, y) arbitrarily close to (a, b); suppose that L is a number.

The function f has a **limit** L at the point (a, b), written

$$\lim_{(x,y) \to (a,b)} f(x, y) = L,$$

if $f(x, y)$ is as close to L as we please whenever the distance from the point (x, y) to the point (a, b) is sufficiently small, but not zero.

We define continuity for functions of two variables in the same way as for functions of one variable:

A function f is **continuous at the point** (a, b) if

$$\lim_{(x,y) \to (a,b)} f(x, y) = f(a, b).$$

A function is **continuous on a region** R in the xy-plane if it is continuous at each point in R.

Thus, if f is continuous at the point (a, b), then f must be defined at (a, b) and the limit, $\lim_{(x,y) \to (a,b)} f(x, y)$, must exist and be equal to the value $f(a, b)$. If a function is defined at a point (a, b) but is not continuous there, then we say that f is *discontinuous* at (a, b).

We now apply the definition of continuity to the functions in Example 1, showing that f is continuous at $(0, 0)$ and that g is discontinuous at $(0, 0)$.

Example 2 Let f and g be the functions in Example 1. Use the definition of the limit to show that:
(a) $\lim_{(x,y) \to (0,0)} f(x, y) = 0$ (b) $\lim_{(x,y) \to (0,0)} g(x, y)$ does not exist.

Solution To investigate these limits of f and g, we consider values of these functions near, but not at, the origin, where they are given by the formulas

$$f(x, y) = \frac{x^2 y}{x^2 + y^2} \qquad g(x, y) = \frac{x^2}{x^2 + y^2}.$$

(a) The graph and contour diagram of f both suggest that $\lim_{(x,y) \to (0,0)} f(x, y) = 0$. To use the definition of the limit, we estimate $|f(x, y) - L|$ with $L = 0$:

$$|f(x, y) - L| = \left| \frac{x^2 y}{x^2 + y^2} - 0 \right| = \left| \frac{x^2}{x^2 + y^2} \right| |y| \le |y| \le \sqrt{x^2 + y^2},$$

Now $\sqrt{x^2 + y^2}$ is the distance from (x, y) to $(0, 0)$. Thus, to make $|f(x, y) - 0| < 0.001$, for example, we need only require (x, y) be within 0.001 of $(0, 0)$. More generally, for any positive number u, no matter how small, we are sure that $|f(x, y) - 0| < u$ whenever (x, y) is no farther than u from $(0, 0)$. This is what we mean by saying that the difference $|f(x, y) - 0|$ can be made as small as we wish by choosing the distance to be sufficiently small. Thus, we conclude that

$$\lim_{(x,y) \to (0,0)} f(x, y) = \lim_{(x,y) \to (0,0)} \frac{x^2 y}{x^2 + y^2} = 0.$$

Notice that since this limit equals $f(0, 0)$, the function f is continuous at $(0, 0)$.

(b) Although the formula defining the function g looks similar to that of f, we saw in Example 1 that g's behavior near the origin is quite different. If we consider points $(x, 0)$ lying along the x-axis near $(0, 0)$, then the values $g(x, 0)$ are equal to 1, while if we consider points $(0, y)$ lying along the y-axis near $(0, 0)$, then the values $g(0, y)$ are equal to 0. Thus, within any distance (no matter how small) from the origin, there are points where $g = 0$ and points where $g = 1$. Therefore the limit $\lim_{(x,y) \to (0,0)} g(x, y)$ does not exist, and thus g is not continuous at $(0, 0)$.

While the notions of limit and continuity look formally the same for one- and two-variable functions, they are somewhat more subtle in the multivariable case. The reason for this is that on the line (1-space), we can approach a point from just two directions (left or right) but in 2-space there are an infinite number of ways to approach a given point.

Exercises and Problems for Section 12.6

Exercises

Are the functions in Exercises 1-6 continuous at all points in the given regions?

1. $\dfrac{1}{x^2 + y^2}$ on the square $-1 \le x \le 1, -1 \le y \le 1$

2. $\dfrac{1}{x^2 + y^2}$ on the square $1 \le x \le 2, 1 \le y \le 2$

3. $\dfrac{y}{x^2 + 2}$ on the disk $x^2 + y^2 \le 1$

4. $\dfrac{e^{\sin x}}{\cos y}$ on the rectangle $-\frac{\pi}{2} \le x \le \frac{\pi}{2}, 0 \le y \le \frac{\pi}{4}$

5. $\tan(xy)$ on the square $-2 \le x \le 2, -2 \le y \le 2$

6. $\sqrt{2x - y}$ on the disk $x^2 + y^2 \le 4$

In Exercises 7–11, find the limits of the functions $f(x, y)$ as $(x, y) \to (0, 0)$. Assume that polynomials, exponentials, logarithmic, and trigonometric functions are continuous.

7. $f(x, y) = e^{-x-y}$
8. $f(x, y) = x^2 + y^2$
9. $f(x, y) = \dfrac{x}{x^2 + 1}$
10. $f(x, y) = \dfrac{x + y}{(\sin y) + 2}$
11. $f(x, y) = \dfrac{\sin(x^2 + y^2)}{x^2 + y^2}$ [Hint: $\lim_{t \to 0} \dfrac{\sin t}{t} = 1$.]

Problems

12. By approaching the origin along the positive x-axis and the positive y-axis, show that the following limit does not exist:
$$\lim_{(x,y) \to (0,0)} \dfrac{2x - y^2}{2x + y^2}.$$

13. By approaching the origin along the positive x-axis and the positive y-axis, show that the following limit does not exist:
$$\lim_{(x,y) \to (0,0)} \dfrac{x + y^2}{2x + y}.$$

14. Is the following function continuous at $(0, 0)$?
$$f(x, y) = \begin{cases} x^2 + y^2 & \text{if } (x, y) \neq (0, 0) \\ 2 & \text{if } (x, y) = (0, 0) \end{cases}$$

15. What value of c makes the following function continuous at $(0, 0)$?
$$f(x, y) = \begin{cases} x^2 + y^2 + 1 & \text{if } (x, y) \neq (0, 0) \\ c & \text{if } (x, y) = (0, 0) \end{cases}$$

16. (a) Use a computer to draw the graph and the contour diagram of the following function:
$$f(x, y) = \begin{cases} \dfrac{xy(x^2 - y^2)}{x^2 + y^2}, & (x, y) \neq (0, 0), \\ 0, & (x, y) = (0, 0). \end{cases}$$
(b) Do your answers to part (a) suggest that f is continuous at $(0, 0)$? Explain your answer.

17. The function f, whose graph and contour diagram are in Figures 12.84 and 12.85, is given by
$$f(x, y) = \begin{cases} \dfrac{xy}{x^2 + y^2}, & (x, y) \neq (0, 0), \\ 0, & (x, y) = (0, 0). \end{cases}$$
(a) Show that $f(0, y)$ and $f(x, 0)$ are each continuous functions of one variable.
(b) Show that rays emanating from the origin are contained in contours of f.
(c) Is f continuous at $(0, 0)$?

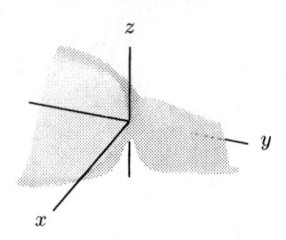

Figure 12.84: Graph of $z = xy/(x^2 + y^2)$

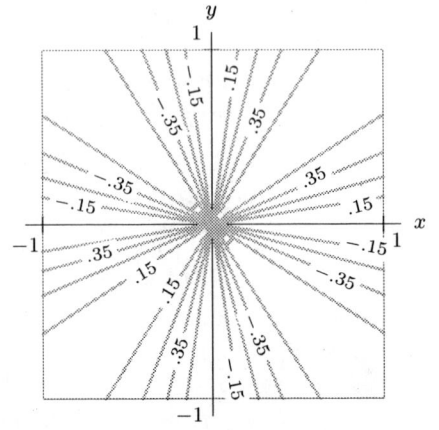

Figure 12.85: Contour diagram of $z = xy/(x^2 + y^2)$

18. Show that the function f does not have a limit at $(0, 0)$ by examining the limits of f as $(x, y) \to (0, 0)$ along the line $y = x$ and along the parabola $y = x^2$:
$$f(x, y) = \dfrac{x^2 y}{x^4 + y^2}, \quad (x, y) \neq (0, 0).$$

19. Show that the function f does not have a limit at $(0, 0)$ by examining the limits of f as $(x, y) \to (0, 0)$ along the curve $y = kx^2$ for different values of k:
$$f(x, y) = \dfrac{x^2}{x^2 + y}, \quad x^2 + y \neq 0.$$

20. Explain why the following function is not continuous along the line $y = 0$:
$$f(x, y) = \begin{cases} 1 - x, & y \geq 0, \\ -2, & y < 0, \end{cases}$$

In Problems 21–22, determine whether there is a value for c making the function continuous everywhere. If so, find it. If not, explain why not.

21. $f(x, y) = \begin{cases} c + y, & x \leq 3, \\ 5 - x, & x > 3. \end{cases}$

22. $f(x, y) = \begin{cases} c + y, & x \leq 3, \\ 5 - y, & x > 3. \end{cases}$

REVIEW EXERCISES AND PROBLEMS FOR CHAPTER TWELVE

CHAPTER SUMMARY (see also Ready Reference at the end of the book)

- **3-Space**
  Cartesian coordinates, x-, y- and z-axes, xy-, xz- and yz-planes, distance formula.
- **Functions of Two Variables**
  Represented by: tables, graphs, formulas, cross-sections (one variable fixed), contours (function value fixed); cylinders (one variable missing).
- **Linear Functions**

  Recognizing linear functions from tables, graphs, contour diagrams, formulas. Converting from one representation to another.
- **Functions of Three Variables**
  Sketching level surfaces (function value fixed) in 3-space; graph of $z = f(x,y)$ is same as level surface $g(x,y,z) = f(x,y) - z = 0$.
- **Continuity**

REVIEW EXERCISES AND PROBLEMS FOR CHAPTER TWELVE

Exercises

Decide if the statements in Exercises 1–5 must be true, might be true, or could not be true. The function $z = f(x,y)$ is defined everywhere.

1. The level curves corresponding to $z = 1$ and $z = -1$ cross at the origin.
2. The level curve $z = 1$ consists of the circle $x^2 + y^2 = 2$ and the circle $x^2 + y^2 = 3$, but no other points.
3. The level curve $z = 1$ consists of two lines which intersect at the origin.
4. If $z = e^{-(x^2+y^2)}$, there is a level curve for every value of z.
5. If $z = e^{-(x^2+y^2)}$, there is a level curve through every point (x,y).

For each of the functions in Exercises 6–9, make a contour plot in the region $-2 < x < 2$ and $-2 < y < 2$. In each case, what is the equation and the shape of the contour lines?

6. $z = 3x - 5y + 1$
7. $z = \sin y$
8. $z = 2x^2 + y^2$
9. $z = e^{-2x^2 - y^2}$

Find equations for the linear functions with the contour diagrams in Exercises 10–11.

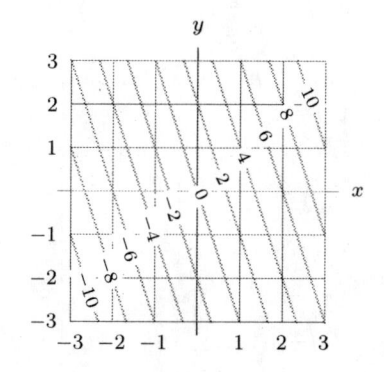

10.

11.

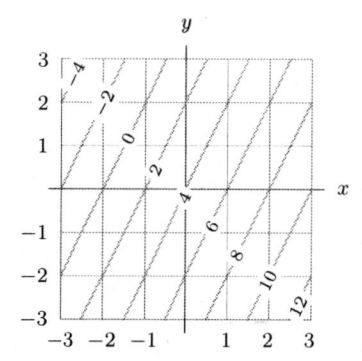

12. Match the functions with the level surfaces in Figure 12.86.

    (a) $f(x,y,z) = x^2 + y^2 + z^2$
    (b) $g(x,y,z) = x^2 + z^2$.

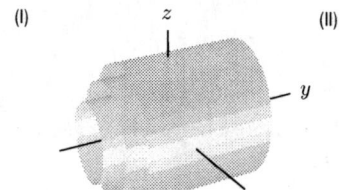

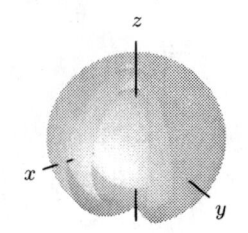

    (I)   (II)

    Figure 12.86

13. Describe the set of points whose x coordinate is 2 and whose y coordinate is 1.

14. Find the equation of the plane through the points $(0,0,2), (0,3,0), (5,0,0)$.

15. Find the center and radius of the sphere with equation $x^2 + 4x + y^2 - 6y + z^2 + 12z = 0$.

Problems

16. Sketch cross-sections of $f(r, h) = \pi r^2 h$, first keeping h fixed, then keeping r fixed.

17. Find a linear function whose graph is the plane that intersects the xy-plane along the line $y = 2x + 2$ and contains the point $(1, 2, 2)$.

18. Figure 12.87 shows the graph of $z = f(x, y)$.

   (a) Suppose y is fixed and positive. Does z increase or decrease as x increases? Graph z against x.
   (b) Suppose x is fixed and positive. Does z increase or decrease as y increases? Graph z against y.

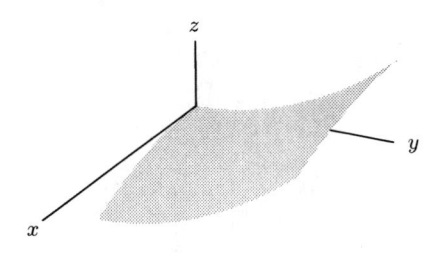

Figure 12.87

19. Match the following descriptions of a company's success with the graphs in Figure 12.88.

   (a) Our success is measured in dollars, plain and simple. More hard work won't hurt, but it also won't help.
   (b) No matter how much money or hard work we put into the company, we just could not make a go of it.
   (c) Although we are not always totally successful, it seems that the amount of money invested does not matter. As long as we put hard work into the company our success will increase.
   (d) The company's success is based on both hard work and investment.

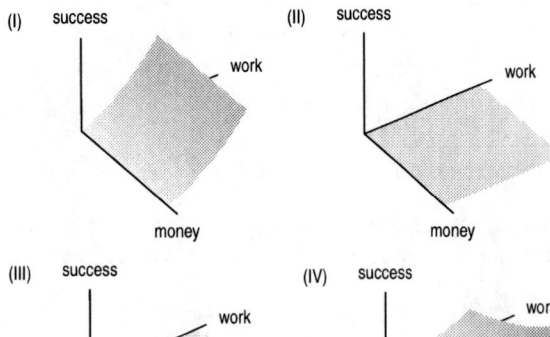

Figure 12.88

20. Figure 12.89 is the contour diagram of $f(x, y)$. Match the functions (a)–(f) with their contour diagrams (I)–(VI). All diagrams are drawn on the same region in the xy-plane.

   (a) $2f(x, y)$ (b) $f(x, y) + 2$ (c) $f(x - 1, y)$
   (d) $f(x, -y)$ (e) $f(2x, y)$ (f) $f(y, x)$

Figure 12.89

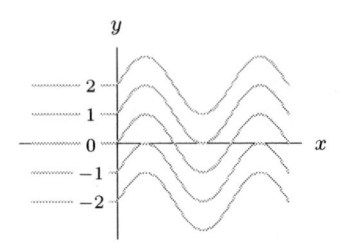

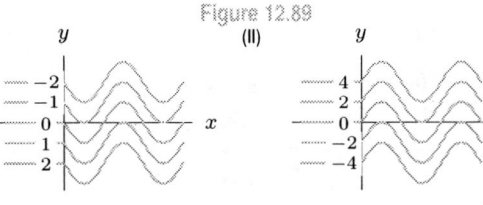

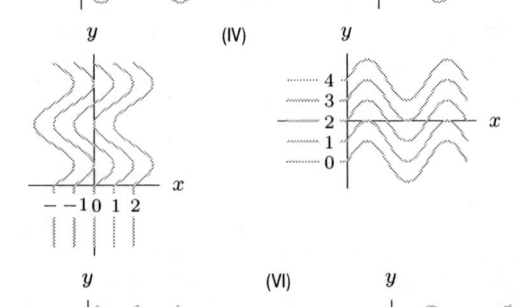

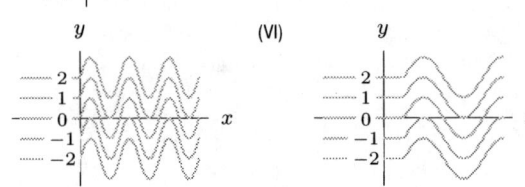

21. Figure 12.90 is the contour diagram of a function $f(x, y)$. Sketch a contour diagram of

   (a) $g(x, y) = (f(x, y))^2$ (b) $h(x, y) = \sin(f(x, y))$.

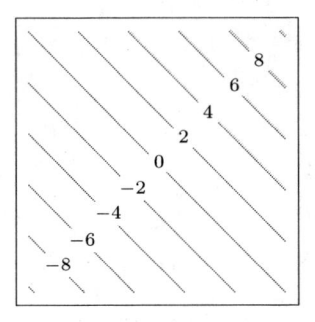

Figure 12.90

22. Values of $f(x,y) = \frac{1}{2}(x+y-2)(x+y-1) + y$ are in Table 12.13.

    (a) Find a pattern in the table. Make a conjecture and use it to complete Table 12.13 without computation. Check by using the formula for f.
    (b) Using the formula, check that the pattern holds for all $x \geq 1$ and $y \geq 1$.

    Table 12.13

	y=1	2	3	4	5	6
x=1	1	3	6	10	15	21
2	2	5	9	14	20	
3	4	8	13	19		
4	7	12	18			
5	11	17				
6	16					

23. You are in a room 30 feet long with a heater at one end. In the morning the room is 65°F. You turn on the heater, which quickly warms up to 85°F. Let $H(x,t)$ be the temperature x feet from the heater, t minutes after the heater is turned on. Figure 12.91 shows the contour diagram for H. How warm is it 10 feet from the heater 5 minutes after it was turned on? 10 minutes after it was turned on?

    Figure 12.91

24. Using the contour diagram in Figure 12.91, sketch the graphs of the one-variable functions $H(x,5)$ and $H(x,20)$. Interpret the two graphs in practical terms, and explain the difference between them.

25. Describe in words the level surfaces of the function $g(x,y,z) = \cos(x+y+z)$.

26. (a) Sketch the level curves of $z = \cos\sqrt{x^2+y^2}$.
    (b) Sketch a cross-section through the surface $z = \cos\sqrt{x^2+y^2}$ in the plane containing the x- and z-axes. Put units on your axes.
    (c) Sketch the cross-section through the surface $z = \cos\sqrt{x^2+y^2}$ in the plane containing the z-axis and the line $y = x$ in the xy-plane.

27. (a) Sketch level curves of $g(x,y) = \sqrt{x^2+y^2} - x$ for $g = 1, 2, 3$.
    (b) For what values of c can level curves of $g = c$ be drawn?

Represent the surfaces in Problems 28–31 as graphs of functions, $f(x,y)$, and as level surfaces of the form $g(x,y,z) = c$. (There are many possible answers.)

28. Paraboloid obtained by shifting $z = x^2 + y^2$ vertically 5 units

29. Plane with intercepts $x = 2, y = 3, z = 4$.

30. Upper half of unit sphere centered at the origin

31. Lower half of sphere of radius 2 centered at $(3,0,0)$.

Problems 32–35 concern a vibrating guitar string. Snapshots of the guitar string at millisecond intervals are shown in Figure 12.92.

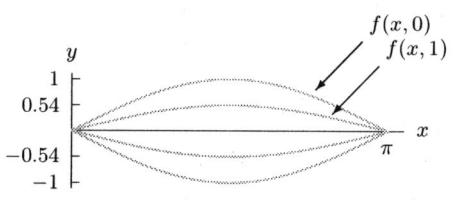

Figure 12.92: A vibrating guitar string: $f(x,t) = \cos t \sin x$ for four t values.

The guitar string is stretched tight along the x-axis from $x = 0$ to $x = \pi$. Each point on the string has an x-value, $0 \leq x \leq \pi$. As the string vibrates, each point on the string moves back and forth on either side of the x-axis. Let $y = f(x,t)$ be the displacement at time t of the point on the string located x units from the left end. A possible formula is

$$y = f(x,t) = \cos t \sin x, \quad 0 \leq x \leq \pi, \quad t \text{ in milliseconds.}$$

32. Explain what the functions $f(x,0)$ and $f(x,1)$ represent in terms of the vibrating string.

33. Explain what the functions $f(0,t)$ and $f(1,t)$ represent in terms of the vibrating string.

34. (a) Sketch graphs of y versus x for fixed t values, $t = 0$, $\pi/4$, $\pi/2$, $3\pi/4$, π.
    (b) Use your graphs to explain why this function could represent a vibrating guitar string.

35. Describe the motion of the guitar strings whose displacements are given by the following:
    (a) $y = g(x,t) = \cos 2t \sin x$
    (b) $y = h(x,t) = \cos t \sin 2x$

For Problems 36–39, use a computer or calculator to sketch the graph of a function with the given shapes. Include the axes and the equation used to generate it in your sketch.

36. A bowl which opens upward and has its vertex at 5 on the z-axis.

37. A plane which has its x, y, and z intercepts all positive.

38. A parabolic cylinder opening upward from along the line $y = x$ in the xy-plane.

39. A cone of circular cross-section opening downward and with its vertex at the origin.

CAS Challenge Problems

40. Let $A = (0, 0, 0)$ and $B = (2, 0, 0)$.
  (a) Find a point C in the xy-plane that is a distance 2 from both A and B.
  (b) Find a point D in 3-space that is a distance 2 from each of A, B, and C.
  (c) Describe the figure obtained by joining A, B, C, and D with straight lines.

41. Let $f(x, y) = 3 + x + 2y$.
  (a) Find formulas for $f(x, f(x, y))$, $f(x, f(x, f(x, y)))$ by hand.
  (b) Consider $f(x, f(x, f(x, f(x, f(x, f(x, y))))))$. Conjecture a formula for this function and check your answer with a computer algebra system.

42. A function $f(x, y, z)$ has the property that $f(1, 0, 1) = 20$, $f(1, 1, 1) = 16$, and $f(1, 1, 2) = 21$.
  (a) Estimate $f(1, 1, 3)$ and $f(1, 2, 1)$, assuming f is a linear function of each variable with the other variables held fixed.
  (b) Suppose in fact that $f(x, y, z) = ax^2 + byz + czx^3 + d2^{x-y}$, for constants a, b, c and d. Which of your estimates in part (a) do you expect to be exact?
  (c) Suppose in addition that $f(0, 0, 1) = 6$. Find an exact formula for f by solving for a, b, c, and d.
  (d) Use the formula in part (c) to evaluate $f(1, 1, 3)$ and $f(1, 2, 1)$ exactly. Do the values confirm your answer to part (b)?

CHECK YOUR UNDERSTANDING

Are the statements in Problems 1–60 true or false? Give reasons for your answer.

1. If $f(x, y)$ is a function of two variables defined for all x and y, then $f(10, y)$ is a function of one variable.

2. The volume V of a box of height h and square base of side length s is a function of h and s.

3. If $H = f(t, d)$ is the function giving the water temperature $H°C$ of a lake at time t hours after midnight and depth d meters, then t is a function of d and H.

4. A table for a function $f(x, y)$ cannot have any values of f appearing twice.

5. The function given by the formula $f(v, w) = e^v / w$ is an increasing function of v when w is a nonzero constant.

6. If $f(x)$ and $g(y)$ are both functions of a single variable, then the product $f(x) \cdot g(y)$ is a function of two variables.

7. Two isotherms representing distinct temperatures on a weather map cannot intersect.

8. On a weather map, there can be two isotherms which represent the same temperature but that do not intersect.

9. A function $f(x, y)$ can be an increasing function of x with y held fixed, and be a decreasing function of y with x held fixed.

10. A function $f(x, y)$ can have the property that $g(x) = f(x, 5)$ is increasing, whereas $h(x) = f(x, 10)$ is decreasing.

11. The plane $x + 2y - 3z = 1$ passes through the origin.

12. The point $(1, 2, 3)$ lies above the plane $z = 2$.

13. The graph of the equation $z = 2$ is a plane parallel to the xz-plane.

14. The points $(1, 0, 1)$ and $(0, -1, 1)$ are the same distance from the origin.

15. The plane $x + y + z = 3$ intersects the x-axis when $x = 3$.

16. The point $(2, -1, 3)$ lies on the graph of the sphere $(x - 2)^2 + (y + 1)^2 + (z - 3)^2 = 25$.

17. There is only one point in the yz-plane that is a distance 3 from the point $(3, 0, 0)$.

18. There is only one point in the yz-plane that is distance 5 from the point $(3, 0, 0)$.

19. The sphere $x^2 + y^2 + z^2 = 10$ intersects the plane $x = 10$.

20. If the point $(0, b, 0)$ has distance 4 from the plane $y = 0$, then b must be 4.

21. The cross-section of the function $f(x, y) = x + y^2$ with $y = 1$ is a line.

22. The function $g(x, y) = 1 - y^2$ has identical parabolas for all cross-sections with $x = c$.

23. The function $g(x, y) = 1 - y^2$ has lines for all cross-sections with $y = c$.

24. The graphs of $f(x, y) = \sin(xy)$ and $g(x, y) = \sin(xy) + 2$ never intersect.

25. The graphs of $f(x, y) = x^2 + y^2$ and $g(x, y) = 1 - x^2 - y^2$ intersect in a circle.

26. If all of the $x = c$ cross-sections of the graph of $f(x, y)$ are lines, then the graph of f is a plane.

27. The only point of intersection of the graphs of $f(x, y)$ and $-f(x, y)$ is the origin.

28. A line parallel to the z-axis can intersect the graph of $f(x, y)$ at most once.

29. A line parallel to the y-axis can intersect the graph of $f(x, y)$ at most once.

30. The point $(0, 0, 10)$ is the highest point on the graph of the function $f(x, y) = 10 - x^2 - y^2$.

31. The contours of the function $f(x, y) = y^2 + (x - 2)^2$ are either circles or a single point.
32. Two contours of $f(x, y)$ with different heights never intersect.
33. If the contours of $g(x, y)$ are concentric circles, then the graph of g is a cone.
34. If the contours for $f(x, y)$ get closer together in a certain direction, then f is increasing in that direction.
35. If all of the contours of $f(x, y)$ are parallel lines, then the graph of f is a plane.
36. The $h = 1$ contour of the function $h(x, y) = xy$ is a hyperbola.
37. If the $f = 10$ contour of the function $f(x, y)$ is identical to the $g = 10$ contour of the function $g(x, y)$, then $f(x, y) = g(x, y)$ for all (x, y).
38. A function $f(x, y)$ has a contour $f = c$ for every value of c.
39. The $f = 5$ contour of the function $f(x, y)$ is identical to the $g = 0$ contour of the function $g(x, y) = f(x, y) - 5$.
40. Contours of $f(x, y) = 3x + 2y$ are lines with slope 3.
41. If the contours of f are all parallel lines, then f is linear.
42. If f is linear, then the contours of f are parallel lines.
43. The function f satisfying $f(0, 0) = 1, f(0, 1) = 4, f(0, 3) = 5$ cannot be linear.
44. The graph of a linear function is always a plane.
45. The cross-section $x = c$ of a linear function $f(x, y)$ is always a line.
46. There is no linear function $f(x, y)$ with a graph parallel to the xy-plane.
47. There is no linear function $f(x, y)$ with a graph parallel to the xz-plane.
48. A linear function $f(x, y) = 2x + 3y - 5$, has exactly one point (a, b) satisfying $f(a, b) = 0$.
49. In a table of values of a linear function, the columns have the same slope as the rows.
50. There is exactly one linear function $f(x, y)$ whose $f = 0$ contour is $y = 2x + 1$.
51. The graph of the function $f(x, y) = x^2 + y^2$ is the same as the level surface $g(x, y, z) = x^2 + y^2 - z = 0$.
52. The graph of $f(x, y) = \sqrt{1 - x^2 - y^2}$ is the same as the level surface $g(x, y, z) = x^2 + y^2 + z^2 = 1$.
53. Any surface which is the graph of a two-variable function $f(x, y)$ can also be represented as the level surface of a three-variable function $g(x, y, z)$.
54. Any surface which is the level surface of a three-variable function $g(x, y, z)$ can also be represented as the graph of a two-variable function $f(x, y)$.
55. The level surfaces of the function $g(x, y, z) = x + 2y + z$ are parallel planes.
56. The level surfaces of $g(x, y, z) = x^2 + y + z^2$ are cylinders with axis along the y-axis.
57. A level surface of a function $g(x, y, z)$ cannot be a single point.
58. If $g(x, y, z) = ax + by + cz + d$, where a, b, c, d are nonzero constants, then the level surfaces of g are planes.
59. If the level surfaces of g are planes, then $g(x, y, z) = ax + by + cz + d$, where a, b, c, d are constants.
60. If the level surfaces $g(x, y, z) = k_1$ and $g(x, y, z) = k_2$ are the same surface, then $k_1 = k_2$.

PROJECTS FOR CHAPTER TWELVE

1. A Heater in a Room

Figure 12.93 shows the contours of the temperature along one wall of a heated room through one winter day, with time indicated as on a 24-hour clock. The room has a heater located at the left-most corner of the wall and one window in the wall. The heater is controlled by a thermostat about 2 feet from the window.

(a) Where is the window?
(b) When is the window open?
(c) When is the heat on?
(d) Draw graphs of the temperature along the wall of the room at 6 am, at 11 am, at 3 pm (15 hours) and at 5 pm (17 hours).
(e) Draw a graph of the temperature as a function of time at the heater, at the window and midway between them.
(f) The temperature at the window at 5 pm (17 hours) is less than at 11 am. Why do you think this might be?
(g) To what temperature do you think the thermostat is set? How do you know?
(h) Where is the thermostat?

648   Chapter Twelve   FUNCTIONS OF SEVERAL VARIABLES

Figure 12.93:

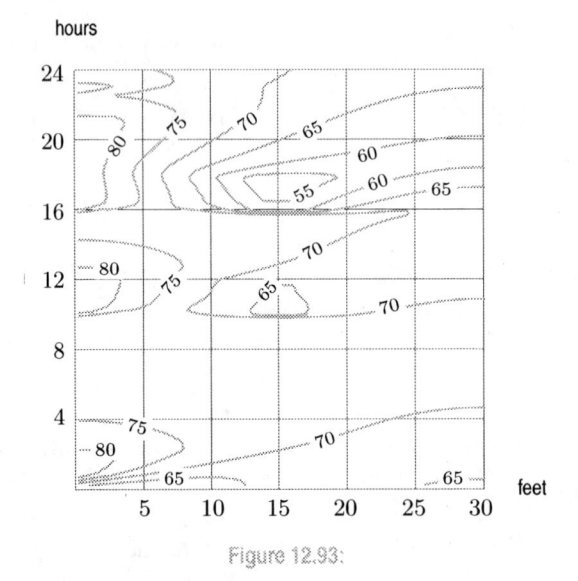

2. Light in a Wave-guide

Figure 12.94 shows the contours of light intensity as a function of location and time in a microscopic wave-guide.

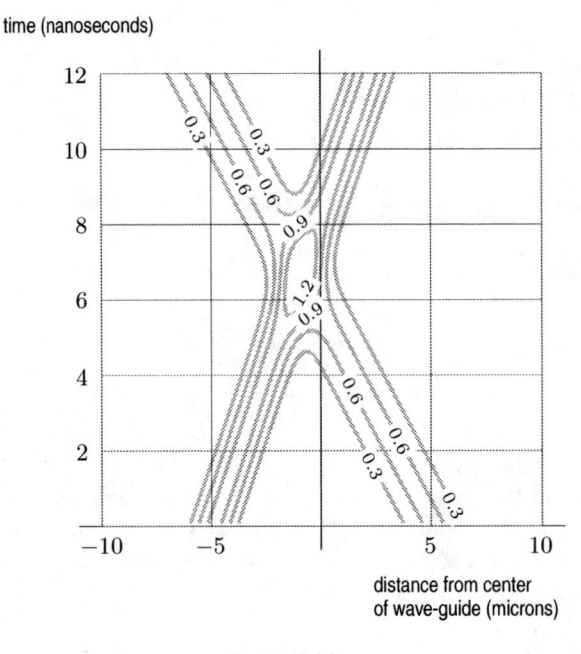

Figure 12.94

(a) Draw graphs showing intensity as a function of location at times 0, 2, 4, 6, 8, and 10 nanoseconds.
(b) If you could create an animation showing how the graph of intensity as a function of location varies as time passes, what would it look like?
(c) Draw a graph of intensity as a function of time at locations -5, 0, and 5 microns from center of wave-guide.
(d) Describe what the light beams are doing in this wave-guide.

Chapter Thirteen

A FUNDAMENTAL TOOL: VECTORS

In one-variable calculus we represented quantities such as velocity by numbers. However, to specify the velocity of a moving object in space, we need to say how fast it is moving and in what direction it is moving. In this chapter *vectors* are used to represent quantities that have direction as well as magnitude.

13.1 DISPLACEMENT VECTORS

Suppose you are a pilot planning a flight from Dallas to Pittsburgh. There are two things you must know: the distance to be traveled (so you have enough fuel to make it) and in what direction to go (so you don't miss Pittsburgh). Both these quantities together specify the displacement or *displacement vector* between the two cities.

> The **displacement vector** from one point to another is an arrow with its tail at the first point and its tip at the second. The **magnitude** (or length) of the displacement vector is the distance between the points, and is represented by the length of the arrow. The **direction** of the displacement vector is the direction of the arrow.

Figure 13.1 shows a map with the displacement vectors from Dallas to Pittsburgh, from Albuquerque to Oshkosh, and from Los Angeles to Buffalo, SD. These displacement vectors have the same length and the same direction. We say that the displacement vectors between the corresponding cities are the same, even though they do not coincide. In other words

> Displacement vectors which point in the same direction and have the same magnitude are considered to be the same, even if they do not coincide.

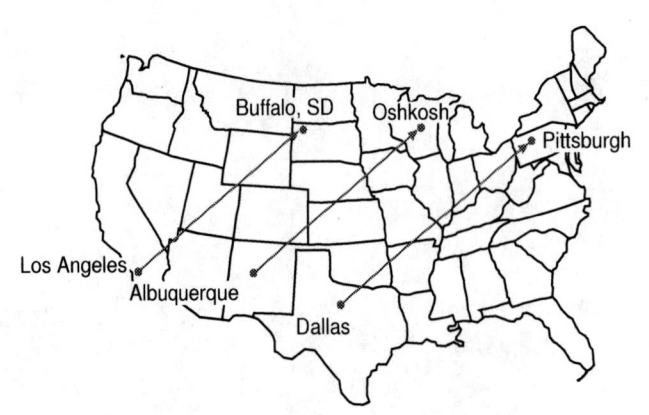

Figure 13.1: Displacement vectors between cities

Notation and Terminology

The displacement vector is our first example of a vector. Vectors have both magnitude and direction; in comparison, a quantity specified only by a number, but no direction, is called a *scalar*.[1] For instance, the time taken by the flight from Dallas to Pittsburgh is a scalar quantity. Displacement is a vector since it requires both distance and direction to specify it.

In this book, vectors are written with an arrow over them, $\vec{v}$, to distinguish them from scalars. Other books use a bold **v** to denote a vector. We use the notation $\overrightarrow{PQ}$ to denote the displacement vector from a point P to a point Q. The magnitude, or length, of a vector $\vec{v}$ is written $\|\vec{v}\|$.

Addition and Subtraction of Displacement Vectors

Suppose NASA commands a robot on Mars to move 75 meters in one direction and then 50 meters in another direction. (See Figure 13.2.) Where does the robot end up? Suppose the displacements are represented by the vectors $\vec{v}$ and $\vec{w}$, respectively. Then the sum $\vec{v} + \vec{w}$ gives the final position.

---

[1] So named by W. R. Hamilton because they are merely numbers on the *scale* from $-\infty$ to ∞.

13.1 DISPLACEMENT VECTORS

> The **sum**, $\vec{v} + \vec{w}$, of two vectors $\vec{v}$ and $\vec{w}$ is the combined displacement resulting from first applying $\vec{v}$ and then $\vec{w}$. (See Figure 13.3.) The sum $\vec{w} + \vec{v}$ gives the same displacement.

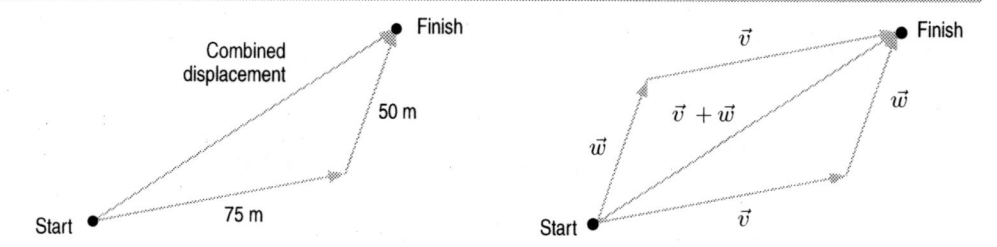

Figure 13.2: Sum of displacements of robots on Mars

Figure 13.3: The sum $\vec{v} + \vec{w} = \vec{w} + \vec{v}$

Suppose two different robots start from the same location. One moves along a displacement vector $\vec{v}$ and the second along a displacement vector $\vec{w}$. What is the displacement vector, $\vec{x}$, from the first robot to the second? (See Figure 13.4.) Since $\vec{v} + \vec{x} = \vec{w}$, we define $\vec{x}$ to be the difference $\vec{x} = \vec{w} - \vec{v}$. In other words, $\vec{w} - \vec{v}$ gets you from $\vec{v}$ to $\vec{w}$.

> The **difference**, $\vec{w} - \vec{v}$, is the displacement vector which, when added to $\vec{v}$, gives $\vec{w}$. That is, $\vec{w} = \vec{v} + (\vec{w} - \vec{v})$. (See Figure 13.4.)

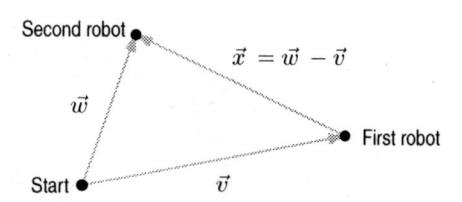

Figure 13.4: The difference $\vec{w} - \vec{v}$

If the robot ends up where it started, then its total displacement vector is the *zero vector*, $\vec{0}$. The zero vector has no direction.

> The **zero vector**, $\vec{0}$, is a displacement vector with zero length.

Scalar Multiplication of Displacement Vectors

If $\vec{v}$ represents a displacement vector, the vector $2\vec{v}$ represents a displacement of twice the magnitude in the same direction as $\vec{v}$. Similarly, $-2\vec{v}$ represents a displacement of twice the magnitude in the opposite direction. (See Figure 13.5.)

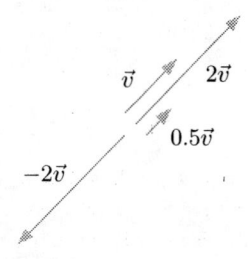

Figure 13.5: Scalar multiples of the vector $\vec{v}$

> If λ is a scalar and $\vec{v}$ is a displacement vector, the **scalar multiple of $\vec{v}$ by λ**, written $\lambda\vec{v}$, is the displacement vector with the following properties:
> - The displacement vector $\lambda\vec{v}$ is parallel to $\vec{v}$, pointing in the same direction if $\lambda > 0$, and in the opposite direction if $\lambda < 0$.
> - The magnitude of $\lambda\vec{v}$ is $|\lambda|$ times the magnitude of $\vec{v}$, that is, $\|\lambda\vec{v}\| = |\lambda|\,\|\vec{v}\|$.

Note that $|\lambda|$ represents the absolute value of the scalar λ while $\|\lambda\vec{v}\|$ represents the magnitude of the vector $\lambda\vec{v}$.

Example 1 Explain why $\vec{w} - \vec{v} = \vec{w} + (-1)\vec{v}$.

Solution The vector $(-1)\vec{v}$ has the same magnitude as $\vec{v}$, but points in the opposite direction. Figure 13.6 shows that the combined displacement $\vec{w} + (-1)\vec{v}$ is the same as the displacement $\vec{w} - \vec{v}$.

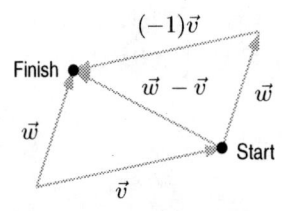

Figure 13.6: Explanation for why $\vec{w} - \vec{v} = \vec{w} + (-1)\vec{v}$

Parallel Vectors

Two vectors $\vec{v}$ and $\vec{w}$ are *parallel* if one is a scalar multiple of the other, that is, if $\vec{w} = \lambda\vec{v}$, for some scalar λ.

Components of Displacement Vectors: The Vectors $\vec{i}$, $\vec{j}$, and $\vec{k}$

Suppose that you live in a city with equally spaced streets running east-west and north-south and that you want to tell someone how to get from one place to another. You'd be likely to tell them how many blocks east-west and how many blocks north-south to go. For example, to get from P to Q in Figure 13.7, we go 4 blocks east and 1 block south. If $\vec{i}$ and $\vec{j}$ are as shown in Figure 13.7, then the displacement vector from P to Q is $4\vec{i} - \vec{j}$.

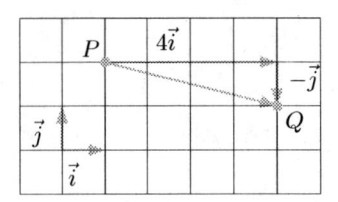

Figure 13.7: The displacement vector from P to Q is $4\vec{i} - \vec{j}$

13.1 DISPLACEMENT VECTORS

We extend the same idea to 3-dimensions. First we choose a Cartesian system of coordinate axes. The three vectors of length 1 shown in Figure 13.8 are the vector $\vec{i}$, which points along the positive x-axis, the vector $\vec{j}$, along the positive y-axis, and the vector $\vec{k}$, along the positive z-axis.

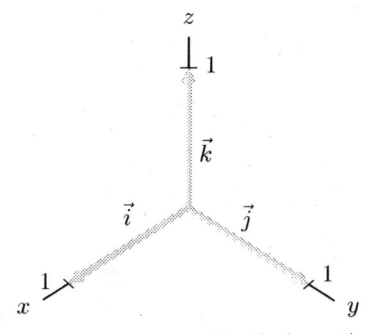

Figure 13.8: The vectors $\vec{i}, \vec{j}$ and $\vec{k}$ in 3-space

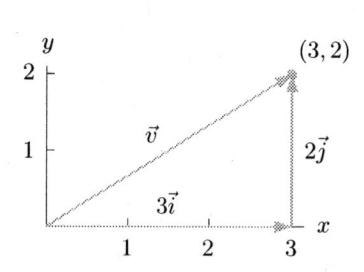

Figure 13.9: We resolve $\vec{v}$ into components by writing $\vec{v} = 3\vec{i} + 2\vec{j}$

Writing Displacement Vectors Using $\vec{i}, \vec{j}, \vec{k}$

Any displacement in 3-space or the plane can be expressed as a combination of displacements in the coordinate directions. For example, Figure 13.9 shows that the displacement vector $\vec{v}$ from the origin to the point $(3, 2)$ can be written as a sum of displacement vectors along the x and y-axes:

$$\vec{v} = 3\vec{i} + 2\vec{j}.$$

This is called *resolving $\vec{v}$ into components*. In general:

> We **resolve** $\vec{v}$ into components by writing $\vec{v}$ in the form
> $$\vec{v} = v_1\vec{i} + v_2\vec{j} + v_3\vec{k}.$$
> We call $v_1\vec{i}, v_2\vec{j}$, and $v_3\vec{k}$ the **components** of $\vec{v}$.

An Alternative Notation for Vectors

Many people write a vector in 3-dimensions as a string of three numbers, that is, as

$$\vec{v} = (v_1, v_2, v_3) \quad \text{instead of} \quad \vec{v} = v_1\vec{i} + v_2\vec{j} + v_3\vec{k}.$$

Since the first notation can be confused with a point and the second cannot, we usually use the second form.

Example 2 Resolve the displacement vector, $\vec{v}$, from the point $P_1 = (2, 4, 10)$ to the point $P_2 = (3, 7, 6)$ into components.

Solution To get from P_1 to P_2, we move 1 unit in the positive x-direction, 3 units in the positive y-direction, and 4 units in the negative z-direction. Hence $\vec{v} = \vec{i} + 3\vec{j} - 4\vec{k}$.

Example 3 Decide whether the vector $\vec{v} = 2\vec{i} + 3\vec{j} + 5\vec{k}$ is parallel to each of the following vectors:
$$\vec{w} = 4\vec{i} + 6\vec{j} + 10\vec{k}, \quad \vec{a} = -\vec{i} - 1.5\vec{j} - 2.5\vec{k}, \quad \vec{b} = 4\vec{i} + 6\vec{j} + 9\vec{k}.$$

Solution Since $\vec{w} = 2\vec{v}$ and $\vec{a} = -0.5\vec{v}$, the vectors $\vec{v}, \vec{w}$, and $\vec{a}$ are parallel. However, $\vec{b}$ is not a multiple of $\vec{v}$ (since, for example, $4/2 \neq 9/5$), so $\vec{v}$ and $\vec{b}$ are not parallel.

Chapter Thirteen A FUNDAMENTAL TOOL: VECTORS

In general, Figure 13.10 shows us how to express the displacement vector between two points in components:

Components of Displacement Vectors

The displacement vector from the point $P_1 = (x_1, y_1, z_1)$ to the point $P_2 = (x_2, y_2, z_2)$ is given in components by

$$\overrightarrow{P_1P_2} = (x_2 - x_1)\vec{i} + (y_2 - y_1)\vec{j} + (z_2 - z_1)\vec{k}.$$

Position Vectors: Displacement of a Point from the Origin

A displacement vector whose tail is at the origin is called a *position vector*. Thus, any point (x_0, y_0, z_0) in space has associated with it the position vector $\vec{r}_0 = x_0\vec{i} + y_0\vec{j} + z_0\vec{k}$. (See Figure 13.11.) In general, a position vector gives the displacement of a point from the origin.

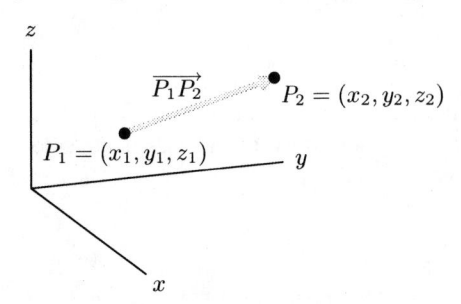

Figure 13.10: The displacement vector $\overrightarrow{P_1P_2} = (x_2 - x_1)\vec{i} + (y_2 - y_1)\vec{j} + (z_2 - z_1)\vec{k}$

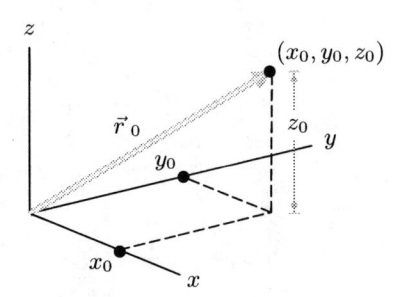

Figure 13.11: The position vector $\vec{r}_0 = x_0\vec{i} + y_0\vec{j} + z_0\vec{k}$

The Components of the Zero Vector

The zero displacement vector has magnitude equal to zero and is written $\vec{0}$. So $\vec{0} = 0\vec{i} + 0\vec{j} + 0\vec{k}$.

The Magnitude of a Vector in Components

For a vector, $\vec{v} = v_1\vec{i} + v_2\vec{j}$, the Pythagorean theorem is used to find its magnitude, $\|\vec{v}\|$. (See Figure 13.12.)

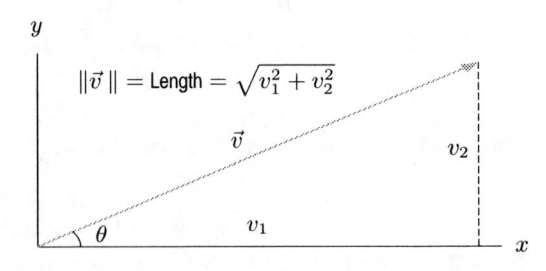

Figure 13.12: Magnitude, $\|\vec{v}\|$, of a 2-dimensional vector, $\vec{v}$

In 3-dimensions, for a vector $\vec{v} = v_1\vec{i} + v_2\vec{j} + v_3\vec{k}$, we have

---

Magnitude of $\vec{v}$ $= \|\vec{v}\| =$ Length of the arrow $= \sqrt{v_1^2 + v_2^2 + v_3^2}.$

For instance, if $\vec{v} = 3\vec{i} - 4\vec{j} + 5\vec{k}$, then $\|\vec{v}\| = \sqrt{3^2 + (-4)^2 + 5^2} = \sqrt{50}$.

Addition and Scalar Multiplication of Vectors in Components

Suppose the vectors $\vec{v}$ and $\vec{w}$ are given in components:

$$\vec{v} = v_1\vec{i} + v_2\vec{j} + v_3\vec{k} \quad \text{and} \quad \vec{w} = w_1\vec{i} + w_2\vec{j} + w_3\vec{k}.$$

Then

$$\vec{v} + \vec{w} = (v_1 + w_1)\vec{i} + (v_2 + w_2)\vec{j} + (v_3 + w_3)\vec{k},$$

and

$$\lambda\vec{v} = \lambda v_1\vec{i} + \lambda v_2\vec{j} + \lambda v_3\vec{k}.$$

Figures 13.13 and 13.14 illustrate these properties in two dimensions. Finally, $\vec{v} - \vec{w} = \vec{v} + (-1)\vec{w}$, so we can write $\vec{v} - \vec{w} = (v_1 - w_1)\vec{i} + (v_2 - w_2)\vec{j} + (v_3 - w_3)\vec{k}$.

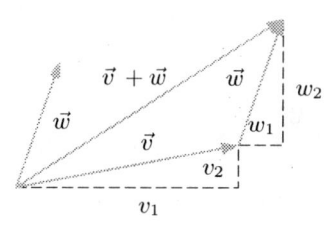

Figure 13.13: Sum $\vec{v} + \vec{w}$ in components

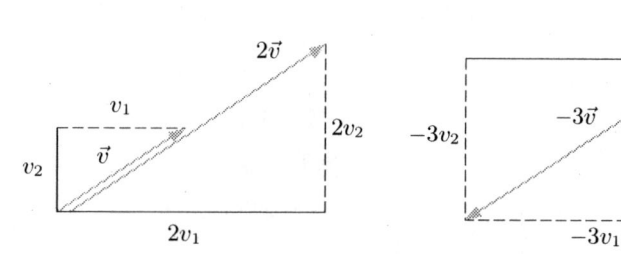

Figure 13.14: Scalar multiples of vectors showing $\vec{v}$, $2\vec{v}$, and $-3\vec{v}$

How to Resolve a Vector into Components

You may wonder how we find the components of a 2-dimensional vector, given its length and direction. Suppose the vector $\vec{v}$ has length v and makes an angle of θ with the x-axis, measured counterclockwise, as in Figure 13.15. If $\vec{v} = v_1\vec{i} + v_2\vec{j}$, Figure 13.15 shows that

$$v_1 = v\cos\theta \quad \text{and} \quad v_2 = v\sin\theta.$$

Thus, we resolve $\vec{v}$ into components by writing

$$\vec{v} = (v\cos\theta)\vec{i} + (v\sin\theta)\vec{j}.$$

Vectors in 3-space are resolved using direction cosines; see Problem 37 on page 681.

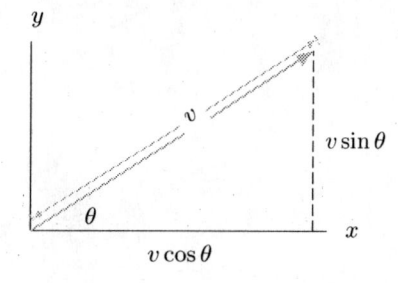

Figure 13.15: Resolving a vector: $\vec{v} = (v\cos\theta)\vec{i} + (v\sin\theta)\vec{j}$

Example 4 Resolve $\vec{v}$ into components if $\|\vec{v}\| = 2$ and $\theta = \pi/6$.

Solution We have $\vec{v} = 2\cos(\pi/6)\vec{i} + 2\sin(\pi/6)\vec{j} = 2(\sqrt{3}/2)\vec{i} + 2(1/2)\vec{j} = \sqrt{3}\vec{i} + \vec{j}$.

Unit Vectors

A *unit vector* is a vector whose magnitude is 1. The vectors $\vec{i}$, $\vec{j}$, and $\vec{k}$ are unit vectors in the directions of the coordinate axes. It is often helpful to find a unit vector in the same direction as a given vector $\vec{v}$. Suppose that $\|\vec{v}\| = 10$; a unit vector in the same direction as $\vec{v}$ is $\vec{v}/10$. In general, a unit vector in the direction of any nonzero vector $\vec{v}$ is

$$\vec{u} = \frac{\vec{v}}{\|\vec{v}\|}.$$

Example 5 Find a unit vector, $\vec{u}$, in the direction of the vector $\vec{v} = \vec{i} + 3\vec{j}$.

Solution If $\vec{v} = \vec{i} + 3\vec{j}$, then $\|\vec{v}\| = \sqrt{1^2 + 3^2} = \sqrt{10}$. Thus, a unit vector in the same direction is given by

$$\vec{u} = \frac{\vec{v}}{\sqrt{10}} = \frac{1}{\sqrt{10}}(\vec{i} + 3\vec{j}) = \frac{1}{\sqrt{10}}\vec{i} + \frac{3}{\sqrt{10}}\vec{j} \approx 0.32\vec{i} + 0.95\vec{j}.$$

Example 6 Find a unit vector at the point (x, y, z) that points radially outward away from the origin.

Solution The vector from the origin to (x, y, z) is the position vector

$$\vec{r} = x\vec{i} + y\vec{j} + z\vec{k}.$$

Thus, if we put its tail at (x, y, z) it will point away from the origin. Its magnitude is

$$\|\vec{r}\| = \sqrt{x^2 + y^2 + z^2},$$

so a unit vector pointing in the same direction is

$$\frac{\vec{r}}{\|\vec{r}\|} = \frac{x\vec{i} + y\vec{j} + z\vec{k}}{\sqrt{x^2 + y^2 + z^2}} = \frac{x}{\sqrt{x^2 + y^2 + z^2}}\vec{i} + \frac{y}{\sqrt{x^2 + y^2 + z^2}}\vec{j} + \frac{z}{\sqrt{x^2 + y^2 + z^2}}\vec{k}.$$

Exercises and Problems for Section 13.1

Exercises

For Exercises 1–4, perform the indicated computation.

1. $(4\vec{i} + 2\vec{j}) - (3\vec{i} - \vec{j})$
2. $(\vec{i} + 2\vec{j}) + (-3)(2\vec{i} + \vec{j})$
3. $-4(\vec{i} - 2\vec{j}) - 0.5(\vec{i} - \vec{k})$
4. $2(0.45\vec{i} - 0.9\vec{j} - 0.01\vec{k}) - 0.5(1.2\vec{i} - 0.1\vec{k})$

Find the length of the vectors in Exercises 5–8.

5. $\vec{v} = \vec{i} - \vec{j} + 3\vec{k}$
6. $\vec{v} = \vec{i} - \vec{j} + 2\vec{k}$
7. $\vec{v} = 1.2\vec{i} - 3.6\vec{j} + 4.1\vec{k}$
8. $\vec{v} = 7.2\vec{i} - 1.5\vec{j} + 2.1\vec{k}$

For Exercises 9–14, perform the indicated operations on the following vectors:

$\vec{a} = 2\vec{j} + \vec{k}$, $\vec{b} = -3\vec{i} + 5\vec{j} + 4\vec{k}$, $\vec{c} = \vec{i} + 6\vec{j}$,
$\vec{x} = -2\vec{i} + 9\vec{j}$, $\vec{y} = 4\vec{i} - 7\vec{j}$, $\vec{z} = \vec{i} - 3\vec{j} - \vec{k}$.

9. $5\vec{b}$
10. $\vec{a} + \vec{z}$
11. $2\vec{c} + \vec{x}$
12. $\|\vec{z}\|$
13. $\|\vec{y}\|$
14. $2\vec{a} + 7\vec{b} - 5\vec{z}$

15. (a) Draw the position vector for $\vec{v} = 5\vec{i} - 7\vec{j}$.
    (b) What is $\|\vec{v}\|$?
    (c) Find the angle between $\vec{v}$ and the positive x-axis.

16. Resolve vector $\vec{v}$ into components if $\|\vec{v}\| = 8$ and the direction of $\vec{v}$ is shown in Figure 13.16.

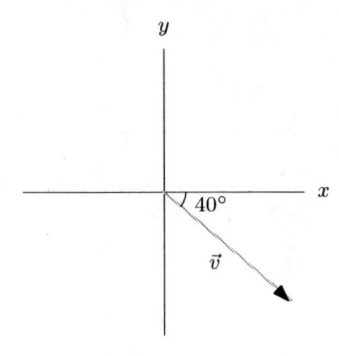

Figure 13.16

13.1 DISPLACEMENT VECTORS

17. Resolve the vectors in Figure 13.17 into components.

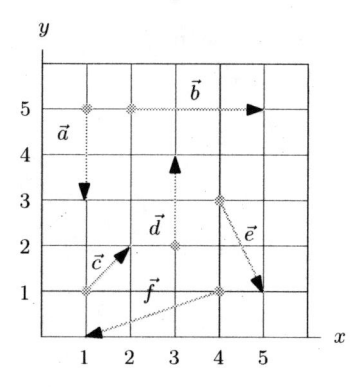

Figure 13.17

Resolve the vectors in Exercises 18–20 into components.

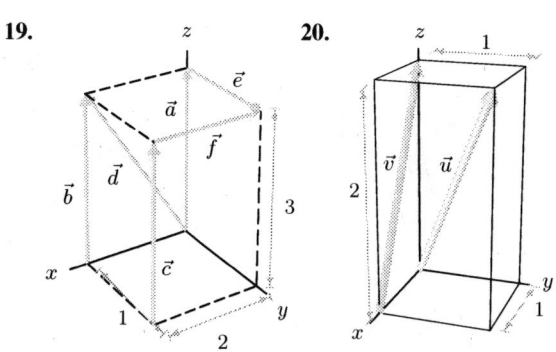

18. A vector starting at the point $Q = (4, 6)$ and ending at the point $P = (1, 2)$.

19.

20.

Problems

21. Which of the following vectors are parallel?

$$\vec{u} = 2\vec{i} + 4\vec{j} - 2\vec{k}, \quad \vec{p} = \vec{i} + \vec{j} + \vec{k},$$
$$\vec{v} = \vec{i} - \vec{j} + 3\vec{k}, \quad \vec{q} = 4\vec{i} - 4\vec{j} + 12\vec{k},$$
$$\vec{w} = -\vec{i} - 2\vec{j} + \vec{k}, \quad \vec{r} = \vec{i} - \vec{j} + \vec{k}.$$

22. Find a vector with length 2 that points in the same direction as $\vec{i} - \vec{j} + 2\vec{k}$.

23. (a) Find a unit vector from the point $P = (1, 2)$ and toward the point $Q = (4, 6)$.
    (b) Find a vector of length 10 pointing in the same direction.

24. Find all vectors $\vec{v}$ in 2 dimensions having $\|\vec{v}\| = 5$ and the $\vec{i}$-component of $\vec{v}$ is $3\vec{i}$.

25. In Figure 13.18, a ship is at point B, a submarine is 6 units below point C, a helicopter is 10 units above point A.
    (a) Find each of the following displacement vectors: from the submarine to the ship, from the helicopter to the ship, and from the submarine to the helicopter.
    (b) Which two objects are closest together?

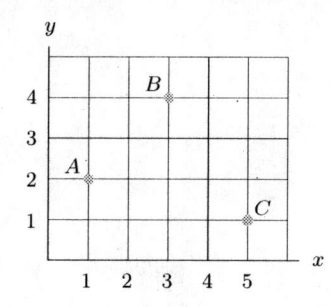

Figure 13.18

A cat on the ground at the point $(1, 4, 0)$ watches a squirrel at the top of a tree. The tree is one unit high with its base at $(2, 4, 0)$. Find the displacement vectors in Problems 26–29.

26. From the origin to the cat.
27. From the bottom of the tree to the squirrel.
28. From the bottom of the tree to the cat.
29. From the cat to the squirrel.
30. Figure 13.19 shows a rectangular box containing several vectors. Are the following statements true or false? Explain.

    (a) $\vec{c} = \vec{f}$ (b) $\vec{a} = \vec{d}$ (c) $\vec{a} = -\vec{b}$
    (d) $\vec{g} = \vec{f} + \vec{a}$ (e) $\vec{e} = \vec{a} - \vec{b}$ (f) $\vec{d} = \vec{g} - \vec{c}$

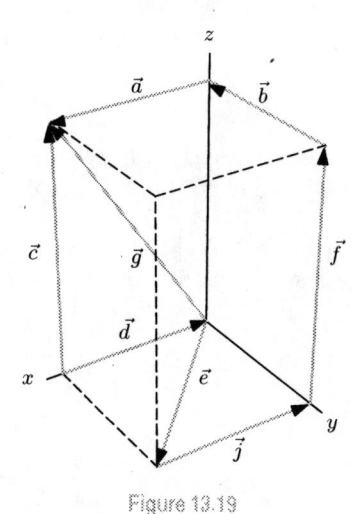

Figure 13.19

31. A truck is traveling due north at 30 km/hr approaching a crossroad. On a perpendicular road a police car is traveling west toward the intersection at 40 km/hr. Both vehicles will reach the crossroad in exactly one hour. Find the vector currently representing the displacement of the truck with respect to the police car.

32. Show that the medians of a triangle intersect at a point $\frac{1}{3}$ of the way along each median from the side it bisects.

33. Show that the lines joining the centroid (the intersection point of the medians) of a face of the tetrahedron and the opposite vertex meet at a point $\frac{1}{4}$ of the way from each centroid to its opposite vertex.

34. Figure 13.20 shows a molecule with four atoms at O, A, B and C. Verify that every atom in the molecule is 2 units away from every other atom.

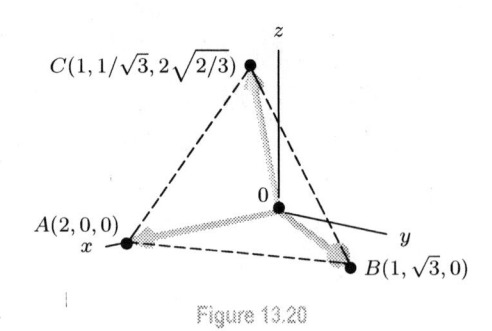

Figure 13.20

13.2 VECTORS IN GENERAL

Besides displacement, there are many quantities that have both magnitude and direction and are added and multiplied by scalars in the same way as displacements. Any such quantity is called a *vector* and is represented by an arrow in the same manner we represent displacements. The length of the arrow is the *magnitude* of the vector, and the direction of the arrow is the direction of the vector.

Velocity Versus Speed

The speed of a moving body tells us how fast it is moving, say 80 km/hr. The speed is just a number; it is therefore a scalar. The velocity, on the other hand, tells us both how fast the body is moving and the direction of motion; it is a vector. For instance, if a car is heading northeast at 80 km/hr, then its velocity is a vector of length 80 pointing northeast.

> The **velocity vector** of a moving object is a vector whose magnitude is the speed of the object and whose direction is the direction of its motion.

Example 1 A car is traveling north at a speed of 100 km/hr, while a plane above is flying horizontally south-west at a speed of 500 km/hr. Draw the velocity vectors of the car and the plane.

Solution Figure 13.21 shows the velocity vectors. The plane's velocity vector is five times as long as the car's, because its speed is five times as great.

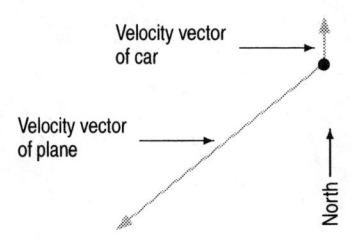

Figure 13.21: Velocity vector of the car is 100 km/hr north and of the plane is 500 km/hr south-west

The next example illustrates that the velocity vectors for two motions add to give the velocity vector for the combined motion, just as displacements do.

13.2 VECTORS IN GENERAL

Example 2 A riverboat is moving with velocity $\vec{v}$ and a speed of 8 km/hr relative to the water. In addition, the river has a current $\vec{c}$ and a speed of 1 km/hr. (See Figure 13.22.) What is the physical significance of the vector $\vec{v} + \vec{c}$?

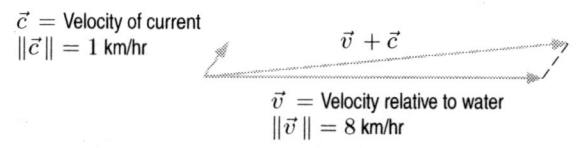

$\vec{c}$ = Velocity of current
$\|\vec{c}\|$ = 1 km/hr

$\vec{v} + \vec{c}$

$\vec{v}$ = Velocity relative to water
$\|\vec{v}\|$ = 8 km/hr

Figure 13.22: Boat's velocity relative to the river bed is the sum, $\vec{v} + \vec{c}$

Solution The vector $\vec{v}$ shows how the boat is moving relative to the water, while $\vec{c}$ shows how the water is moving relative to the riverbed. During an hour, imagine that the boat first moves 8 km relative to the water, which remains still; this displacement is represented by $\vec{v}$. Then imagine the water moving 1 km while the boat remains stationary relative to the water; this displacement is represented by $\vec{c}$. The combined displacement is represented by $\vec{v} + \vec{c}$. Thus, the vector $\vec{v} + \vec{c}$ is the velocity of the boat relative to the riverbed.

Note that the effective speed of the boat is not necessarily 9 km/hr unless the boat is moving in the direction of the current. Although we add the velocity vectors, we do not necessarily add their lengths.

Scalar multiplication also makes sense for velocity vectors. For example, if $\vec{v}$ is a velocity vector, then $-2\vec{v}$ represents a velocity of twice the magnitude in the opposite direction.

Example 3 A ball is moving with velocity $\vec{v}$ when it hits a wall at a right angle and bounces straight back, with its speed reduced by 20%. Express its new velocity in terms of the old one.

Solution The new velocity is $-0.8\vec{v}$, where the negative sign expresses the fact that the new velocity is in the direction opposite to the old.

We can represent velocity vectors in components in the same way we did on page 655.

Example 4 Represent the velocity vectors of the car and the plane in Example 1 using components. Take north to be the positive y-axis, east to be the positive x-axis, and upward to be the positive z-axis.

Solution The car is traveling north at 100 km/hr, so the y-component of its velocity is $100\vec{j}$ and the x-component is $0\vec{i}$. Since it is traveling horizontally, the z-component is $0\vec{k}$. So we have

$$\text{Velocity of car} = 0\vec{i} + 100\vec{j} + 0\vec{k} = 100\vec{j}.$$

The plane's velocity vector also has $\vec{k}$ component equal to zero. Since it is traveling southwest, its $\vec{i}$ and $\vec{j}$ components have negative coefficients (north and east are positive). Since the plane is traveling at 500 km/hr, in one hour it is displaced $500/\sqrt{2} \approx 354$ km to the west and 354 km to the south. (See Figure 13.23.) Thus,

Distance traveled by the car in one hour → 100
Distance traveled by the plane in one hour →
500
45°
$500/\sqrt{2}$
$500/\sqrt{2} \approx 354$
North

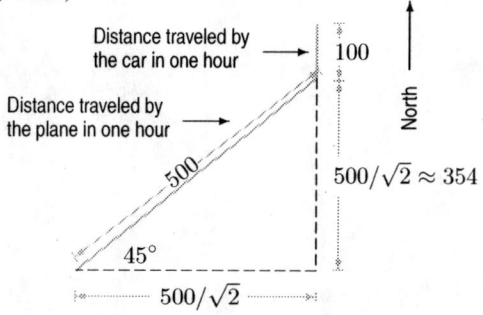

Figure 13.23: Distance traveled by the plane and car in one hour

Chapter Thirteen A FUNDAMENTAL TOOL: VECTORS

$$\text{Velocity of plane} = -(500\cos 45°)\vec{i} - (500\sin 45°)\vec{j} \approx -354\vec{i} - 354\vec{j}.$$

Of course, if the car were climbing a hill or if the plane were descending for a landing, then the $\vec{k}$ component would not be zero.

Acceleration

Another example of a vector quantity is acceleration. Acceleration, like velocity, is specified by both a magnitude and a direction — for example, the acceleration due to gravity is 9.81 m/sec^2 vertically downward.

Force

Force is another example of a vector quantity. Suppose you push on an open door. The result depends both on how hard you push and in what direction. Thus, to specify a force we must give its magnitude (or strength) and the direction in which it is acting. For example, the gravitational force exerted on an object by the earth is a vector pointing from the object toward the center of the earth; its magnitude is the strength of the gravitational force.

Example 5 The earth travels around the sun in an ellipse. The gravitational force on the earth and the velocity of the earth are governed by the following laws:
Newton's Law of Gravitation: The magnitude of the gravitational attraction, F, between two masses m_1 and m_2 at a distance r apart is given by $F = Gm_1m_2/r^2$, where G is a constant. The force vector lies along the line between the masses.
Kepler's Second Law: The line joining a planet to the sun sweeps out equal areas in equal times.
(a) Sketch vectors representing the gravitational force of the sun on the earth at two different positions in the earth's orbit.
(b) Sketch the velocity vector of the earth at two points in its orbit.

Solution (a) Figure 13.24 shows the earth orbiting the sun. Note that the gravitational force vector always points toward the sun and is larger when the earth is closer to the sun because of the r^2 term in the denominator. (In fact, the real orbit looks much more like a circle than we have shown here.)
(b) The velocity vector points in the direction of motion of the earth. Thus, the velocity vector is tangent to the ellipse. See Figure 13.25. Furthermore, the velocity vector is longer at points of the orbit where the planet is moving quickly, because the magnitude of the velocity vector is the speed. Kepler's Second Law enables us to determine when the earth is moving quickly and when it is moving slowly. Over a fixed period of time, say one month, the line joining the earth to the sun sweeps out a sector having a certain area. Figure 13.25 shows two sectors swept out in two different one-month time-intervals. Kepler's law says that the areas of the two sectors are the same. Thus, the earth must move farther in a month when it is close to the sun than when it is far from the sun. Therefore, the earth moves faster when it is closer to the sun and slower when it is farther away.

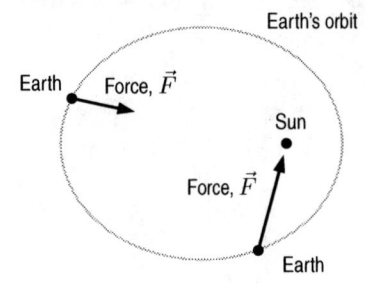

Figure 13.24: Gravitational force, $\vec{F}$, exerted by the sun on the earth: Greater magnitude closer to sun

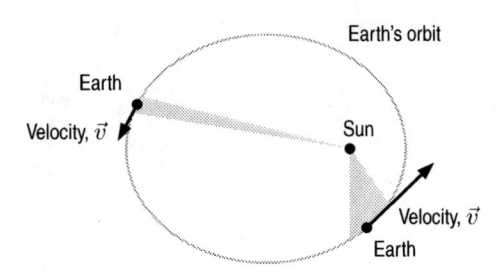

Figure 13.25: The velocity vector, $\vec{v}$, of the earth: Greater magnitude closer to the sun

Properties of Addition and Scalar Multiplication

In general, vectors add, subtract, and are multiplied by scalars in the same way as displacement vectors. Thus, for any vectors $\vec{u}, \vec{v}$, and $\vec{w}$ and any scalars α and β, we have the following properties:

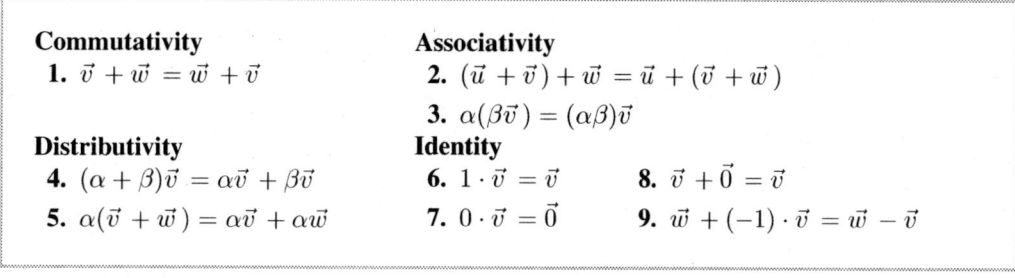

Commutativity
1. $\vec{v} + \vec{w} = \vec{w} + \vec{v}$

Associativity
2. $(\vec{u} + \vec{v}) + \vec{w} = \vec{u} + (\vec{v} + \vec{w})$
3. $\alpha(\beta\vec{v}) = (\alpha\beta)\vec{v}$

Distributivity
4. $(\alpha + \beta)\vec{v} = \alpha\vec{v} + \beta\vec{v}$
5. $\alpha(\vec{v} + \vec{w}) = \alpha\vec{v} + \alpha\vec{w}$

Identity
6. $1 \cdot \vec{v} = \vec{v}$
7. $0 \cdot \vec{v} = \vec{0}$
8. $\vec{v} + \vec{0} = \vec{v}$
9. $\vec{w} + (-1) \cdot \vec{v} = \vec{w} - \vec{v}$

Problems 24–30 at the end of this section ask for a justification of these results in terms of displacement vectors.

Using Components

Example 6 A plane, heading due east at an airspeed of 600 km/hr, experiences a wind of 50 km/hr blowing toward the northeast. Find the plane's direction and ground speed.

Solution We choose a coordinate system with the x-axis pointing east and the y-axis pointing north. See Figure 13.26.

The airspeed tells us the speed of the plane relative to still air. Thus, the plane is moving due east with velocity $\vec{v} = 600\vec{i}$ relative to still air. In addition, the air is moving with a velocity $\vec{w}$. Writing $\vec{w}$ in components, we have

$$\vec{w} = (50\cos 45°)\vec{i} + (50\sin 45°)\vec{j} = 35.4\vec{i} + 35.4\vec{j}.$$

The vector $\vec{v} + \vec{w}$ represents the displacement of the plane in one hour relative to the ground. Therefore, $\vec{v} + \vec{w}$ is the velocity of the plane relative to the ground. In components, we have

$$\vec{v} + \vec{w} = 600\vec{i} + \left(35.4\vec{i} + 35.4\vec{j}\right) = 635.4\vec{i} + 35.4\vec{j}.$$

The direction of the plane's motion relative to the ground is given by the angle θ in Figure 13.26, where

$$\tan\theta = \frac{35.4}{635.4}$$

so

$$\theta = \arctan\left(\frac{35.4}{635.4}\right) = 3.2°.$$

The ground speed is the speed of the plane relative to the ground, so

$$\text{Groundspeed} = \sqrt{635.4^2 + 35.4^2} = 636.4 \text{ km/hr}.$$

Thus, the speed of the plane relative to the ground has been increased slightly by the wind. (This is as we would expect, as the wind has a positive component in the direction in which the plane is traveling.) The angle θ shows how far the plane is blown off course by the wind.

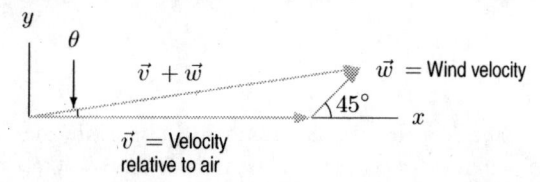

Figure 13.26: Plane's velocity relative to the ground is the sum $\vec{v} + \vec{w}$

Vectors in n Dimensions

Using the alternative notation $\vec{v} = (v_1, v_2, v_3)$ for a vector in 3-space, we can define a vector in n dimensions as a string of n numbers. Thus, a vector in n dimensions can be written as

$$\vec{c} = (c_1, c_2, \ldots, c_n).$$

Addition and scalar multiplication are defined by the formulas

$$\vec{v} + \vec{w} = (v_1, v_2, \ldots, v_n) + (w_1, w_2, \ldots, w_n) = (v_1 + w_1, v_2 + w_2, \ldots, v_n + w_n)$$

and

$$\lambda \vec{v} = \lambda(v_1, v_2, \ldots, v_n) = (\lambda v_1, \lambda v_2, \ldots, \lambda v_n).$$

Why Do We Want Vectors in n Dimensions?

Vectors in two and three dimensions can be used to model displacement, velocities, or forces. But what about vectors in n dimensions? There is another interpretation of 3-dimensional vectors (or 3-vectors) which is useful: they can be thought of as listing 3 different quantities — for example, the displacements parallel to the x, y, and z axes. Similarly, the n-vector

$$\vec{c} = (c_1, c_2, \ldots, c_n)$$

can be thought of as a way of keeping n different quantities organized. For example, a *population* vector $\vec{N}$ shows the number of children and adults in a population:

$$\vec{N} = (\text{Number of children, Number of adults}),$$

or, if we are interested in a more detailed breakdown of ages, we might give the number in each ten-year age bracket in the population (up to age 110) in the form

$$\vec{N} = (N_1, N_2, N_3, N_4, \ldots, N_{10}, N_{11}),$$

where N_1 is the population aged 0–9, and N_2 is the population aged 10–19, and so on.

A *consumption* vector,

$$\vec{q} = (q_1, q_2, \ldots, q_n)$$

shows the quantities $q_1, q_2, \ldots, q_n$ consumed of each of n different goods. A *price* vector

$$\vec{p} = (p_1, p_2, \ldots, p_n)$$

contains the prices of n different items.

In 1907, Hermann Minkowski used vectors with four components when he introduced *space-time coordinates*, whereby each event is assigned a vector position $\vec{v}$ with four coordinates, three for its position in space and one for time:

$$\vec{v} = (x, y, z, t).$$

Example 7 Suppose the vector $\vec{I}$ represents the number of copies, in thousands, made by each of four copy centers in the month of December and $\vec{J}$ represents the number of copies made at the same four copy centers during the previous eleven months (the "year-to-date"). If $\vec{I} = (25, 211, 818, 642)$, and $\vec{J} = (331, 3227, 1377, 2570)$, compute $\vec{I} + \vec{J}$. What does this sum represent?

Solution   The sum is

$$\vec{I} + \vec{J} = (25 + 331, 211 + 3227, 818 + 1377, 642 + 2570) = (356, 3438, 2195, 3212).$$

Each term in $\vec{I} + \vec{J}$ represents the sum of the number of copies made in December plus those in the previous eleven months, that is, the total number of copies made during the entire year at that particular copy center.

Example 8 The price vector $\vec{p} = (p_1, p_2, p_3)$ represents the prices in dollars of three goods. Write a vector which gives the prices of the same goods in cents.

Solution   The prices in cents are $100p_1$, $100p_2$, and $100p_3$ respectively, so the new price vector is

$$(100p_1, 100p_2, 100p_3) = 100\vec{p}.$$

Exercises and Problems for Section 13.2

Exercises

In Exercises 1–5, say whether the given quantity is a vector or a scalar.

1. The population of the US.
2. The distance from Seattle to St. Louis.
3. The magnetic field at a point on the earth's surface.
4. The temperature at a point on the earth's surface.
5. The populations of each of the 50 states.
6. Which is traveling faster, a car whose velocity vector is $21\vec{i} + 35\vec{j}$, or a car whose velocity vector is $40\vec{i}$, assuming that the units are the same for both directions?
7. Give the components of the velocity vector of a boat which is moving at 40 km/hr in a direction 20° south of west. (Assume north is in the positive y-direction.)

8. Give the components of the velocity vector for wind blowing at 10 km/hr toward the southeast. (Assume north is the positive y-direction.)
9. A car is traveling at a speed of 50 km/hr. The positive y-axis is north and the positive x-axis is east. Resolve the car's velocity vector (in 2-space) into components if the car is traveling in each of the following directions:

    (a) East   (b) South
    (c) Southeast (d) Northwest.

10. What angle does a force of $\vec{F} = 15\vec{i} + 18\vec{j}$ make with the x-axis?

Problems

11. Shortly after takeoff, a plane is climbing northwest through still air at an airspeed of 200 km/hr, and rising at a rate of 300 m/min. Resolve its velocity vector into components. The x-axis points east, the y-axis points north, and the z-axis points up.

12. A boat is heading due east at 25 km/hr (relative to the water). The current is moving toward the southwest at 10 km/hr.

    (a) Give the vector representing the actual movement of the boat.
    (b) How fast is the boat going, relative to the ground?
    (c) By what angle does the current push the boat off of its due east course?

13. An airplane is flying at an airspeed of 500 km/hr in a wind blowing at 60 km/hr toward the southeast. In what direction should the plane head to end up going due east? What is the airplane's speed relative to the ground?

14. A plane is heading due east and climbing at the rate of 80 km/hr. If its airspeed is 480 km/hr and there is a wind blowing 100 km/hr to the northeast, what is the ground speed of the plane?

15. An airplane is flying at an airspeed of 600 km/hr in a cross-wind that is blowing from the northeast at a speed of 50 km/hr. In what direction should the plane head to end up going due east?

16. A model rocket is shot into the air at an angle with the earth of about 60°. The rocket is going fast initially but slows down as it reaches its highest point. It picks up speed again as it falls to earth.

    (a) Sketch a graph showing the path of the rocket. Draw several velocity vectors on your graph.
    (b) A second rocket has a parachute that deploys as it begins its descent. How do the velocity vectors from part (a) change for this rocket?

17. A car drives clockwise around the track in Figure 13.27, slowing down at the curves and speeding up along the straight portions. Sketch velocity vectors at the points P, Q, and R.

Figure 13.27

18. A racing car drives clockwise around the track shown in Figure 13.27 at a constant speed. At what point on the track does the car have the longest acceleration vector, and in roughly what direction is it pointing? (Recall that acceleration is the rate of change of velocity.)

19. There are five students in a class. Their scores on the midterm (out of 100) are given by the vector $\vec{v} = (73, 80, 91, 65, 84)$. Their scores on the final (out of 100) are given by $\vec{w} = (82, 79, 88, 70, 92)$. If the final counts twice as much as the midterm, find a vector giving the total scores (as a percentage) of the students.

20. The price vector of beans, rice, and tofu is $(0.30, 0.20, 0.50)$ in dollars per pound. Express it in dollars per ounce.

21. Two forces, represented by the vectors $\vec{F}_1 = 8\vec{i} - 6\vec{j}$ and $\vec{F}_2 = 3\vec{i} + 2\vec{j}$, are acting on an object. Give a vector representing the force that must be applied to the object if it is to remain stationary.

22. One force is pushing an object in a direction 50° south of east with a force of 25 newtons. A second force is simultaneously pushing the object in a direction 70° north of west with a force of 60 newtons. If the object is to remain stationary, give the direction and magnitude of the third force which must be applied to the object to counterbalance the first two.

23. An object is to be moved vertically upward by a crane. As the crane cannot get directly above the object, three ropes are attached to guide the object. One rope is pulled parallel to the ground with a force of 100 newtons in a direction 30° north of east. The second rope is pulled parallel to the ground with a force of 70 newtons in a direction 80° south of east. If the crane is attached to the third rope and can pull with a total force of 3000 newtons, find the force vector for the crane. What is the resulting (total) force on the object? (Assume vector $\vec{i}$ points east, vector $\vec{j}$ points north, and vector $\vec{k}$ points vertically up.)

Use the geometric definition of addition and scalar multiplication to explain each of the properties in Problems 24–31.

24. $\vec{w} + \vec{v} = \vec{v} + \vec{w}$

25. $(\alpha + \beta)\vec{v} = \alpha\vec{v} + \beta\vec{v}$

26. $\alpha(\vec{v} + \vec{w}) = \alpha\vec{v} + \alpha\vec{w}$

27. $\alpha(\beta\vec{v}) = (\alpha\beta)\vec{v}$

28. $\vec{v} + \vec{0} = \vec{v}$

29. $1\vec{v} = \vec{v}$

30. $\vec{v} + (-1)\vec{w} = \vec{v} - \vec{w}$

31. $(\vec{u} + \vec{v}) + \vec{w} = \vec{u} + (\vec{v} + \vec{w})$

13.3 THE DOT PRODUCT

We have seen how to add vectors; can we multiply two vectors together? In the next two sections we will see two different ways of doing so: the *scalar product* (or *dot product*) which produces a scalar, and the *vector product* (or *cross product*), which produces a vector.

Definition of the Dot Product

The dot product links geometry and algebra. We already know how to calculate the length of a vector from its components; the dot product gives us a way of computing the angle between two vectors. For any two vectors $\vec{v} = v_1\vec{i} + v_2\vec{j} + v_3\vec{k}$ and $\vec{w} = w_1\vec{i} + w_2\vec{j} + w_3\vec{k}$, shown in Figure 13.28, we define a scalar as follows:

> The following two definitions of the **dot product**, or **scalar product**, $\vec{v} \cdot \vec{w}$, are equivalent:
> - **Geometric definition**
> $\vec{v} \cdot \vec{w} = \|\vec{v}\|\|\vec{w}\|\cos\theta$ where θ is the angle between $\vec{v}$ and $\vec{w}$ and $0 \leq \theta \leq \pi$.
> - **Algebraic definition**
> $\vec{v} \cdot \vec{w} = v_1w_1 + v_2w_2 + v_3w_3$.
> Notice that the dot product of two vectors is a *number*.

Why don't we give just one definition of $\vec{v} \cdot \vec{w}$? The reason is that both definitions are equally important; the geometric definition gives us a picture of what the dot product means and the algebraic definition gives us a way of calculating it.

How do we know the two definitions are equivalent — that is, they really do define the same thing? First, we observe that the two definitions give the same result in a particular example. Then we show why they are equivalent in general.

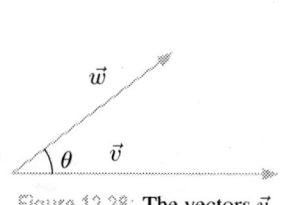

Figure 13.28: The vectors $\vec{v}$ and $\vec{w}$

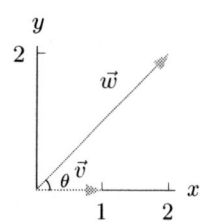

Figure 13.29: Calculating the dot product of the vectors $v = \vec{i}$ and $\vec{w} = 2\vec{i} + 2\vec{j}$ geometrically and algebraically gives the same result

Example 1 Suppose $\vec{v} = \vec{i}$ and $\vec{w} = 2\vec{i} + 2\vec{j}$. Compute $\vec{v} \cdot \vec{w}$ both geometrically and algebraically.

Solution To use the geometric definition, see Figure 13.29. The angle between the vectors is $\pi/4$, or $45°$, and the lengths of the vectors are given by

$$\|\vec{v}\| = 1 \quad \text{and} \quad \|\vec{w}\| = 2\sqrt{2}.$$

Thus,

$$\vec{v} \cdot \vec{w} = \|\vec{v}\|\|\vec{w}\|\cos\theta = 1 \cdot 2\sqrt{2}\cos\left(\frac{\pi}{4}\right) = 2.$$

Using the algebraic definition, we get the same result:

$$\vec{v} \cdot \vec{w} = 1 \cdot 2 + 0 \cdot 2 = 2.$$

Why the Two Definitions of the Dot Product Give the Same Result

In the previous example, the two definitions give the same value for the dot product. To show that the geometric and algebraic definitions of the dot product always give the same result, we must show that, for any vectors $\vec{v} = v_1\vec{i} + v_2\vec{j} + v_3\vec{k}$ and $\vec{w} = w_1\vec{i} + w_2\vec{j} + w_3\vec{k}$ with an angle θ between them:

$$\|\vec{v}\|\|\vec{w}\|\cos\theta = v_1w_1 + v_2w_2 + v_3w_3.$$

One method follows; a method which does not use trigonometry is given in Problem 46 on page 672.

Using the Law of Cosines. Suppose that $0 < \theta < \pi$, so that the vectors $\vec{v}$ and $\vec{w}$ form a triangle. (See Figure 13.30.) By the Law of Cosines, we have

$$\|\vec{v} - \vec{w}\|^2 = \|\vec{v}\|^2 + \|\vec{w}\|^2 - 2\|\vec{v}\|\|\vec{w}\|\cos\theta.$$

This result is also true for $\theta = 0$ and $\theta = \pi$. We calculate the lengths using components:

$$\|\vec{v}\|^2 = v_1^2 + v_2^2 + v_3^2$$
$$\|\vec{w}\|^2 = w_1^2 + w_2^2 + w_3^2$$
$$\|\vec{v} - \vec{w}\|^2 = (v_1 - w_1)^2 + (v_2 - w_2)^2 + (v_3 - w_3)^2$$
$$= v_1^2 - 2v_1w_1 + w_1^2 + v_2^2 - 2v_2w_2 + w_2^2 + v_3^2 - 2v_3w_3 + w_3^2.$$

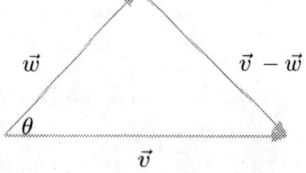

Figure 13.30: Triangle used in the justification of $\|\vec{v}\|\|\vec{w}\|\cos\theta = v_1w_1 + v_2w_2 + v_3w_3$

Substituting into the Law of Cosines and canceling, we see that
$$-2v_1w_1 - 2v_2w_2 - 2v_3w_3 = -2\|\vec{v}\|\|\vec{w}\|\cos\theta.$$

Therefore we have the result we wanted, namely that:
$$v_1w_1 + v_2w_2 + v_3w_3 = \|\vec{v}\|\|\vec{w}\|\cos\theta.$$

Properties of the Dot Product

The following properties of the dot product can be justified using the algebraic definition; see Problem 39 on page 672. For a geometric interpretation of Property 3, see Problem 43.

> **Properties of the Dot Product.** For any vectors $\vec{u}$, $\vec{v}$, and $\vec{w}$ and any scalar λ,
> 1. $\vec{v} \cdot \vec{w} = \vec{w} \cdot \vec{v}$
> 2. $\vec{v} \cdot (\lambda\vec{w}) = \lambda(\vec{v} \cdot \vec{w}) = (\lambda\vec{v}) \cdot \vec{w}$
> 3. $(\vec{v} + \vec{w}) \cdot \vec{u} = \vec{v} \cdot \vec{u} + \vec{w} \cdot \vec{u}$

Perpendicularity, Magnitude, and Dot Products

Two vectors are perpendicular if the angle between them is $\pi/2$ or $90°$. Since $\cos(\pi/2) = 0$, if $\vec{v}$ and $\vec{w}$ are perpendicular, then $\vec{v} \cdot \vec{w} = 0$. Conversely, provided that $\vec{v} \cdot \vec{w} = 0$, then $\cos\theta = 0$, so $\theta = \pi/2$ and the vectors are perpendicular. Thus, we have the following result:

> Two nonzero vectors $\vec{v}$ and $\vec{w}$ are **perpendicular**, or **orthogonal**, if and only if
> $$\vec{v} \cdot \vec{w} = 0.$$

For example: $\vec{i} \cdot \vec{j} = 0, \vec{j} \cdot \vec{k} = 0, \vec{i} \cdot \vec{k} = 0$.

If we take the dot product of a vector with itself, then $\theta = 0$ and $\cos\theta = 1$. For any vector $\vec{v}$:

> Magnitude and dot product are related as follows:
> $$\vec{v} \cdot \vec{v} = \|\vec{v}\|^2.$$

For example: $\vec{i} \cdot \vec{i} = 1, \vec{j} \cdot \vec{j} = 1, \vec{k} \cdot \vec{k} = 1$.

Using the Dot Product

Depending on the situation, one definition of the dot product may be more convenient to use than the other. In Example 2 which follows, the geometric definition is the only one which can be used because we are not given components. In Example 3, the algebraic definition is used.

Example 2 Suppose the vector $\vec{b}$ is fixed and has length 2; the vector $\vec{a}$ is free to rotate and has length 3. What are the maximum and minimum values of the dot product $\vec{a} \cdot \vec{b}$ as the vector $\vec{a}$ rotates through all possible positions? What positions of $\vec{a}$ and $\vec{b}$ lead to these values?

Solution The geometric definition gives $\vec{a} \cdot \vec{b} = \|\vec{a}\|\|\vec{b}\|\cos\theta = 3 \cdot 2\cos\theta = 6\cos\theta$. Thus, the maximum value of $\vec{a} \cdot \vec{b}$ is 6, and it occurs when $\cos\theta = 1$ so $\theta = 0$, that is, when $\vec{a}$ and $\vec{b}$ point in the same direction. The minimum value of $\vec{a} \cdot \vec{b}$ is -6, and it occurs when $\cos\theta = -1$ so $\theta = \pi$, that is, when $\vec{a}$ and $\vec{b}$ point in opposite directions. (See Figure 13.31.)

13.3 THE DOT PRODUCT 667

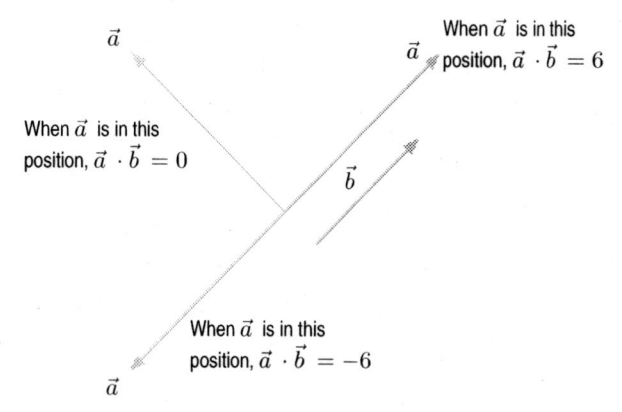

Figure 13.31: Maximum and minimum values of $\vec{a} \cdot \vec{b}$ obtained from a fixed vector $\vec{b}$ of length 2 and rotating vector $\vec{a}$ of length 3

Example 3 Which pairs from the following list of 3-dimensional vectors are perpendicular to one another?

$$\vec{u} = \vec{i} + \sqrt{3}\,\vec{k}, \quad \vec{v} = \vec{i} + \sqrt{3}\,\vec{j}, \quad \vec{w} = \sqrt{3}\,\vec{i} + \vec{j} - \vec{k}.$$

Solution The geometric definition tells us that two vectors are perpendicular if and only if their dot product is zero. Since the vectors are given in components, we calculate dot products using the algebraic definition:

$$\vec{v} \cdot \vec{u} = (\vec{i} + \sqrt{3}\,\vec{j} + 0\vec{k}) \cdot (\vec{i} + 0\vec{j} + \sqrt{3}\,\vec{k}) = 1 \cdot 1 + \sqrt{3} \cdot 0 + 0 \cdot \sqrt{3} = 1,$$
$$\vec{v} \cdot \vec{w} = (\vec{i} + \sqrt{3}\,\vec{j} + 0\vec{k}) \cdot (\sqrt{3}\,\vec{i} + \vec{j} - \vec{k}) = 1 \cdot \sqrt{3} + \sqrt{3} \cdot 1 + 0(-1) = 2\sqrt{3},$$
$$\vec{w} \cdot \vec{u} = (\sqrt{3}\,\vec{i} + \vec{j} - \vec{k}) \cdot (\vec{i} + 0\vec{j} + \sqrt{3}\,\vec{k}) = \sqrt{3} \cdot 1 + 1 \cdot 0 + (-1) \cdot \sqrt{3} = 0.$$

So the only two vectors which are perpendicular are $\vec{w}$ and $\vec{u}$.

Normal Vectors and the Equation of a Plane

In Section 12.4 we wrote the equation of a plane given its x-slope, y-slope and z-intercept. Now we write the equation of a plane using a vector and a point lying on the plane. A *normal vector* to a plane is a vector that is perpendicular to the plane, that is, it is perpendicular to every displacement vector between any two points in the plane. Let $\vec{n} = a\vec{i} + b\vec{j} + c\vec{k}$ be a normal vector to the plane, let $P_0 = (x_0, y_0, z_0)$ be a fixed point in the plane, and let $P = (x, y, z)$ be any other point in the plane. Then $\overrightarrow{P_0P} = (x - x_0)\vec{i} + (y - y_0)\vec{j} + (z - z_0)\vec{k}$ is a vector whose head and tail both lie in the plane. (See Figure 13.32.) Thus, the vectors $\vec{n}$ and $\overrightarrow{P_0P}$ are perpendicular, so $\vec{n} \cdot \overrightarrow{P_0P} = 0$. The algebraic definition of the dot product gives $\vec{n} \cdot \overrightarrow{P_0P} = a(x - x_0) + b(y - y_0) + c(z - z_0)$, so we obtain the following result:

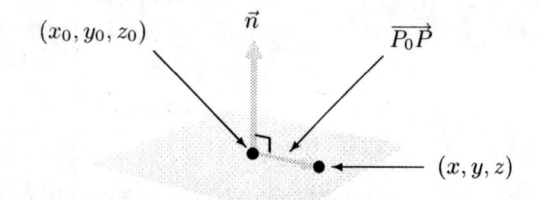

Figure 13.32: Plane with normal $\vec{n}$ and containing a fixed point (x_0, y_0, z_0)

> The **equation of the plane** with normal vector $\vec{n} = a\vec{i} + b\vec{j} + c\vec{k}$ and containing the point $P_0 = (x_0, y_0, z_0)$ is
> $$a(x - x_0) + b(y - y_0) + c(z - z_0) = 0.$$
> Letting $d = ax_0 + by_0 + cz_0$ (a constant), we can write the equation of the plane in the form
> $$ax + by + cz = d.$$

Example 4 Find the equation of the plane perpendicular to $-\vec{i} + 3\vec{j} + 2\vec{k}$ and passing through the point $(1, 0, 4)$.

Solution The equation of the plane is
$$-(x - 1) + 3(y - 0) + 2(z - 4) = 0,$$
which simplifies to
$$-x + 3y + 2z = 7.$$

Example 5 Find a normal vector to the plane with equation (a) $x - y + 2z = 5$ (b) $z = 0.5x + 1.2y$.

Solution
(a) Since the coefficients of $\vec{i}$, $\vec{j}$, and $\vec{k}$ in a normal vector are the coefficients of x, y, z in the equation of the plane, a normal vector is $\vec{n} = \vec{i} - \vec{j} + 2\vec{k}$.
(b) Before we can find a normal vector, we rewrite the equation of the plane in the form
$$0.5x + 1.2y - z = 0.$$
Thus, a normal vector is $\vec{n} = 0.5\vec{i} + 1.2\vec{j} - \vec{k}$.

The Dot Product in n Dimensions

The algebraic definition of the dot product can be extended to vectors in higher dimensions.

> If $\vec{u} = (u_1, \ldots, u_n)$ and $\vec{v} = (v_1, \ldots, v_n)$ then the dot product of $\vec{u}$ and $\vec{v}$ is the **scalar**
> $$\vec{u} \cdot \vec{v} = u_1 v_1 + \ldots + u_n v_n.$$

Example 6 A video store sells videos, tapes, CDs, and computer games. We define the quantity vector $\vec{q} = (q_1, q_2, q_3, q_4)$, where q_1, q_2, q_3, q_4 denote the quantities sold of each of the items, and the price vector $\vec{p} = (p_1, p_2, p_3, p_4)$, where p_1, p_2, p_3, p_4 denote the price per unit of each item. What does the dot product $\vec{p} \cdot \vec{q}$ represent?

Solution The dot product is $\vec{p} \cdot \vec{q} = p_1 q_1 + p_2 q_2 + p_3 q_3 + p_4 q_4$. The quantity $p_1 q_1$ represents the revenue received by the store for the videos, $p_2 q_2$ represents the revenue for the tapes, and so on. The dot product represents the total revenue received by the store for the sale of these four items.

Resolving a Vector into Components: Projections

In Section 13.1, we resolved a vector into components parallel to the axes. Now we see how to resolve a vector, $\vec{v}$, into components, called $\vec{v}_{\text{parallel}}$ and $\vec{v}_{\text{perp}}$, which are parallel and perpendicular, respectively, to a given nonzero vector, $\vec{u}$. (See Figure 13.33.)

13.3 THE DOT PRODUCT

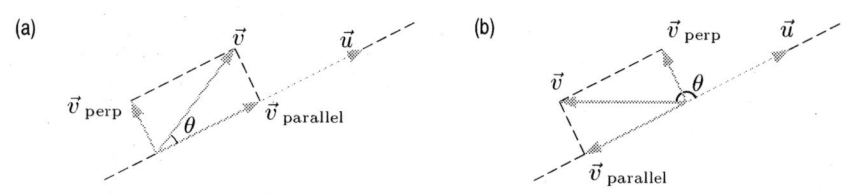

Figure 13.33: Resolving $\vec{v}$ into components parallel and perpendicular to $\vec{u}$
(a) $0 < \theta < \pi/2$ (b) $\pi/2 < \theta < \pi$

The projection of $\vec{v}$ on $\vec{u}$, written $\vec{v}_{\text{parallel}}$, measures (in some sense) how much the vector $\vec{v}$ is aligned with the vector $\vec{u}$. The length of $\vec{v}_{\text{parallel}}$ is the length of the shadow cast by $\vec{v}$ on a line in the direction of $\vec{u}$.

To compute $\vec{v}_{\text{parallel}}$, we assume $\vec{u}$ is a unit vector. (If not, create one by dividing by its length.) Then Figure 13.33(a) shows that, if $0 \leq \theta \leq \pi/2$:

$$\|\vec{v}_{\text{parallel}}\| = \|\vec{v}\| \cos\theta = \vec{v} \cdot \vec{u} \quad \text{(since } \|\vec{u}\| = 1\text{)}.$$

Now $\vec{v}_{\text{parallel}}$ is a scalar multiple of $\vec{u}$, and since $\vec{u}$ is a unit vector,

$$\vec{v}_{\text{parallel}} = (\|\vec{v}\| \cos\theta)\vec{u} = (\vec{v} \cdot \vec{u})\vec{u}.$$

A similar argument shows that if $\pi/2 < \theta \leq \pi$, as in Figure 13.33(b), this formula for $\vec{v}_{\text{parallel}}$ still holds. The vector $\vec{v}_{\text{perp}}$ is specified by

$$\vec{v}_{\text{perp}} = \vec{v} - \vec{v}_{\text{parallel}}.$$

Thus, we have the following results:

Projection of $\vec{v}$ on the Line in the Direction of the Unit Vector $\vec{u}$

If $\vec{v}_{\text{parallel}}$ and $\vec{v}_{\text{perp}}$ are components of $\vec{v}$ which are parallel and perpendicular, respectively, to $\vec{u}$, then

$$\text{Projection of } \vec{v} \text{ on to } \vec{u} = \vec{v}_{\text{parallel}} = (\vec{v} \cdot \vec{u})\vec{u} \quad \text{provided } \|\vec{u}\| = 1$$

and $\quad \vec{v} = \vec{v}_{\text{parallel}} + \vec{v}_{\text{perp}} \quad$ so $\quad \vec{v}_{\text{perp}} = \vec{v} - \vec{v}_{\text{parallel}}.$

Example 7 Figure 13.34 shows the force the wind exerts on the sail of a sailboat. Find the component of the force in the direction in which the sailboat is traveling.

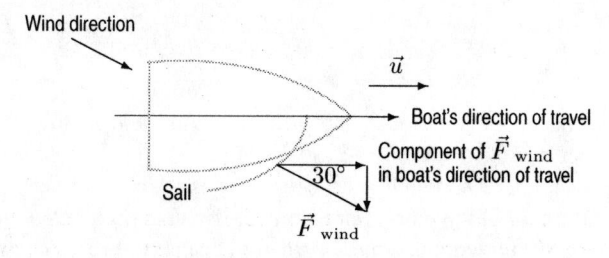

Figure 13.34: Wind moving a sailboat

Solution
Let $\vec{u}$ be a unit vector in the direction of travel. The force of the wind on the sail makes an angle of 30° with $\vec{u}$. Thus, the component of this force in the direction of $\vec{u}$ is

$$\vec{F}_{\text{parallel}} = (\vec{F} \cdot \vec{u})\vec{u} = \|\vec{F}\|(\cos 30°)\vec{u} = 0.87\|\vec{F}\|\vec{u}.$$

Thus, the boat is being pushed forward with about 87% of the total force due to the wind. (In fact, the interaction of wind and sail is much more complex than this model suggests.)

A Physical Interpretation of the Dot Product: Work

In physics, the word "work" has a slightly different meaning from its everyday meaning. In physics, when a force of magnitude F acts on an object through a distance d, we say the *work*, W, done by the force is

$$W = Fd,$$

provided the force and the displacement are in the same direction. For example, if a 1 kg body falls 10 meters under the force of gravity, which is 9.8 newtons, then the work done by gravity is

$$W = (9.8 \text{ newtons}) \cdot (10 \text{ meters}) = 98 \text{ joules}.$$

What if the force and the displacement are not in the same direction? Suppose a force $\vec{F}$ acts on an object as it moves along a displacement vector $\vec{d}$. Let θ be the angle between $\vec{F}$ and $\vec{d}$. First, we assume $0 \le \theta \le \pi/2$. Figure 13.35 shows how we can resolve $\vec{F}$ into components that are parallel and perpendicular to $\vec{d}$:

$$\vec{F} = \vec{F}_{\text{parallel}} + \vec{F}_{\text{perp}},$$

Then the work done by $\vec{F}$ is defined to be

$$W = \|\vec{F}_{\text{parallel}}\| \|\vec{d}\|.$$

We see from Figure 13.35 that $\vec{F}_{\text{parallel}}$ has magnitude $\|\vec{F}\| \cos \theta$. So the work is given by the dot product:

$$W = (\|\vec{F}\| \cos \theta)\|\vec{d}\| = \|\vec{F}\|\|\vec{d}\| \cos \theta = \vec{F} \cdot \vec{d}.$$

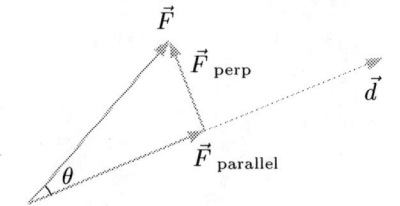

Figure 13.35: Resolving the force $\vec{F}$ into two forces, one parallel to $\vec{d}$, one perpendicular to $\vec{d}$

The formula $W = \vec{F} \cdot \vec{d}$ holds when $\pi/2 < \theta \le \pi$ also. In that case, the work done by the force is negative and the object is moving against the force. Thus, we have the following definition:

The **work**, W, done by a force $\vec{F}$ acting on an object through a displacement $\vec{d}$ is given by

$$W = \vec{F} \cdot \vec{d}.$$

Notice that if the vectors $\vec{F}$ and $\vec{d}$ are parallel and in the same direction, with magnitudes F and d, then $\cos \theta = \cos 0 = 1$, so $W = \|\vec{F}\|\|\vec{d}\| = Fd$, which is the original definition. When the vectors are perpendicular, $\cos \theta = \cos \frac{\pi}{2} = 0$, so $W = 0$ and no work is done in the technical definition of the word. For example, if you carry a heavy box across the room at the same horizontal height, no work is done by gravity because the force of gravity is vertical but the motion is horizontal.

Exercises and Problems for Section 13.3

Exercises

For Exercises 1–9, perform the following operations on the given 3-dimensional vectors.

$\vec{a} = 2\vec{j} + \vec{k}$ $\vec{b} = -3\vec{i} + 5\vec{j} + 4\vec{k}$ $\vec{c} = \vec{i} + 6\vec{j}$

$\vec{y} = 4\vec{i} - 7\vec{j}$ $\vec{z} = \vec{i} - 3\vec{j} - \vec{k}$

1. $\vec{a} \cdot \vec{y}$
2. $\vec{c} \cdot \vec{y}$
3. $\vec{a} \cdot \vec{b}$
4. $\vec{a} \cdot \vec{z}$
5. $\vec{a} \cdot (\vec{c} + \vec{y})$
6. $\vec{c} \cdot \vec{a} + \vec{a} \cdot \vec{y}$
7. $(\vec{a} \cdot \vec{b})\vec{a}$
8. $(\vec{a} \cdot \vec{y})(\vec{c} \cdot \vec{z})$
9. $((\vec{c} \cdot \vec{c})\vec{a}) \cdot \vec{a}$

In Exercises 10–13, find a normal vector to the plane.

10. $2x + y - z = 5$
11. $z = 3x + 4y - 7$
12. $z - 5(x - 2) = 3(5 - y)$
13. $2(x - z) = 3(x + y)$

Problems

14. Let θ be the angle between $\vec{v}$ and $\vec{w}$, with $0 < \theta < \pi/2$. What is the effect on $\vec{v} \cdot \vec{w}$ of increasing each of the following quantities? Does $\vec{v} \cdot \vec{w}$ increase or decrease?
    (a) $\|\vec{v}\|$
    (b) θ

15. (a) Give a vector that is parallel to, but not equal to, $\vec{v} = 4\vec{i} + 3\vec{j}$.
    (b) Give a vector that is perpendicular to $\vec{v}$.

16. Compute the angle between the vectors $\vec{i} + \vec{j} + \vec{k}$ and $\vec{i} - \vec{j} - \vec{k}$.

17. Which pairs of the vectors $\sqrt{3}\vec{i} + \vec{j}, 3\vec{i} + \sqrt{3}\vec{j}, \vec{i} - \sqrt{3}\vec{j}$ are parallel and which are perpendicular?

18. For what values of t are $\vec{u} = t\vec{i} - \vec{j} + \vec{k}$ and $\vec{v} = t\vec{i} + t\vec{j} - 2\vec{k}$ perpendicular? Are there values of t for which $\vec{u}$ and $\vec{v}$ are parallel?

19. List any vectors which are parallel to each other and any vectors which are perpendicular to each other:

    $\vec{v}_1 = \vec{i} - 2\vec{j}$, $\vec{v}_2 = 2\vec{i} + 4\vec{j}$, $\vec{v}_3 = 3\vec{i} + 1.5\vec{j}$,
    $\vec{v}_4 = -1.2\vec{i} + 2.4\vec{j}$, $\vec{v}_5 = -5\vec{i} - 2.5\vec{j}$, $\vec{v}_6 = 12\vec{i} - 12\vec{j}$,
    $\vec{v}_7 = 4\vec{i} + 2\vec{j}$, $\vec{v}_8 = 3\vec{i} - 6\vec{j}$, $\vec{v}_9 = 0.70\vec{i} - 0.35\vec{j}$

20. Which pairs (if any) of vectors from the following list
    (a) Are perpendicular?
    (b) Are parallel?
    (c) Have an angles less than $\pi/2$ between them?
    (d) Have an angle of more than $\pi/2$ between them?

    $\vec{a} = \vec{i} - 3\vec{j} - \vec{k}$ $\vec{b} = \vec{i} + \vec{j} + 2\vec{k}$
    $\vec{c} = -2\vec{i} - \vec{j} + \vec{k}$ $\vec{d} = -\vec{i} - \vec{j} + \vec{k}$

21. Write $\vec{a} = 3\vec{i} + 2\vec{j} - 6\vec{k}$ as the sum of two vectors, one parallel, and one perpendicular, to $\vec{d} = 2\vec{i} - 4\vec{j} + \vec{k}$.

In Problems 22–26, find an equation of a plane that satisfies the given conditions.

22. Perpendicular to the vector $-\vec{i} + 2\vec{j} + \vec{k}$ and passing through the point $(1, 0, 2)$.

23. Perpendicular to the vector $5\vec{i} + \vec{j} - 2\vec{k}$ and passing through the point $(0, 1, -1)$.

24. Perpendicular to the vector $2\vec{i} - 3\vec{j} + 7\vec{k}$ and passing through the point $(1, -1, 2)$.

25. Parallel to the plane $2x + 4y - 3z = 1$ and through the point $(1, 0, -1)$.

26. Through $(-2, 3, 2)$ and parallel to $3x + y + z = 4$.

27. A plane has equation $z = 5x - 2y + 7$.
    (a) Find a value of λ making the vector $\lambda\vec{i} + \vec{j} + 0.5\vec{k}$ normal to the plane.
    (b) Find a value of a so that the point $(a + 1, a, a - 1)$ lies on the plane.

28. Find the points where the plane $z = 5x - 4y + 3$ intersects each of the coordinate axes. Find the lengths of the sides and the angles of the triangle formed by these points.

29. Find the angle between the planes $5(x - 1) + 3(y + 2) + 2z = 0$ and $x + 3(y - 1) + 2(z + 4) = 0$.

30. Find angle BAC if $A = (2, 2, 2)$, $B = (4, 2, 1)$, and $C = (2, 3, 1)$.

31. Let $A = (0, 4)$, $B = (-1, -3)$, and $C = (-5, 1)$. Draw triangle ABC and find each of its interior angles.

32. Let S be the triangle with vertices $A = (2, 2, 2)$, $B = (4, 2, 1)$, and $C = (2, 3, 1)$.
    (a) Find the length of the shortest side of S.
    (b) Find the cosine of the angle BAC at vertex A.

33. A 100-meter dash is run on a track in the direction of the vector $\vec{v} = 2\vec{i} + 6\vec{j}$. The wind velocity $\vec{w}$ is $5\vec{i} + \vec{j}$ km/hr. The rules say that a legal wind speed measured in the direction of the dash must not exceed 5 km/hr. Will the race results be disqualified due to an illegal wind? Justify your answer.

34. A basketball gymnasium is 25 meters high, 80 meters wide and 200 meters long. For a half time stunt, the cheerleaders want to run two strings, one from each of the two corners above one basket to the diagonally opposite corners of the gym floor. What is the cosine of the angle made by the strings as they cross?

35. A course has four exams, weighted 10%, 15%, 25%, 50%, respectively. The class average on each of these exams is 75%, 91%, 84%, 87%, respectively. What do the vectors $\vec{a} = (0.75, 0.91, 0.84, 0.87)$ and $\vec{w} = (0.1, 0.15, 0.25, 0.5)$ represent, in terms of the course? Calculate the dot product $\vec{w} \cdot \vec{a}$. What does it represent, in terms of the course?

36. A street vendor sells six items, with prices p_1 dollars per unit, p_2 dollars per unit, and so on. The vendor's price vector is $\vec{p} = (p_1, p_2, p_3, p_4, p_5, p_6) = (1.00, 3.50, 4.00, 2.75, 5.00, 3.00)$. The vendor sells q_1 units of the first item, q_2 units of the second item, and so on. The vendor's quantity vector is $\vec{q} = (q_1, q_2, q_3, q_4, q_5, q_6) = (43, 57, 12, 78, 20, 35)$. Find $\vec{p} \cdot \vec{q}$, give its units, and explain its significance to the vendor.

37. A consumption vector of three goods is defined by $\vec{x} = (x_1, x_2, x_3)$, where x_1, x_2 and x_3 are the quantities consumed of the three goods. A budget constraint is represented by the equation $\vec{p} \cdot \vec{x} = k$, where $\vec{p}$ is the price vector of the three goods and k is a constant. Show that the difference between two consumption vectors corresponding to points satisfying the same budget constraint is perpendicular to the price vector $\vec{p}$.

38. Consider a point P and plane through the point P_0 with normal vector $\vec{n}$.

    (a) What do all the points satisfying the following condition have in common:
    $$\vec{n} \cdot \overrightarrow{P_0 P} > 0?$$
    What about all the points satisfying $\vec{n} \cdot \overrightarrow{P_0 P} < 0$?

    (b) Express the conditions $\vec{n} \cdot \overrightarrow{P_0 P} > 0$ and $\vec{n} \cdot \overrightarrow{P_0 P} < 0$ in terms of coordinates.

    (c) Use parts (a) and (b) to decide which of the points
    $$P = (-1, -1, 1), \quad Q = (-1, -1, -1)$$
    $$R = (1, 1, 1)$$
    are on the same side of the plane $2x - 3y + 4z = 4$.

39. Show why each of the properties of the dot product in the box on page 666 follows from the algebraic definition of the dot product:
    $$\vec{v} \cdot \vec{w} = v_1 w_1 + v_2 w_2 + v_3 w_3$$

40. What does Property 2 of the dot product in the box on page 666 say geometrically?

41. Show that the vectors $(\vec{b} \cdot \vec{c})\vec{a} - (\vec{a} \cdot \vec{c})\vec{b}$ and $\vec{c}$ are perpendicular.

42. Show that if $\vec{u}$ and $\vec{v}$ are two vectors such that
    $$\vec{u} \cdot \vec{w} = \vec{v} \cdot \vec{w}$$
    for every vector $\vec{w}$, then
    $$\vec{u} = \vec{v}.$$

43. Figure 13.36 shows that, given three vectors $\vec{u}$, $\vec{v}$, and $\vec{w}$, the sum of the components of $\vec{v}$ and $\vec{w}$ in the direction of $\vec{u}$ is the component of $\vec{v} + \vec{w}$ in the direction of $\vec{u}$. (Although the figure is drawn in two dimensions, this result is also true in three dimensions.) Use this figure to explain why the geometric definition of the dot product satisfies $(\vec{v} + \vec{w}) \cdot \vec{u} = \vec{v} \cdot \vec{u} + \vec{w} \cdot \vec{u}$.

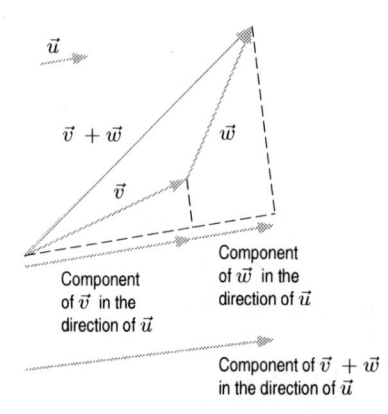

Figure 13.36: Component of $\vec{v} + \vec{w}$ in the direction of $\vec{u}$ is the sum of the components of $\vec{v}$ and $\vec{w}$ in that direction

44. (a) Using the geometric definition of the dot product, show that
    $$\vec{u} \cdot (-\vec{v}) = -(\vec{u} \cdot \vec{v}).$$
    [Hint: What happens to the angle when you multiply $\vec{v}$ by -1?]

    (b) Using the geometric definition of the dot product, show that for any negative scalar λ
    $$\vec{u} \cdot (\lambda \vec{v}) = \lambda(\vec{u} \cdot \vec{v})$$
    $$(\lambda \vec{u}) \cdot \vec{v} = \lambda(\vec{u} \cdot \vec{v}).$$

45. The Law of Cosines for a triangle with side lengths a, b, and c, and with angle C opposite side c, says
    $$c^2 = a^2 + b^2 - 2ab \cos C.$$
    On page 665, we used the Law of Cosines to show that the two definitions of the dot product are equivalent. In this problem, use the geometric definition of the dot product and its properties in the box on page 666 to prove the Law of Cosines. [Hint: Let $\vec{u}$ and $\vec{v}$ be the displacement vectors from C to the other two vertices, and express c^2 in terms of $\vec{u}$ and $\vec{v}$.]

46. Use the following steps and the results of Problems 43–44 to show (without trigonometry) that the geometric and algebraic definitions of the dot product are equivalent.
    Let $\vec{u} = u_1 \vec{i} + u_2 \vec{j} + u_3 \vec{k}$ and $\vec{v} = v_1 \vec{i} + v_2 \vec{j} + v_3 \vec{k}$ be any vectors. Write $(\vec{u} \cdot \vec{v})_{\text{geom}}$ for the result of the dot product computed geometrically. Substitute $\vec{u} = u_1 \vec{i} + u_2 \vec{j} + u_3 \vec{k}$ and use Problems 43–44 to expand $(\vec{u} \cdot \vec{v})_{\text{geom}}$. Substitute for $\vec{v}$ and expand. Then calculate the dot products $\vec{i} \cdot \vec{i}$, $\vec{i} \cdot \vec{j}$, etc. geometrically.

47. For any vectors $\vec{v}$ and $\vec{w}$, consider the following function of t:

$$q(t) = (\vec{v} + t\vec{w}) \cdot (\vec{v} + t\vec{w}).$$

(a) Explain why $q(t) \geq 0$ for all real t.

(b) Expand $q(t)$ as a quadratic polynomial in t using the properties on page 666.

(c) Using the discriminant of the quadratic, show that,

$$|\vec{v} \cdot \vec{w}| \leq \|\vec{v}\| \|\vec{w}\|.$$

13.4 THE CROSS PRODUCT

In the previous section we combined two vectors to get a number, the dot product. In this section we see another way of combining two vectors, this time to get a vector, the *cross product*. Any two vectors in 3-space form a parallelogram. We define the cross product using this parallelogram.

The Area of a Parallelogram

Consider the parallelogram formed by the vectors $\vec{v}$ and $\vec{w}$ with an angle of θ between them. Then Figure 13.37 shows

$$\text{Area of parallelogram} = \text{Base} \cdot \text{Height} = \|\vec{v}\| \|\vec{w}\| \sin\theta.$$

How would we compute the area of the parallelogram if we were given $\vec{v}$ and $\vec{w}$ in components, $\vec{v} = v_1\vec{i} + v_2\vec{j} + v_3\vec{k}$ and $\vec{w} = w_1\vec{i} + w_2\vec{j} + w_3\vec{k}$? Project 1 on page 682 shows that if $\vec{v}$ and $\vec{w}$ are in the xy-plane, so $v_3 = w_3 = 0$, then

$$\text{Area of parallelogram} = |v_1w_2 - v_2w_1|.$$

What if $\vec{v}$ and $\vec{w}$ do not lie in the xy-plane? The cross product will enable us to compute the area of the parallelogram formed by any two vectors.

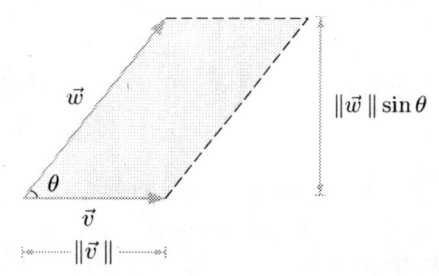

Figure 13.37: Parallelogram formed by $\vec{v}$ and $\vec{w}$ has
Area $= \|\vec{v}\| \|\vec{w}\| \sin\theta$

Definition of the Cross Product

We define the cross product of the vectors $\vec{v}$ and $\vec{w}$, written $\vec{v} \times \vec{w}$, to be a vector perpendicular to both $\vec{v}$ and $\vec{w}$. The magnitude of this vector is the area of the parallelogram formed by the two vectors. The direction of $\vec{v} \times \vec{w}$ is given by the normal vector, $\vec{n}$, to the plane defined by $\vec{v}$ and $\vec{w}$. If we require that $\vec{n}$ be a unit vector, there are two choices for $\vec{n}$, pointing out of the plane in opposite directions. We pick one by the following rule (see Figure 13.38):

> **The right-hand rule:** Place $\vec{v}$ and $\vec{w}$ so that their tails coincide and curl the fingers of your right hand through the smaller of the two angles from $\vec{v}$ to $\vec{w}$; your thumb points in the direction of the normal vector, $\vec{n}$.

Like the dot product, there are two equivalent definitions of the cross product:

The following two definitions of the **cross product** or **vector product** $\vec{v} \times \vec{w}$ are equivalent:

- **Geometric definition**

  If $\vec{v}$ and $\vec{w}$ are not parallel, then

  $$\vec{v} \times \vec{w} = \begin{pmatrix} \text{Area of parallelogram} \\ \text{with edges } \vec{v} \text{ and } \vec{w} \end{pmatrix} \vec{n} = (\|\vec{v}\|\|\vec{w}\|\sin\theta)\vec{n},$$

  where $0 \leq \theta \leq \pi$ is the angle between $\vec{v}$ and $\vec{w}$ and $\vec{n}$ is the unit vector perpendicular to $\vec{v}$ and $\vec{w}$ pointing in the direction given by the right-hand rule. If $\vec{v}$ and $\vec{w}$ are parallel, then $\vec{v} \times \vec{w} = \vec{0}$.

- **Algebraic definition**

  $$\vec{v} \times \vec{w} = (v_2 w_3 - v_3 w_2)\vec{i} + (v_3 w_1 - v_1 w_3)\vec{j} + (v_1 w_2 - v_2 w_1)\vec{k}$$

  where $\vec{v} = v_1\vec{i} + v_2\vec{j} + v_3\vec{k}$ and $\vec{w} = w_1\vec{i} + w_2\vec{j} + w_3\vec{k}$.

Notice that the magnitude of the $\vec{k}$ component is the area of a 2-dimensional parallelogram and the other components have a similar form. Problems 33 and 31 at the end of this section show that the geometric and algebraic definitions of the cross product give the same result.

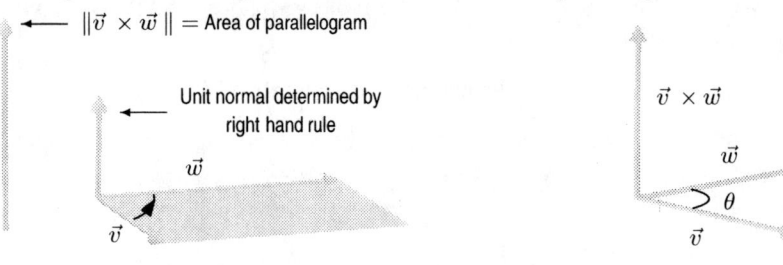

Figure 13.38: Area of parallelogram $= \|\vec{v} \times \vec{w}\|$

Figure 13.39: The cross product $\vec{v} \times \vec{w}$

Unlike the dot product, the cross product is only defined for three-dimensional vectors. The geometric definition shows us that the cross product is *rotation invariant*. Imagine the two vectors $\vec{v}$ and $\vec{w}$ as two metal rods welded together. Attach a third rod whose direction and length correspond to $\vec{v} \times \vec{w}$. (See Figure 13.39.) Then, no matter how we turn this set of rods, the third will still be the cross product of the first two.

The algebraic definition is more easily remembered by writing it as a 3×3 determinant. (See Appendix D.)

$$\vec{v} \times \vec{w} = \begin{vmatrix} \vec{i} & \vec{j} & \vec{k} \\ v_1 & v_2 & v_3 \\ w_1 & w_2 & w_3 \end{vmatrix} = (v_2 w_3 - v_3 w_2)\vec{i} + (v_3 w_1 - v_1 w_3)\vec{j} + (v_1 w_2 - v_2 w_1)\vec{k}.$$

Example 1 Find $\vec{i} \times \vec{j}$ and $\vec{j} \times \vec{i}$.

Solution The vectors $\vec{i}$ and $\vec{j}$ both have magnitude 1 and the angle between them is $\pi/2$. By the right-hand rule, the vector $\vec{i} \times \vec{j}$ is in the direction of $\vec{k}$, so $\vec{n} = \vec{k}$ and we have

$$\vec{i} \times \vec{j} = \left(\|\vec{i}\|\|\vec{j}\|\sin\frac{\pi}{2}\right)\vec{k} = \vec{k}.$$

Similarly, the right-hand rule says that the direction of $\vec{j} \times \vec{i}$ is $-\vec{k}$, so

$$\vec{j} \times \vec{i} = (\|\vec{j}\|\|\vec{i}\|\sin\frac{\pi}{2})(-\vec{k}) = -\vec{k}.$$

Similar calculations show that $\vec{j} \times \vec{k} = \vec{i}$ and $\vec{k} \times \vec{i} = \vec{j}$.

Example 2 For any vector $\vec{v}$, find $\vec{v} \times \vec{v}$.

Solution Since $\vec{v}$ is parallel to itself, $\vec{v} \times \vec{v} = \vec{0}$.

Example 3 Find the cross product of $\vec{v} = 2\vec{i} + \vec{j} - 2\vec{k}$ and $\vec{w} = 3\vec{i} + \vec{k}$ and check that the cross product is perpendicular to both $\vec{v}$ and $\vec{w}$.

Solution Writing $\vec{v} \times \vec{w}$ as a determinant and expanding it into three two-by-two determinants, we have

$$\vec{v} \times \vec{w} = \begin{vmatrix} \vec{i} & \vec{j} & \vec{k} \\ 2 & 1 & -2 \\ 3 & 0 & 1 \end{vmatrix} = \vec{i} \begin{vmatrix} 1 & -2 \\ 0 & 1 \end{vmatrix} - \vec{j} \begin{vmatrix} 2 & -2 \\ 3 & 1 \end{vmatrix} + \vec{k} \begin{vmatrix} 2 & 1 \\ 3 & 0 \end{vmatrix}$$

$$= \vec{i}\,(1(1) - 0(-2)) - \vec{j}\,(2(1) - 3(-2)) + \vec{k}\,(2(0) - 3(1))$$
$$= \vec{i} - 8\vec{j} - 3\vec{k}.$$

To check that $\vec{v} \times \vec{w}$ is perpendicular to $\vec{v}$, we compute the dot product:
$$\vec{v} \cdot (\vec{v} \times \vec{w}) = (2\vec{i} + \vec{j} - 2\vec{k}) \cdot (\vec{i} - 8\vec{j} - 3\vec{k}) = 2 - 8 + 6 = 0.$$

Similarly,
$$\vec{w} \cdot (\vec{v} \times \vec{w}) = (3\vec{i} + 0\vec{j} + \vec{k}) \cdot (\vec{i} - 8\vec{j} - 3\vec{k}) = 3 + 0 - 3 = 0.$$

Thus, $\vec{v} \times \vec{w}$ is perpendicular to both $\vec{v}$ and $\vec{w}$.

Properties of the Cross Product

The right-hand rule tells us that $\vec{v} \times \vec{w}$ and $\vec{w} \times \vec{v}$ point in opposite directions. The magnitudes of $\vec{v} \times \vec{w}$ and $\vec{w} \times \vec{v}$ are the same, so $\vec{w} \times \vec{v} = -(\vec{v} \times \vec{w})$. (See Figure 13.40.)

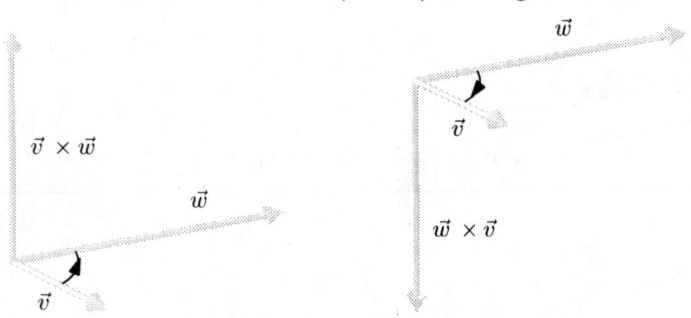

Figure 13.40: Diagram showing $\vec{v} \times \vec{w} = -(\vec{w} \times \vec{v})$

This explains the first of the following properties. The other two are derived in Problems 25, 28, and 33 at the end of this section.

Properties of the Cross Product

For vectors $\vec{u}, \vec{v}, \vec{w}$ and scalar λ
1. $\vec{w} \times \vec{v} = -(\vec{v} \times \vec{w})$
2. $(\lambda \vec{v}) \times \vec{w} = \lambda(\vec{v} \times \vec{w}) = \vec{v} \times (\lambda \vec{w})$
3. $\vec{u} \times (\vec{v} + \vec{w}) = \vec{u} \times \vec{v} + \vec{u} \times \vec{w}$.

676 Chapter Thirteen A FUNDAMENTAL TOOL: VECTORS

The Equivalence of the Two Definitions of the Cross Product

Problem 33 on page 678 uses geometric arguments to show that the cross product distributes over addition. Problem 31 then shows how the formula in the algebraic definition of the cross product can be derived from the geometric definition.

The Equation of a Plane Through Three Points

The equation of a plane is determined by a point $P_0 = (x_0, y_0, z_0)$ on the plane, and a normal vector, $\vec{n} = a\vec{i} + b\vec{j} + c\vec{k}$:

$$a(x - x_0) + b(y - y_0) + c(z - z_0) = 0.$$

However, a plane can also be determined by three points on it (provided they do not lie on a line). In that case we can find an equation of the plane by first determining two vectors in the plane and then finding a normal vector using the cross product, as in the following example.

Example 4 Find an equation of the plane containing the points $P = (1, 3, 0)$, $Q = (3, 4, -3)$, and $R = (3, 6, 2)$.

Solution Since the points P and Q are in the plane, the displacement vector between them, $\overrightarrow{PQ}$, is in the plane, where

$$\overrightarrow{PQ} = (3-1)\vec{i} + (4-3)\vec{j} + (-3-0)\vec{k} = 2\vec{i} + \vec{j} - 3\vec{k}.$$

The displacement vector $\overrightarrow{PR}$ is also in the plane, where

$$\overrightarrow{PR} = (3-1)\vec{i} + (6-3)\vec{j} + (2-0)\vec{k} = 2\vec{i} + 3\vec{j} + 2\vec{k}.$$

Thus, a normal vector, $\vec{n}$, to the plane is given by

$$\vec{n} = \overrightarrow{PQ} \times \overrightarrow{PR} = \begin{vmatrix} \vec{i} & \vec{j} & \vec{k} \\ 2 & 1 & -3 \\ 2 & 3 & 2 \end{vmatrix} = 11\vec{i} - 10\vec{j} + 4\vec{k}.$$

Since the point $(1, 3, 0)$ is on the plane, the equation of the plane is

$$11(x-1) - 10(y-3) + 4(z-0) = 0,$$

which simplifies to

$$11x - 10y + 4z = -19.$$

You should check that P, Q, and R satisfy the equation of the plane.

Areas and Volumes Using the Cross Product and Determinants

We can use the cross product to calculate the area of the parallelogram with sides $\vec{v}$ and $\vec{w}$. We say that $\vec{v} \times \vec{w}$ is the *area vector* of the parallelogram. The geometric definition of the cross product tells us that $\vec{v} \times \vec{w}$ is normal to the parallelogram and gives us the following result:

Area of a parallelogram with edges $\vec{v} = v_1\vec{i} + v_2\vec{j} + v_3\vec{k}$ and $\vec{w} = w_1\vec{i} + w_2\vec{j} + w_3\vec{k}$ is given by

$$\text{Area} = \|\vec{v} \times \vec{w}\|, \quad \text{where} \quad \vec{v} \times \vec{w} = \begin{vmatrix} \vec{i} & \vec{j} & \vec{k} \\ v_1 & v_2 & v_3 \\ w_1 & w_2 & w_3 \end{vmatrix}.$$

13.4 THE CROSS PRODUCT 677

Example 5 Find the area of the parallelogram with edges $\vec{v} = 2\vec{i} + \vec{j} - 3\vec{k}$ and $\vec{w} = \vec{i} + 3\vec{j} + 2\vec{k}$.

Solution We calculate the cross product:

$$\vec{v} \times \vec{w} = \begin{vmatrix} \vec{i} & \vec{j} & \vec{k} \\ 2 & 1 & -3 \\ 1 & 3 & 2 \end{vmatrix} = (2+9)\vec{i} - (4+3)\vec{j} + (6-1)\vec{k} = 11\vec{i} - 7\vec{j} + 5\vec{k}.$$

The area of the parallelogram with edges $\vec{v}$ and $\vec{w}$ is the magnitude of the vector $\vec{v} \times \vec{w}$:

$$\text{Area} = \|\vec{v} \times \vec{w}\| = \sqrt{11^2 + (-7)^2 + 5^2} = \sqrt{195}.$$

Volume of a Parallelepiped

Consider the parallelepiped with sides formed by $\vec{a}, \vec{b}$, and $\vec{c}$. (See Figure 13.41.) Since the base is formed by the vectors $\vec{b}$ and $\vec{c}$, we have

$$\text{Area of base of parallelepiped} = \|\vec{b} \times \vec{c}\|.$$

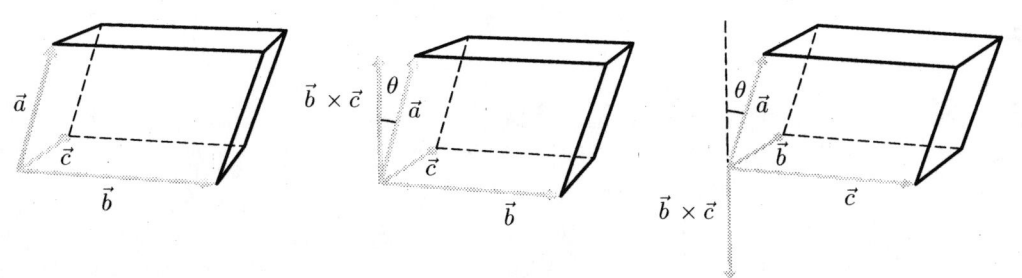

Figure 13.41: Volume of a Parallelepiped

Figure 13.42: The vectors $\vec{a}, \vec{b}, \vec{c}$ are called a right-handed set

Figure 13.43: The vectors $\vec{a}, \vec{b}, \vec{c}$ are called a left-handed set

The vectors $\vec{a}, \vec{b}$, and $\vec{c}$ can be arranged either as in Figure 13.42 or as in Figure 13.43. In either case,

$$\text{Height of parallelepiped} = \|\vec{a}\| \cos \theta,$$

where θ is the angle shown in the figures. In Figure 13.42 the angle θ is less than $\pi/2$, so the product, $(\vec{b} \times \vec{c}) \cdot \vec{a}$, called the *triple product*, is positive. Thus, in this case

$$\text{Volume of parallelepiped} = \text{Base} \cdot \text{Height} = \|\vec{b} \times \vec{c}\| \cdot \|\vec{a}\| \cos \theta = (\vec{b} \times \vec{c}) \cdot \vec{a}.$$

In Figure 13.43, the angle, $\pi - \theta$, between $\vec{a}$ and $\vec{b} \times \vec{c}$ is more than $\pi/2$, so the product $(\vec{b} \times \vec{c}) \cdot \vec{a}$ is negative. Thus, in this case we have

$$\text{Volume} = \text{Base} \cdot \text{Height} = \|\vec{b} \times \vec{c}\| \cdot \|\vec{a}\| \cos \theta = -\|\vec{b} \times \vec{c}\| \cdot \|\vec{a}\| \cos(\pi - \theta)$$
$$= -(\vec{b} \times \vec{c}) \cdot \vec{a} = \left|(\vec{b} \times \vec{c}) \cdot \vec{a}\right|.$$

Therefore, in both cases the volume is given by $\left|(\vec{b} \times \vec{c}) \cdot \vec{a}\right|$. Using determinants, we can write

Volume of a parallelepiped with edges $\vec{a}, \vec{b}, \vec{c}$ is given by

$$\text{Volume} = \left|(\vec{b} \times \vec{c}) \cdot \vec{a}\right| = \text{Absolute value of the determinant} \begin{vmatrix} a_1 & a_2 & a_3 \\ b_1 & b_2 & b_3 \\ c_1 & c_2 & c_3 \end{vmatrix}.$$

Exercises and Problems for Section 13.4

Exercises

In Exercises 1–6, use the algebraic definition to find $\vec{v} \times \vec{w}$.

1. $\vec{v} = \vec{k}, \vec{w} = \vec{j}$
2. $\vec{v} = -\vec{i}, \vec{w} = \vec{j} + \vec{k}$
3. $\vec{v} = \vec{i} + \vec{k}, \vec{w} = \vec{i} + \vec{j}$
4. $\vec{v} = \vec{i} + \vec{j} + \vec{k}, \vec{w} = \vec{i} + \vec{j} + -\vec{k}$
5. $\vec{v} = 2\vec{i} - 3\vec{j} + \vec{k}, \vec{w} = \vec{i} + 2\vec{j} - \vec{k}$
6. $\vec{v} = 2\vec{i} - \vec{j} - \vec{k}, \vec{w} = -6\vec{i} + 3\vec{j} + 3\vec{k}$

In Exercises 7–8, use the properties on page 675 to find:

7. $\left((\vec{i} + \vec{j}) \times \vec{i}\right) \times \vec{j}$
8. $(\vec{i} + \vec{j}) \times (\vec{i} \times \vec{j})$

Use the geometric definition in Exercises 9–10 to find:

9. $2\vec{i} \times (\vec{i} + \vec{j})$
10. $(\vec{i} + \vec{j}) \times (\vec{i} - \vec{j})$

11. If $\vec{v} = 3\vec{i} - 2\vec{j} + 4\vec{k}$ and $\vec{w} = \vec{i} + 2\vec{j} - \vec{k}$, find $\vec{v} \times \vec{w}$ and $\vec{w} \times \vec{v}$. What is the relation between the two answers?

12. For $\vec{a} = 3\vec{i} + \vec{j} - \vec{k}$ and $\vec{b} = \vec{i} - 4\vec{j} + 2\vec{k}$, find $\vec{a} \times \vec{b}$ and check that it is perpendicular to both $\vec{a}$ and $\vec{b}$.

Find an equation for the plane through the points in Exercises 13–14.

13. $(1, 0, 0), (0, 1, 0), (0, 0, 1)$.
14. $(3, 4, 2), (-2, 1, 0), (0, 2, 1)$.

Problems

15. Suppose $\vec{v} \cdot \vec{w} = 5$ and $\|\vec{v} \times \vec{w}\| = 3$, and the angle between $\vec{v}$ and $\vec{w}$ is θ. Find
    (a) $\tan \theta$
    (b) θ.

16. If $\vec{v} \times \vec{w} = 2\vec{i} - 3\vec{j} + 5\vec{k}$, and $\vec{v} \cdot \vec{w} = 3$, find $\tan \theta$ where θ is the angle between $\vec{v}$ and $\vec{w}$.

17. If $\vec{v}$ and $\vec{w}$ are both parallel to the xy-plane, what can you conclude about $\vec{v} \times \vec{w}$? Explain.

18. Let $A = (-1, 3, 0), B = (3, 2, 4),$ and $C = (1, -1, 5)$.
    (a) Find an equation for the plane that passes through these three points.
    (b) Find the area of the triangle determined by these three points.

19. Let $P = (0, 1, 0), \quad Q = (-1, 1, 2), \quad R = (2, 1, -1)$. Find
    (a) The area of the triangle PQR.
    (b) The equation for a plane that contains $P, Q,$ and R.

20. Find a vector parallel to the line of intersection of the two planes $4x - 3y + 2z = 12$ and $x + 5y - z = 25$.

21. Find a vector parallel to the intersection of the planes $2x - 3y + 5z = 2$ and $4x + y - 3z = 7$.

22. Find the equation of the plane through the origin which is perpendicular to the line of intersection of the planes in Problem 21.

23. Find the equation of the plane through the point $(4, 5, 6)$ which is perpendicular to the line of intersection of the planes in Problem 21.

24. Find the equation of a plane through the origin and perpendicular to $x - y + z = 5$ and $2x + y - 2z = 7$.

25. Use the algebraic definition to check that
    $$\vec{a} \times (\vec{b} + \vec{c}) = (\vec{a} \times \vec{b}) + (\vec{a} \times \vec{c}).$$

26. In this problem, we arrive at the algebraic definition for the cross product by a different route. Let $\vec{a} = a_1\vec{i} + a_2\vec{j} + a_3\vec{k}$ and $\vec{b} = b_1\vec{i} + b_2\vec{j} + b_3\vec{k}$. We seek a vector $\vec{v} = x\vec{i} + y\vec{j} + z\vec{k}$ which is perpendicular to both $\vec{a}$ and $\vec{b}$. Use this requirement to construct two equations for $x, y,$ and z. Eliminate x and solve for y in terms of z. Then eliminate y and solve for x in terms of z. Since z can be any value whatsoever (the direction of $\vec{v}$ is unaffected), select the value for z which eliminates the denominator in the equation you obtained. How does the resulting expression for $\vec{v}$ compare to the formula we derived on page 674?

27. Suppose $\vec{a}$ and $\vec{b}$ are vectors in the xy-plane, such that $\vec{a} = a_1\vec{i} + a_2\vec{j}$ and $\vec{b} = b_1\vec{i} + b_2\vec{j}$ with $0 < a_2 < a_1$ and $0 < b_1 < b_2$.
    (a) Sketch $\vec{a}$ and $\vec{b}$ and the vector $\vec{c} = -a_2\vec{i} + a_1\vec{j}$. Shade the parallelogram formed by $\vec{a}$ and $\vec{b}$.
    (b) What is the relation between $\vec{a}$ and $\vec{c}$? [Hint: Find $\vec{c} \cdot \vec{a}$ and $\vec{c} \cdot \vec{c}$.]
    (c) Find $\vec{c} \cdot \vec{b}$.
    (d) Explain why $\vec{c} \cdot \vec{b}$ gives the area of the parallelogram formed by $\vec{a}$ and $\vec{b}$.
    (e) Check that in this case $\vec{a} \times \vec{b} = (a_1 b_2 - a_2 b_1)\vec{k}$.

28. If $\vec{v}$ and $\vec{w}$ are nonzero vectors, use the geometric definition of the cross product to explain why
    $$(\lambda \vec{v}) \times \vec{w} = \lambda(\vec{v} \times \vec{w}) = \vec{v} \times (\lambda \vec{w}).$$
    Consider the cases $\lambda > 0$, and $\lambda = 0$, and $\lambda < 0$ separately.

29. Use a parallelepiped to show that $\vec{a} \cdot (\vec{b} \times \vec{c}) = (\vec{a} \times \vec{b}) \cdot \vec{c}$ for any vectors $\vec{a}, \vec{b},$ and $\vec{c}$.

30. Show that $\|\vec{a} \times \vec{b}\|^2 = \|\vec{a}\|^2 \|\vec{b}\|^2 - (\vec{a} \cdot \vec{b})^2$.

31. Use the fact that $\vec{i} \times \vec{i} = \vec{0}, \vec{i} \times \vec{j} = \vec{k}, \vec{i} \times \vec{k} = -\vec{j}$, and so on, together with the properties on page 675 to derive the algebraic definition for the cross product.

32. For vectors $\vec{a}$ and $\vec{b}$, let $\vec{c} = \vec{a} \times (\vec{b} \times \vec{a})$.

(a) Show that $\vec{c}$ lies in the plane containing $\vec{a}$ and $\vec{b}$.
(b) Use Problems 29 and 30 to show that $\vec{a} \cdot \vec{c} = 0$ and $\vec{b} \cdot \vec{c} = \|\vec{a}\|^2\|\vec{b}\|^2 - (\vec{a} \cdot \vec{b})^2$.
(c) Show that
$$\vec{a} \times (\vec{b} \times \vec{a}) = \|\vec{a}\|^2\vec{b} - (\vec{a} \cdot \vec{b})\vec{a}.$$

33. Use the result of Problem 29 to show that the cross product distributes over addition. First, use distributivity for the dot product to show that for any vector $\vec{d}$,
$$[(\vec{a}+\vec{b}) \times \vec{c}] \cdot \vec{d} = [(\vec{a} \times \vec{c}) + (\vec{b} \times \vec{c})] \cdot \vec{d}.$$
Next, show that for any vector $\vec{d}$,
$$[((\vec{a}+\vec{b}) \times \vec{c}) - (\vec{a} \times \vec{c}) - (\vec{b} \times \vec{c})] \cdot \vec{d} = 0.$$
Finally, explain why you can conclude that
$$(\vec{a}+\vec{b}) \times \vec{c} = (\vec{a} \times \vec{c}) + (\vec{b} \times \vec{c}).$$

34. Figure 13.44 shows the tetrahedron determined by three vectors $\vec{a}, \vec{b}, \vec{c}$. The *area vector* of a face is a vector perpendicular to the face, pointing outward, whose magnitude is the area of the face. Show that the sum of the four outward pointing area vectors of the faces equals the zero vector.

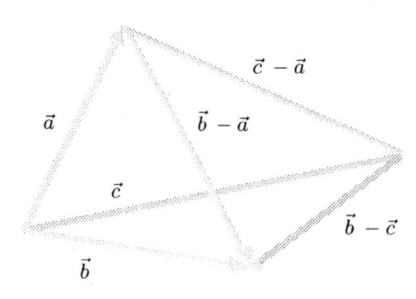

Figure 13.44

CHAPTER SUMMARY (see also Ready Reference at the end of the book)

- **Vectors**
  Geometric definition of vector addition, subtraction and scalar multiplication, resolving into $\vec{i}, \vec{j}$, and $\vec{k}$ components, magnitude of a vector, algebraic properties of addition and scalar multiplication.

- **Dot Product**
  Geometric and algebraic definition, algebraic properties, using dot products to find angles and determine perpendicularity, the equation of a plane with given normal vector passing through a given point, projection of a vector in a direction given by a unit vector.

- **Cross Product**
  Geometric and algebraic definition, algebraic properties, cross product and volume, finding the equation of a plane through three points.

REVIEW EXERCISES AND PROBLEMS FOR CHAPTER THIRTEEN

Exercises

In Exercises 1–10, use $\vec{v} = 2\vec{i} + 3\vec{j} - \vec{k}$ and $\vec{w} = \vec{i} - \vec{j} + 2\vec{k}$ to calculate the given quantities.

1. $\vec{v} + 2\vec{w}$
2. $3\vec{v} - \vec{w} - \vec{v}$
3. $\vec{v} \cdot \vec{w}$
4. $\vec{v} \times \vec{w}$
5. $\vec{v} \times \vec{v}$
6. $\|\vec{v} + \vec{w}\|$
7. $(\vec{v} \cdot \vec{w})\vec{v}$
8. $(\vec{v} \times \vec{w}) \times \vec{w}$
9. $(\vec{v} \times \vec{w}) \cdot \vec{w}$
10. $(\vec{v} \times \vec{w}) \times (\vec{v} \times \vec{w})$

11. Let $\vec{v} = 3\vec{i} + 2\vec{j} - 2\vec{k}$ and $\vec{w} = 4\vec{i} - 3\vec{j} + \vec{k}$. Find each of the following:

(a) $\vec{v} \cdot \vec{w}$
(b) $\vec{v} \times \vec{w}$
(c) A vector of length 5 parallel to vector $\vec{v}$
(d) The angle between vectors $\vec{v}$ and $\vec{w}$
(e) The component of vector $\vec{v}$ in the direction of vector $\vec{w}$
(f) A vector perpendicular to vector $\vec{v}$
(g) A vector perpendicular to both vectors $\vec{v}$ and $\vec{w}$

In Exercises 12–15, find a vector with the given property.

12. Length 10, parallel to $2\vec{i} + 3\vec{j} - \vec{k}$.
13. Unit vector perpendicular to $\vec{i} + \vec{j}$ and $\vec{i} - \vec{j} - \vec{k}$
14. Unit vector in the xy-plane perpendicular to $3\vec{i} - 2\vec{j}$.
15. The vector obtained from $4\vec{i} + 3\vec{j}$ by rotating it 90° counterclockwise.

Problems

16. For what values of t are the following pairs of vectors parallel?

(a) $2\vec{i} + (t^2 + \frac{2}{3}t + 1)\vec{j} + t\vec{k}$, $6\vec{i} + 8\vec{j} + 3\vec{k}$
(b) $t\vec{i} + \vec{j} + (t-1)\vec{k}$, $2\vec{i} - 4\vec{j} + \vec{k}$
(c) $2t\vec{i} + t\vec{j} + t\vec{k}$, $6\vec{i} + 3\vec{j} + 3\vec{k}$.

17. Suppose $\vec{v} \cdot \vec{w} = 8$ and $\vec{v} \times \vec{w} = 12\vec{i} - 3\vec{j} + 4\vec{k}$ and that the angle between $\vec{v}$ and $\vec{w}$ is θ. Find

(a) $\tan\theta$ (b) θ

18. Find a vector normal to $4(x-1) + 6(z+3) = 12$.

19. Find the equation of the plane through the origin which is parallel to $z = 4x - 3y + 8$.

20. Consider the plane $5x - y + 7z = 21$.

(a) Find a point on the x-axis on this plane.
(b) Find two other points on the plane.
(c) Find a vector perpendicular to the plane.
(d) Find a vector parallel to the plane.

21. Find the area of the triangle with vertices $P = (-2, 2, 0)$, $Q = (1, 3, -1)$, and $R = (-4, 2, 1)$.

22. A plane is drawn through the points $A = (2, 1, 0)$, $B = (0, 1, 3)$ and $C = (1, 0, 1)$. Find

(a) Two vectors lying in the plane.
(b) A vector perpendicular to the plane.
(c) The equation of the plane.

23. Given the points $P = (1, 2, 3)$, $Q = (3, 5, 7)$, and $R = (2, 5, 3)$, find:

(a) A unit vector perpendicular to a plane containing P, Q, R.
(b) The angle between PQ and PR.
(c) The area of the triangle PQR.
(d) The distance from R to the line through P and Q.

24. Find the distance from the point $P = (2, -1, 3)$ to the plane $2x + 4y - z = -1$.

25. Find an equation of the plane passing through the three points $(1, 1, 1), (1, 4, 5), (-3, -2, 0)$. Find the distance from the origin to the plane.

26. An airplane heads northeast at an airspeed of 700 km/hr, but there is a wind blowing from the west at 60 km/hr. In what direction does the plane end up flying? What is its speed relative to the ground?

27. A man wishes to row the shortest possible distance from north to south across a river which is flowing at 4 km/hr from the east. He can row at 5 km/hr.

(a) In which direction should he steer?
(b) If there is a wind of 10 km/hr from the southwest, in which direction should he steer to try and go directly across the river? What happens?

28. A particle moving with speed v hits a barrier at an angle of 60° and bounces off at an angle of 60° in the opposite direction with speed reduced by 20 percent. See Figure 13.45. Find the velocity vector of the object after impact.

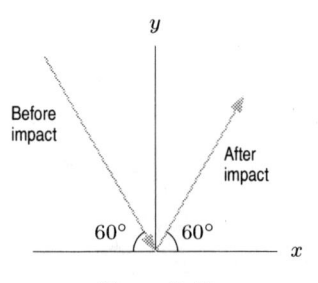

Figure 13.45

29. An airport is at the point $(200, 10, 0)$ and an approaching plane is at the point $(550, 60, 4)$. Assume that the xy-plane is horizontal, with the x-axis pointing eastward and the y-axis pointing northward. Also assume that the z-axis is upward and that all distances are measured in kilometers. The plane flies due west at a constant altitude at a speed of 500 km/hr for half an hour. It then descends at 200 km/hr, heading straight for the airport.

(a) Find the velocity vector of the plane while it is flying at constant altitude.
(b) Find the coordinates of the point at which the plane starts to descend.
(c) Find a vector representing the velocity of the plane when it is descending.

30. The earth is at the origin, the moon is at the point $(384, 0)$, and a spaceship is at $(280, 90)$, where distance is in thousands of kilometers.

(a) What is the displacement vector of the moon relative to the earth? Of the spaceship relative to the earth? Of the spaceship relative to the moon?
(b) How far is the spaceship from the earth? From the moon?
(c) The gravitational force on the spaceship from the earth is 461 newtons, and force from the moon is 26 newtons. What is the resulting force?

31. A large ship is being towed by two tugs. The larger tug exerts a force which is 25% greater than the smaller tug and at an angle of 30 degrees north of east. Which direction must the smaller tug pull to ensure that the ship travels due east?

32. The current in a river is pushing a boat in direction 25° north of east with a speed of 12 km/hr. The wind is pushing the same boat in a direction 80° south of east with a speed of 7 km/hr. Find the velocity vector of the boat's engine (relative to the water) if the boat actually moves due east at a speed of 40 km/hr relative to the ground.

33. Three people are trying to hold a ferocious lion still for the veterinarian. The lion, in the center, is wearing a collar with three ropes attached to it and each person has hold of a rope. Charlie is pulling in the direction 62° west of north with a force of 175 newtons and Sam is pulling in the direction 43° east of north with a force of 200 newtons. What is the direction and magnitude of the force which must be exerted by Alice on the third rope to counterbalance Sam and Charlie?

34. An object is moving counterclockwise at a constant speed around the circle $x^2 + y^2 = 1$, where x and y are measured in meters. It completes one revolution every minute.

   (a) What is its speed?
   (b) What is its velocity vector 30 seconds after it passes the point $(1, 0)$? Does your answer change if the object is moving clockwise? Explain.

35. An object is attached by a string to a fixed point and rotates 30 times per minute in a horizontal plane. Show that the speed of the object is constant but the velocity is not. What does this imply about the acceleration?

36. Find the vector $\vec{v}$ with all of the following properties:
   - Magnitude 10
   - Angle of 45° with positive x-axis
   - Angle of 75° with positive y-axis
   - Positive $\vec{k}$-component.

37. (a) A vector $\vec{v}$ of magnitude v makes an angle α with the positive x-axis, angle β with the positive y-axis, and angle γ with the positive z-axis. Show that
$$\vec{v} = v\cos\alpha \vec{i} + v\cos\beta \vec{j} + v\cos\gamma \vec{k}.$$

   (b) $\cos\alpha$, $\cos\beta$, and $\cos\gamma$ are called *direction cosines*. Show that
$$\cos^2\alpha + \cos^2\beta + \cos^2\gamma = 1.$$

38. Two lines in space are skew if they are not parallel and do not intersect. Determine the minimum distance between two such lines.

CAS Challenge Problems

39. Let $\vec{a} = x\vec{i} + y\vec{j} + z\vec{k}$, $\vec{b} = u\vec{i} + v\vec{j} + w\vec{k}$, and $\vec{c} = m\vec{a} + n\vec{b}$. Compute $(\vec{a} \times \vec{b}) \cdot \vec{c}$ and $(\vec{a} \times \vec{b}) \times (\vec{a} \times \vec{c})$, and explain the geometric meaning of your answers.

40. Let $\vec{a} = x\vec{i} + y\vec{j} + z\vec{k}$, $\vec{b} = u\vec{i} + v\vec{j} + w\vec{k}$ and $\vec{c} = r\vec{i} + s\vec{j} + t\vec{k}$. Show that the parallelepiped with edges $\vec{a}, \vec{b}, \vec{c}$ has the same volume as the parallelepiped with edges $\vec{a}, \vec{b}, 2\vec{a} - \vec{b} + \vec{c}$. Explain this result geometrically.

41. Let $\vec{a} = \vec{i} + 2\vec{j} + 3\vec{k}$, $\vec{b} = 2\vec{i} + \vec{j} + 2\vec{k}$ and let θ be the angle between $\vec{a}$ and $\vec{b}$.

   (a) For $\vec{c} = x\vec{i} + y\vec{j} + z\vec{k}$, write the following conditions as equations in x, y, z and solve them:
$$\vec{a} \cdot \vec{c} = 0, \quad \vec{b} \cdot \vec{c} = 0, \quad \|\vec{c}\|^2 = \|\vec{a}\|^2 \|\vec{b}\|^2 \sin^2\theta.$$
   [Hint: Use the dot product to find $\sin^2\theta$.]
   (b) Compute the cross product $\vec{a} \times \vec{b}$ and compare with your answer in part (a). What do you notice? Explain.

42. Let $A = (0, 0, 0)$, $B = (2, 0, 0)$, $C = (1, \sqrt{3}, 0)$ and $D = (1, 1/\sqrt{3}, 2\sqrt{2/3})$.

   (a) Show that A, B, C, D are all the same distance from each other.
   (b) Find the point $P = (x, y, z)$ which is equidistant from A, B, C and D by setting up and solving three equations in $x, y,$ and z.
   (c) Use the dot product to find the angle APB. (In chemistry, this angle is often approximated by 109.5°. A methane molecule can be represented by four hydrogen atoms at points A, B, C and D, and a carbon atom at P.)

43. Let $P = (x, y, z)$, $Q = (u, v, w)$ and $R = (r, s, t)$ be points on the plane $ax + by + cz = d$.

   (a) What is the relation between $\vec{PQ} \times \vec{PR}$ and the normal vector to the plane, $a\vec{i} + b\vec{j} + c\vec{k}$?
   (b) Express $\vec{PQ} \times \vec{PR}$ in terms of $x, y, z, u, v, w, r, s, t$.
   (c) Use the equation for the plane to eliminate $z, w,$ and t from the expression you obtained in part (b), and simplify. Does your answer agree with what you said in part (a)?

CHECK YOUR UNDERSTANDING

Are the statements in Problems 1–30 true or false? Give reasons for your answer.

1. There is exactly one unit vector parallel to a given nonzero vector $\vec{v}$.

2. The vector $\frac{1}{\sqrt{3}}\vec{i} + \frac{-1}{\sqrt{3}}\vec{j} + \frac{2}{\sqrt{3}}\vec{k}$ is a unit vector.

3. The length of the vector $2\vec{v}$ is twice the length of the vector $\vec{v}$.

4. If $\vec{v}$ and $\vec{w}$ are any two vectors, then $\|\vec{v} + \vec{w}\| = \|\vec{v}\| + \|\vec{w}\|$.

5. If $\vec{v}$ and $\vec{w}$ are any two vectors, then $\|\vec{v} - \vec{w}\| = \|\vec{v}\| - \|\vec{w}\|$.

6. The vectors $2\vec{i} - \vec{j} + \vec{k}$ and $\vec{i} - 2\vec{j} + \vec{k}$ are parallel.
7. The vector $\vec{u} + \vec{v}$ is always larger in magnitude than both $\vec{u}$ and $\vec{v}$.
8. For any scalar c and vector $\vec{v}$ we have $\|c\vec{v}\| = c\|\vec{v}\|$.
9. The displacement vector from $(1,1,1)$ to $(1,2,3)$ is $-\vec{j} - 2\vec{k}$.
10. The displacement vector from (a,b) to (c,d) is the same as the displacement vector from (c,d) to (a,b).
11. The quantity $\vec{u} \cdot \vec{v}$ is a vector.
12. The plane $x + 2y - 3z = 5$ has normal vector $\vec{i} + 2\vec{j} - 3\vec{k}$.
13. If $\vec{u} \cdot \vec{v} < 0$ then the angle between $\vec{u}$ and $\vec{v}$ is greater than $\pi/2$.
14. An equation of the plane with normal vector $\vec{i} + \vec{j} + \vec{k}$ containing the point $(1,2,3)$ is $z = x + y$.
15. The triangle in 3-space with vertices $(1,1,0), (0,1,0)$ and $(0,1,1)$ has a right angle.
16. The dot product $\vec{v} \cdot \vec{v}$ is never negative.
17. If $\vec{u} \cdot \vec{v} = 0$ then either $\vec{u} = 0$ or $\vec{v} = 0$.
18. If $\vec{u}, \vec{v}$ and $\vec{w}$ are all nonzero, and $\vec{u} \cdot \vec{v} = \vec{u} \cdot \vec{w}$, then $\vec{v} = \vec{w}$.
19. For any vectors $\vec{u}$ and $\vec{v}$: $(\vec{u} + \vec{v}) \cdot (\vec{u} - \vec{v}) = \|\vec{u}\|^2 - \|\vec{v}\|^2$.
20. If $\|\vec{u}\| = 1$, then the vector $\vec{v} - (\vec{v} \cdot \vec{u})\vec{u}$ is perpendicular to $\vec{u}$.
21. $\vec{u} \times \vec{v}$ is a vector.
22. $\vec{u} \times \vec{v}$ has direction parallel to both $\vec{u}$ and $\vec{v}$.
23. $\|\vec{u} \times \vec{v}\| = \|\vec{u}\|\|\vec{v}\|$.
24. $(\vec{i} \times \vec{j}) \cdot \vec{k} = \vec{i} \cdot (\vec{j} \times \vec{k})$.
25. If $\vec{v}$ is a nonzero vector and $\vec{v} \times \vec{u} = \vec{v} \times \vec{w}$, then $\vec{u} = \vec{w}$.
26. The value of $\vec{v} \cdot (\vec{v} \times \vec{w})$ is always 0.
27. The value of $\vec{v} \times \vec{w}$ is never the same as $\vec{v} \cdot \vec{w}$.
28. The area of the triangle with two sides given by $\vec{i} + \vec{j}$ and $\vec{j} + 2\vec{k}$ is $3/2$.
29. Given a nonzero vector $\vec{v}$ in 3-space, there is a nonzero vector $\vec{w}$ such that $\vec{v} \times \vec{w} = \vec{0}$.
30. It is never true that $\vec{v} \times \vec{w} = \vec{w} \times \vec{v}$.

PROJECTS FOR CHAPTER THIRTEEN

1. **Cross Product of Vectors in the Plane**
   Let $\vec{a} = a_1\vec{i} + a_2\vec{j}$ and $\vec{b} = b_1\vec{i} + b_2\vec{j}$ be two nonparallel vectors in 2-space, as in Figure 13.46.

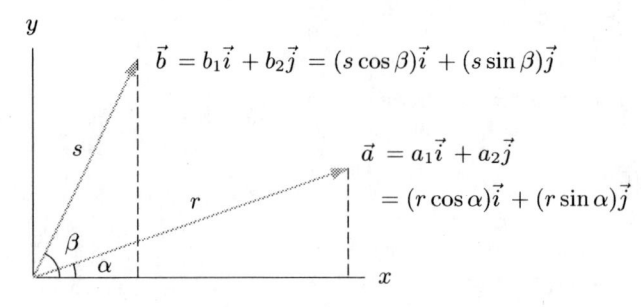

   Figure 13.46

   (a) Use the identity $\sin(\beta - \alpha) = (\sin\beta\cos\alpha - \cos\beta\sin\alpha)$ to derive the formula for the area of the parallelogram formed by $\vec{a}$ and $\vec{b}$:

   $$\text{Area of parallelogram} = |a_1 b_2 - a_2 b_1|.$$

   (b) Show that $a_1 b_2 - a_2 b_1$ is positive when the rotation from $\vec{a}$ to $\vec{b}$ is counterclockwise, and negative when it is clockwise.
   (c) Use parts (a) and (b) to show that the geometric and algebraic definitions of $\vec{a} \times \vec{b}$ give the same result.

2. The Dot Product in Genetics[2]

We define[3] the angle between two n-dimensional vectors, $\vec{v}$ and $\vec{w}$, using the dot product:

$$\cos\theta = \frac{\vec{v}\cdot\vec{w}}{\|\vec{v}\|\|\vec{w}\|} = \frac{v_1w_1 + v_2w_2 + \cdots + v_nw_n}{\|\vec{v}\|\|\vec{w}\|}, \qquad \text{provided } \|\vec{v}\|, \|\vec{w}\| \neq 0.$$

We use this idea of angle to measure how close two populations are to one another genetically. The table shows the relative frequencies of four alleles (variants of a gene) in four populations.

Allele	Eskimo	Bantu	English	Korean
A_1	0.29	0.10	0.20	0.22
A_2	0.00	0.08	0.06	0.00
B	0.03	0.12	0.06	0.20
O	0.67	0.69	0.66	0.57

Let $\vec{a}_1$ be the 4-vector showing the relative frequencies of the alleles in the Eskimo population. Let $\vec{a}_2, \vec{a}_3, \vec{a}_4$ be the corresponding vectors for the Bantu, English, and Korean populations, respectively. The genetic distance between two populations is defined as the angle between the corresponding vectors.

(a) Using this definition, is the English population closer genetically to the Bantus or to the Koreans? Explain.

(b) Is the English population closer to a half Eskimo, half Bantu population than to the Bantu population alone?

(c) Among all possible populations that are a mix of Eskimo and Bantu, find the mix that is closest to the English population.

---

[2] Adapted from Cavalli-Sforza and Edwards, "Models and Estimation Procedures," Am J. Hum. Genet., Vol. 19 (1967), pp. 223-57.

[3] The result of Problem 47 on page 673 shows that the quantity on the right-hand side of this equation is between -1 and 1, so this definition makes sense.

Chapter Fourteen

DIFFERENTIATING FUNCTIONS OF SEVERAL VARIABLES

For a function of one variable, $y = f(x)$, the derivative $dy/dx = f'(x)$ gives the rate of change of y with respect to x. For a function of two variables, $z = f(x, y)$, there is no such thing as *the* rate of change, since x and y can each vary while the other is held fixed or both can vary at once. However, we can consider the rate of change with respect to each one of the independent variables. This chapter introduces these *partial derivatives* and several ways they can be used to get a complete picture of the way the function varies.

14.1 THE PARTIAL DERIVATIVE

The derivative of a one-variable function measures its rate of change. In this section we see how a two variable function has two rates of change: one as x changes (with y held constant) and one as y changes (with x held constant).

Rate of Change of Temperature in a Metal Rod: a One-Variable Problem

Imagine an unevenly heated metal rod lying along the x-axis, with its left end at the origin and x measured in meters. (See Figure 14.1.) Let $u(x)$ be the temperature (in °C) of the rod at the point x. Table 14.1 gives values of $u(x)$. We see that the temperature increases as we move along the rod, reaching its maximum at $x = 4$, after which it starts to decrease.

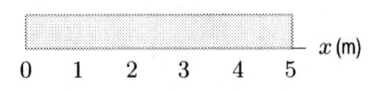

Figure 14.1: Unevenly heated metal rod

Table 14.1 *Temperature $u(x)$ of the rod*

x (m)	0	1	2	3	4	5
$u(x)$ (°C)	125	128	135	160	175	160

Example 1 Estimate the derivative $u'(2)$ using Table 14.1 and explain what the answer means in terms of temperature.

Solution The derivative $u'(2)$ is defined as a limit of difference quotients:

$$u'(2) = \lim_{h \to 0} \frac{u(2+h) - u(2)}{h}.$$

Choosing $h = 1$ so that we can use the data in Table 14.1, we get

$$u'(2) \approx \frac{u(2+1) - u(2)}{1} = \frac{160 - 135}{1} = 25.$$

This means that the temperature increases at a rate of approximately 25°C per meter as we go from left to right, past $x = 2$.

Rate of Change of Temperature in a Metal Plate

Imagine an unevenly heated thin rectangular metal plate lying in the xy-plane with its lower left corner at the origin and x and y measured in meters. The temperature (in °C) at the point (x, y) is $T(x, y)$. See Figure 14.2 and Table 14.2. How does T vary near the point $(2, 1)$? We consider the horizontal line $y = 1$ containing the point $(2, 1)$. The temperature along this line is the cross-section, $T(x, 1)$, of the function $T(x, y)$ with $y = 1$. Suppose we write $u(x) = T(x, 1)$.

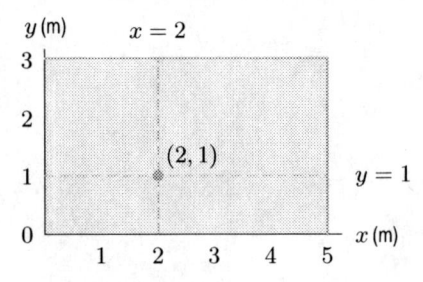

Figure 14.2: Unevenly heated metal plate

Table 14.2 *Temperature (°C) of a metal plate*

y (m) \ x (m)	0	1	2	3	4	5
3	85	90	110	135	155	180
2	100	110	120	145	190	170
1	125	128	135	160	175	160
0	120	135	155	160	160	150

What is the meaning of the derivative $u'(2)$? It is the rate of change of temperature T in the x-direction at the point $(2, 1)$, keeping y fixed. Denote this rate of change by $T_x(2, 1)$, so that

$$T_x(2, 1) = u'(2) = \lim_{h \to 0} \frac{u(2+h) - u(2)}{h} = \lim_{h \to 0} \frac{T(2+h, 1) - T(2, 1)}{h}.$$

We call $T_x(2,1)$ the *partial derivative of T with respect to x at the point* $(2,1)$. Taking $h = 1$, we can read values of T from the row with $y = 1$ in Table 13.2, giving

$$T_x(2,1) \approx \frac{T(3,1) - T(2,1)}{1} = \frac{160 - 135}{1} = 25°\text{C/m}.$$

The fact that $T_x(2,1)$ is positive means that the temperature of the plate is increasing as we move past the point $(2,1)$ in the direction of increasing x (that is, horizontally from left to right in Figure 14.2).

Example 2 Estimate the rate of change of T in the y-direction at the point $(2,1)$.

Solution The temperature along the line $x = 2$ is the cross-section of T with $x = 2$, that is, the function $v(y) = T(2, y)$. If we denote the rate of change of T in the y-direction at $(2,1)$ by $T_y(2,1)$, then

$$T_y(2,1) = v'(1) = \lim_{h \to 0} \frac{v(1+h) - v(1)}{h} = \lim_{h \to 0} \frac{T(2, 1+h) - T(2,1)}{h}.$$

We call $T_y(2,1)$ the *partial derivative of T with respect to y at the point* $(2,1)$. Taking $h = 1$ so that we can use the column with $x = 2$ in Table 13.2, we get

$$T_y(2,1) \approx \frac{T(2, 1+1) - T(2,1)}{1} = \frac{120 - 135}{1} = -15°\text{C/m}.$$

The fact that $T_y(2,1)$ is negative means that the temperature decreases as y increases.

Definition of the Partial Derivative

We study the influence of x and y separately on the value of the function $f(x, y)$ by holding one fixed and letting the other vary. This leads to the following definitions.

Partial Derivatives of f With Respect to x and y

For all points at which the limits exist, we define the **partial derivatives at the point** (a, b) by

$$f_x(a,b) = \begin{array}{c}\text{Rate of change of } f \text{ with respect to } x \\ \text{at the point } (a,b)\end{array} = \lim_{h \to 0} \frac{f(a+h, b) - f(a,b)}{h},$$

$$f_y(a,b) = \begin{array}{c}\text{Rate of change of } f \text{ with respect to } y \\ \text{at the point } (a,b)\end{array} = \lim_{h \to 0} \frac{f(a, b+h) - f(a,b)}{h}.$$

If we let a and b vary, we have the **partial derivative functions** $f_x(x, y)$ and $f_y(x, y)$.

Just as with ordinary derivatives, there is an alternative notation:

Alternative Notation for Partial Derivatives

If $z = f(x, y)$, we can write

$$f_x(x, y) = \frac{\partial z}{\partial x} \quad \text{and} \quad f_y(x, y) = \frac{\partial z}{\partial y},$$

$$f_x(a, b) = \left.\frac{\partial z}{\partial x}\right|_{(a,b)} \quad \text{and} \quad f_y(a, b) = \left.\frac{\partial z}{\partial y}\right|_{(a,b)}.$$

Chapter Fourteen / DIFFERENTIATING FUNCTIONS OF SEVERAL VARIABLES

We use the symbol ∂ to distinguish partial derivatives from ordinary derivatives. In cases where the independent variables have names different from x and y, we adjust the notation accordingly. For example, the partial derivatives of $f(u, v)$ are denoted by f_u and f_v.

Visualizing Partial Derivatives on a Graph

The ordinary derivative of a one-variable function is the slope of its graph. How do we visualize the partial derivative $f_x(a, b)$? The graph of the one-variable function $f(x, b)$ is the curve where the vertical plane $y = b$ cuts the graph of $f(x, y)$. (See Figure 14.3.) Thus, $f_x(a, b)$ is the slope of the tangent line to this curve at $x = a$.

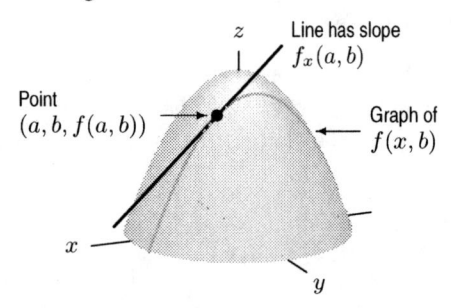

Figure 14.3: The curve $z = f(x, b)$ on the graph of f has slope $f_x(a, b)$ at $x = a$

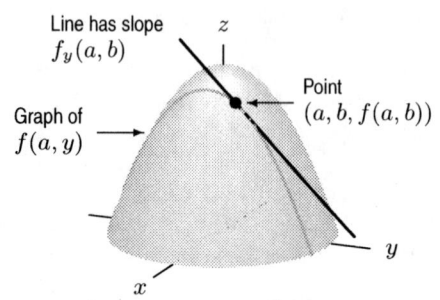

Figure 14.4: The curve $z = f(a, y)$ on the graph of f has slope $f_y(a, b)$ at $y = b$

Similarly, the graph of the function $f(a, y)$ is the curve where the vertical plane $x = a$ cuts the graph of f, and the partial derivative $f_y(a, b)$ is the slope of this curve at $y = b$. (See Figure 14.4.)

Example 3 At each point labeled on the graph of the surface $z = f(x, y)$ in Figure 14.5, say whether each partial derivative is positive or negative.

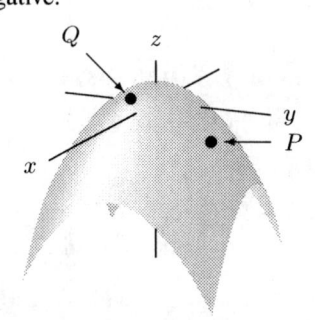

Figure 14.5: Decide the signs of f_x and f_y at P and Q

Solution The positive x-axis points out of the page. Imagine heading off in this direction from the point marked P; we descend steeply. So the partial derivative with respect to x is negative at P, with quite a large absolute value. The same is true for the partial derivative with respect to y at P, since there is also a steep descent in the positive y-direction.

At the point marked Q, heading in the positive x-direction results in a gentle descent, whereas heading in the positive y-direction results in a gentle ascent. Thus, the partial derivative f_x at Q is negative but small (that is, near zero), and the partial derivative f_y is positive but small.

Estimating Partial Derivatives From a Contour Diagram

The graph of a function $f(x, y)$ often makes clear the sign of the partial derivatives. However, numerical estimates of these derivatives are more easily made from a contour diagram than a surface graph. If we move parallel to one of the axes on a contour diagram, the partial derivative is the rate of change of the value of the function on the contours. For example, if the values on the contours are increasing as we move in the positive direction, then the partial derivative must be positive.

14.1 THE PARTIAL DERIVATIVE

Example 4 Figure 14.6 shows the contour diagram for the temperature $H(x, t)$ (in °C) in a room as a function of distance x (in meters) from a heater and time t (in minutes) after the heater has been turned on. What are the signs of $H_x(10, 20)$ and $H_t(10, 20)$? Estimate these partial derivatives and explain the answers in practical terms.

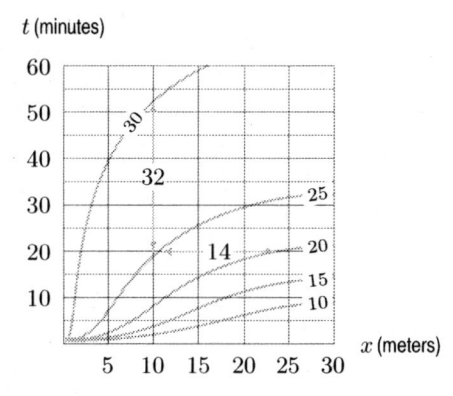

Figure 14.6: Temperature in a heated room: Heater at $x = 0$ is turned on at $t = 0$

Solution The point $(10, 20)$ is nearly on the $H = 25$ contour. As x increases, we move toward the $H = 20$ contour, so H is decreasing and $H_x(10, 20)$ is negative. This makes sense because the $H = 30$ contour is to the left: As we move further from the heater, the temperature drops. On the other hand, as t increases, we move toward the $H = 30$ contour, so H is increasing; as t decreases H decreases. Thus, $H_t(10, 20)$ is positive. This says that as time passes, the room warms up.

To estimate the partial derivatives, use a difference quotient. Looking at the contour diagram, we see there is a point on the $H = 20$ contour about 14 units to the right of the point $(10, 20)$. Hence, H decreases by 5 when x increases by 14, so we find

$$\text{Rate of change of } H \text{ with respect to } x = H_x(10, 20) \approx \frac{-5}{14} \approx -0.36°\text{C/meter}.$$

This means that near the point 10 m from the heater, after 20 minutes the temperature drops about 0.36, or one third, of a degree, for each meter we move away from the heater.

To estimate $H_t(10, 20)$, we notice that the $H = 30$ contour is about 32 units directly above the point $(10, 20)$. So H increases by 5 when t increases by 32. Hence,

$$\text{Rate of change of } H \text{ with respect to } t = H_t(10, 20) \approx \frac{5}{32} = 0.16°\text{C/minute}.$$

This means that after 20 minutes the temperature is going up about 0.16, or 1/6, of a degree each minute at the point 10 m from the heater.

Using Units to Interpret Partial Derivatives

The meaning of a partial derivative can often be explained using units.

Example 5 Suppose that your weight w in pounds is a function $f(c, n)$ of the number c of calories you consume daily and the number n of minutes you exercise daily. Using the units for w, c and n, interpret in everyday terms the statements

$$\frac{\partial w}{\partial c}(2000, 15) = 0.02 \quad \text{and} \quad \frac{\partial w}{\partial n}(2000, 15) = -0.025.$$

Solution The units of $\partial w / \partial c$ are pounds per calorie. The statement

$$\frac{\partial w}{\partial c}(2000, 15) = 0.02$$

means that if you are presently consuming 2000 calories daily and exercising 15 minutes daily, you will weigh 0.02 pounds more for each extra calorie you consume daily, or about 2 pounds for each extra 100 calories per day. The units of $\partial w/\partial n$ are pounds per minute. The statement

$$\frac{\partial w}{\partial n}(2000, 15) = -0.025$$

means that for the same calorie consumption and number of minutes of exercise, you will weigh 0.025 pounds less for each extra minute you exercise daily, or about 1 pound less for each extra 40 minutes per day. So if you eat an extra 100 calories each day and exercise about 80 minutes more each day, your weight should remain roughly steady.

Exercises and Problems for Section 14.1

Exercises

1. Using difference quotients, estimate $f_x(3, 2)$ and $f_y(3, 2)$ for the function given by

$$f(x, y) = \frac{x^2}{y+1}.$$

[Recall: A difference quotient is an expression of the form $(f(a + h, b) - f(a, b))/h$.]

2. Use difference quotients with $\Delta x = 0.1$ and $\Delta y = 0.1$ to estimate $f_x(1, 3)$ and $f_y(1, 3)$ where

$$f(x, y) = e^{-x} \sin y.$$

Then give better estimates by using $\Delta x = 0.01$ and $\Delta y = 0.01$.

3. The quantity, Q, of beef purchased at a store, in kilograms per week, is a function of the price of beef, b, and the price of chicken, c, both in dollars per kilogram.

   (a) Do you expect $\partial Q/\partial b$ to be positive or negative? Explain.
   (b) Do you expect $\partial Q/\partial c$ to be positive or negative? Explain.
   (c) Interpret the statement $\partial Q/\partial b = -213$ in terms of quantity of beef purchased.

4. The sales of a product, $S = f(p, a)$, is a function of the price, p, of the product (in dollars per unit) and the amount, a, spent on advertising (in thousands of dollars).

   (a) Do you expect f_p to be positive or negative? Why?
   (b) Explain the meaning of the statement $f_a(8, 12) = 150$ in terms of sales.

5. A drug is injected into a patient's blood vessel. The function $c = f(x, t)$ represents the concentration of the drug at a distance x mm in the direction of the blood flow measured from the point of injection and at time t seconds since the injection. What are the units of the following partial derivatives? What are their practical interpretations? What do you expect their signs to be?

   (a) $\partial c/\partial x$ (b) $\partial c/\partial t$

6. The monthly mortgage payment in dollars, P, for a house is a function of three variables

$$P = f(A, r, N),$$

where A is the amount borrowed in dollars, r is the interest rate, and N is the number of years before the mortgage is paid off.

   (a) $f(92000, 14, 30) = 1090.08$. What does this tell you, in financial terms?
   (b) $\left.\dfrac{\partial P}{\partial r}\right|_{(92000,14,30)} = 72.82$. What is the financial significance of the number 72.82?
   (c) Would you expect $\partial P/\partial A$ to be positive or negative? Why?
   (d) Would you expect $\partial P/\partial N$ to be positive or negative? Why?

7. Suppose that x is the average price of a new car and that y is the average price of a gallon of gasoline. Then q_1, the number of new cars bought in a year, depends on both x and y, so $q_1 = f(x, y)$. Similarly, if q_2 is the quantity of gas bought in a year, then $q_2 = g(x, y)$.

   (a) What do you expect the signs of $\partial q_1/\partial x$ and $\partial q_2/\partial y$ to be? Explain.
   (b) What do you expect the signs of $\partial q_1/\partial y$ and $\partial q_2/\partial x$ to be? Explain.

8. Figure 14.7 is a contour diagram of $f(x, y)$. In each of the following cases, list the marked points in the diagram (there may be none or more than one) at which

   (a) $f_x < 0$ (b) $f_y > 0$
   (c) $f_{xx} > 0$ (d) $f_{yy} < 0$

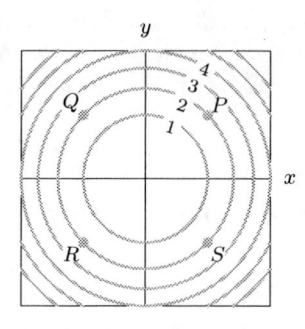

Figure 14.7

9. Figure 14.8 is a contour diagram for $z = f(x, y)$. Is f_x positive or negative? Is f_y positive or negative? Estimate $f(2, 1)$, $f_x(2, 1)$, and $f_y(2, 1)$.

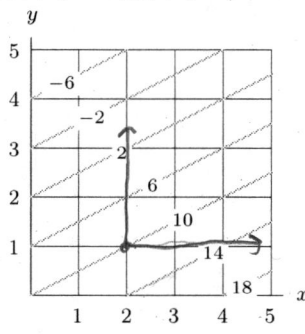

Figure 14.8

For Exercises 10–12, refer to Table 12.2 on page 611 giving the temperature adjusted for wind-chill, C, in °F, as a function $f(w, T)$ of the wind speed, w, in mph, and the temperature, T, in °F. The temperature adjusted for wind-chill tells you how cold it feels, as a result of the combination of wind and temperature.

10. Estimate $f_w(10, 25)$. What does your answer mean in practical terms?

11. Estimate $f_T(5, 20)$. What does your answer mean in practical terms?

12. From Table 12.2 you can see that when the temperature is 20°F, the temperature adjusted for wind-chill drops by an average of about 0.8°F with every 1 mph increase in wind speed from 5 mph to 10 mph. Which partial derivative is this telling you about?

Problems

13. Figure 14.9 gives a contour diagram for the number n of foxes per square kilometer in southwestern England. Estimate $\partial n/\partial x$ and $\partial n/\partial y$ at the points A, B, and C, where x is kilometers east and y is kilometers north.

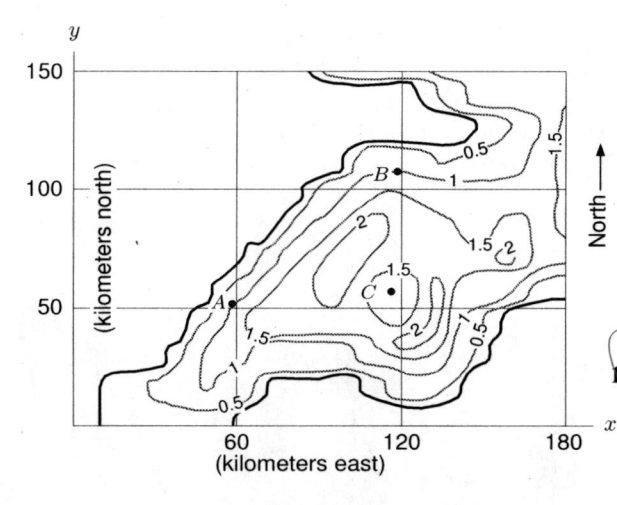

Figure 14.9

14. The cardiac output, represented by c, is the volume of blood flowing through a person's heart, per unit time. The systemic vascular resistance (SVR), represented by s, is the resistance to blood flowing through veins and arteries. Let p be a person's blood pressure. Then p is a function of c and s, so $p = f(c, s)$.

(a) What does $\partial p/\partial c$ represent?

Suppose now that $p = kcs$, where k is a constant.

(b) Sketch the level curves of p. What do they represent? Label your axes.
(c) For a person with a weak heart, it is desirable to have the heart pumping against less resistance, while maintaining the same blood pressure. Such a person may be given the drug nitroglycerine to decrease the SVR and the drug Dopamine to increase the cardiac output. Represent this on a graph showing level curves. Put a point A on the graph representing the person's state before drugs are given and a point B for after.
(d) Right after a heart attack, a patient's cardiac output drops, thereby causing the blood pressure to drop. A common mistake made by medical residents is to get the patient's blood pressure back to normal by using drugs to increase the SVR, rather than by increasing the cardiac output. On a graph of the level curves of p, put a point D representing the patient before the heart attack, a point E representing the patient right after the heart attack, and a third point F representing the patient after the resident has given the drugs to increase the SVR.

15. Figure 14.10 shows a contour diagram for the monthly payment P as a function of the interest rate, $r\%$, and the amount, L, of a 5-year loan. Estimate $\partial P/\partial r$ and $\partial P/\partial L$ at the following points. In each case, give the units and the everyday meaning of your answer.

(a) $r = 8, L = 4000$ (b) $r = 8, L = 6000$
(c) $r = 13, L = 7000$

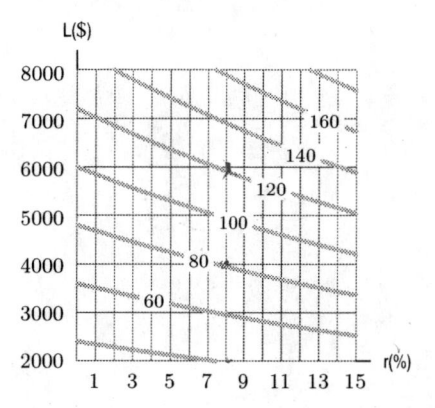

Figure 14.10

16. An experiment to measure the toxicity of formaldehyde yielded the data in Table 14.3. The values show the percent, $P = f(t, c)$, of rats surviving an exposure to formaldehyde at a concentration of c (in parts per million, ppm) after t months. Estimate $f_t(18, 6)$ and $f_c(18, 6)$. Interpret your answers in terms of formaldehyde toxicity.

Table 14.3

		Time t (months)					
		14	16	18	20	22	24
Conc. c (ppm)	0	100	100	100	99	97	95
	2	100	99	98	97	95	92
	6	96	95	93	90	86	80
	15	96	93	82	70	58	36

17. The surface $z = f(x, y)$ is shown in Figure 14.11. The points A and B are in the xy-plane.

(a) What is the sign of

(i) $f_x(A)$? (ii) $f_y(A)$?

(b) The point P in the xy-plane moves along a straight line from A to B. How does the sign of $f_x(P)$ change? How does the sign of $f_y(P)$ change?

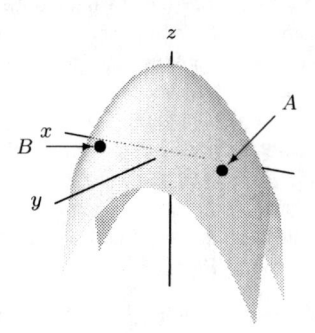

Figure 14.11

18. Figure 14.12 shows the saddle-shaped surface $z = f(x, y)$.

(a) What is the sign of $f_x(0, 5)$?
(b) What is the sign of $f_y(0, 5)$?

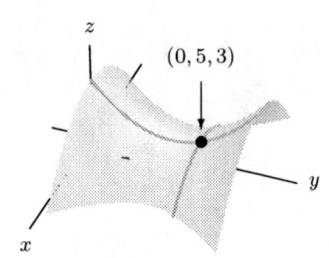

Figure 14.12

19. Figure 14.13 shows contours of $f(x, y)$ with values of f on the contours omitted. If $f_x(P) > 0$, find the sign of

(a) $f_y(P)$ (b) $f_y(Q)$ (c) $f_x(Q)$

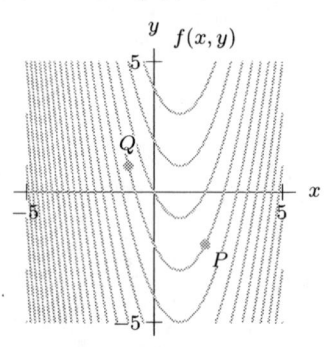

Figure 14.13

20. In each case, give a possible contour diagram for the function $f(x, y)$ if

(a) $f_x > 0$ and $f_y > 0$ (b) $f_x > 0$ and $f_y < 0$
(c) $f_x < 0$ and $f_y > 0$ (d) $f_x < 0$ and $f_y < 0$

An airport can be cleared of fog by heating the air. The amount of heat required depends on the air temperature and the wetness of the fog. Problems 21–23 involve Figure 14.14 which shows the heat $H(T, w)$ required (in calories per cubic meter of fog) as a function of the temperature T (in degrees Celsius) and the water content w (in grams per cubic meter of fog). Note that Figure 14.14 is not a contour diagram, but shows cross-sections of H with w fixed at 0.1, 0.2, 0.3, and 0.4.

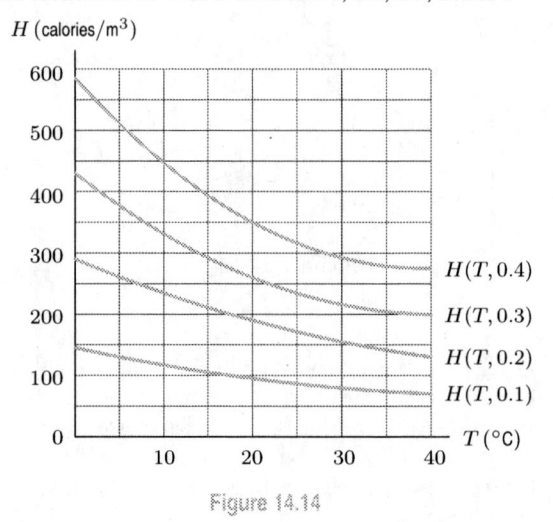

Figure 14.14

21. Use Figure 14.14 to estimate $H_T(10, 0.1)$. Interpret the partial derivative in practical terms.

22. Make a table of values for $H(T, w)$ from Figure 14.14, and use it to estimate $H_T(T, w)$ for $T = 10, 20,$ and 30 and $w = 0.1, 0.2,$ and 0.3.

23. Repeat Problem 22 for $H_w(T, w)$ at $T = 10, 20,$ and 30 and $w = 0.1, 0.2,$ and 0.3. What is the practical meaning of these partial derivatives?

14.2 COMPUTING PARTIAL DERIVATIVES ALGEBRAICALLY

Since the partial derivative $f_x(x, y)$ is the ordinary derivative of the function $f(x, y)$ with y held constant and $f_y(x, y)$ is the ordinary derivative of $f(x, y)$ with x held constant, we can use all the differentiation formulas from one-variable calculus to find partial derivatives.

Example 1 Let $f(x, y) = \dfrac{x^2}{y + 1}$. Find $f_x(3, 2)$ algebraically.

Solution We use the fact that $f_x(3, 2)$ equals the derivative of $f(x, 2)$ at $x = 3$. Since

$$f(x, 2) = \frac{x^2}{2+1} = \frac{x^2}{3},$$

differentiating with respect to x, we have

$$f_x(x, 2) = \frac{\partial}{\partial x}\left(\frac{x^2}{3}\right) = \frac{2x}{3}, \quad \text{and so} \quad f_x(3, 2) = 2.$$

Example 2 Compute the partial derivatives with respect to x and with respect to y for the following functions.
(a) $f(x, y) = y^2 e^{3x}$
(b) $z = (3xy + 2x)^5$
(c) $g(x, y) = e^{x+3y} \sin(xy)$

Solution (a) This is the product of a function of x (namely e^{3x}) and a function of y (namely y^2). When we differentiate with respect to x, we think of the function of y as a constant, and vice versa. Thus,

$$f_x(x, y) = y^2 \frac{\partial}{\partial x}\left(e^{3x}\right) = 3y^2 e^{3x},$$

$$f_y(x, y) = e^{3x} \frac{\partial}{\partial y}(y^2) = 2y e^{3x}.$$

(b) Here we use the chain rule:

$$\frac{\partial z}{\partial x} = 5(3xy + 2x)^4 \frac{\partial}{\partial x}(3xy + 2x) = 5(3xy + 2x)^4(3y + 2),$$

$$\frac{\partial z}{\partial y} = 5(3xy + 2x)^4 \frac{\partial}{\partial y}(3xy + 2x) = 5(3xy + 2x)^4 3x = 15x(3xy + 2x)^4.$$

(c) Since each function in the product is a function of both x and y, we need to use the product rule for each partial derivative:

$$g_x(x, y) = \left(\frac{\partial}{\partial x}(e^{x+3y})\right)\sin(xy) + e^{x+3y}\frac{\partial}{\partial x}(\sin(xy)) = e^{x+3y}\sin(xy) + e^{x+3y}y\cos(xy),$$

$$g_y(x, y) = \left(\frac{\partial}{\partial y}(e^{x+3y})\right)\sin(xy) + e^{x+3y}\frac{\partial}{\partial y}(\sin(xy)) = 3e^{x+3y}\sin(xy) + e^{x+3y}x\cos(xy).$$

For functions of three or more variables, we find partial derivatives by the same method: Differentiate with respect to one variable, regarding the other variables as constants. For a function $f(x, y, z)$, the partial derivative $f_x(a, b, c)$ gives the rate of change of f with respect to x along the line $y = b, z = c$.

Example 3 Find all the partial derivatives of $f(x, y, z) = \dfrac{x^2 y^3}{z}$.

Solution To find $f_x(x, y, z)$, for example, we consider y and z as fixed, giving

$$f_x(x, y, z) = \frac{2xy^3}{z}, \quad \text{and} \quad f_y(x, y, z) = \frac{3x^2 y^2}{z}, \quad \text{and} \quad f_z(x, y, z) = -\frac{x^2 y^3}{z^2}.$$

Interpretation of Partial Derivatives

Example 4 A vibrating guitar string, originally at rest along the x-axis, is shown in Figure 14.15. Let x be the distance in meters from the left end of the string. At time t seconds the point x has been displaced $y = f(x, t)$ meters vertically from its rest position, where

$$y = f(x, t) = 0.003 \sin(\pi x) \sin(2765t).$$

Evaluate $f_x(0.3, 1)$ and $f_t(0.3, 1)$ and explain what each means in practical terms.

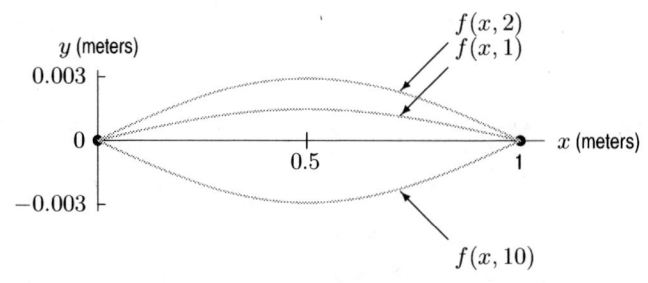

Figure 14.15: The position of a vibrating guitar string at several different times: Graph of $f(x, t)$ for $t = 1, 2, 10$.

Solution Differentiating $f(x, t) = 0.003 \sin(\pi x) \sin(2765t)$ with respect to x, we have

$$f_x(x, t) = 0.003\pi \cos(\pi x) \sin(2765t).$$

In particular, substituting $x = 0.3$ and $t = 1$ gives

$$f_x(0.3, 1) = 0.003\pi \cos(\pi(0.3)) \sin(2765) \approx 0.002.$$

To see what $f_x(0.3, 1)$ means, think about the function $f(x, 1)$. The graph of $f(x, 1)$ in Figure 14.16 is a snapshot of the string at the time $t = 1$. Thus, the derivative $f_x(0.3, 1)$ is the slope of the string at the point $x = 0.3$ at the instant when $t = 1$.

Similarly, taking the derivative of $f(x, t) = 0.003 \sin(\pi x) \sin(2765t)$ with respect to t, we get

$$f_t(x, t) = (0.003)(2765) \sin(\pi x) \cos(2765t) = 8.3 \sin(\pi x) \cos(2765t).$$

Since $f(x, t)$ is in meters and t is in seconds, the derivative $f_t(0.3, 1)$ is in m/sec. Thus, substituting $x = 0.3$ and $t = 1$,

$$f_t(0.3, 1) = 8.3 \sin(\pi(0.3)) \cos(2765(1)) \approx 6 \text{ m/sec}.$$

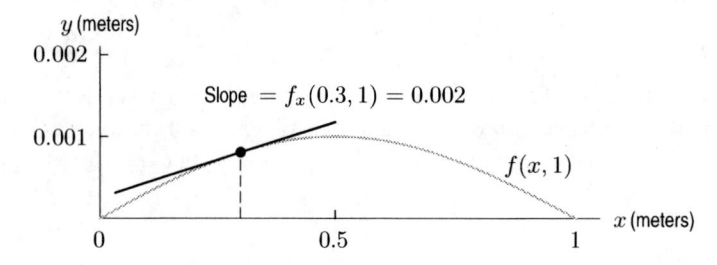

Figure 14.16: Graph of $f(x, 1)$: Snapshot of the shape of the string at $t = 1$ sec

To see what $f_t(0.3, 1)$ means, think about the function $f(0.3, t)$. The graph of $f(0.3, t)$ is a position versus time graph that tracks the up and down movement of the point on the string where $x = 0.3$.

(See Figure 14.17.) The derivative $f_t(0.3, 1) = 6$ m/sec is the velocity of that point on the string at time $t = 1$. The fact that $f_t(0.3, 1)$ is positive indicates that the point is moving upward when $t = 1$.

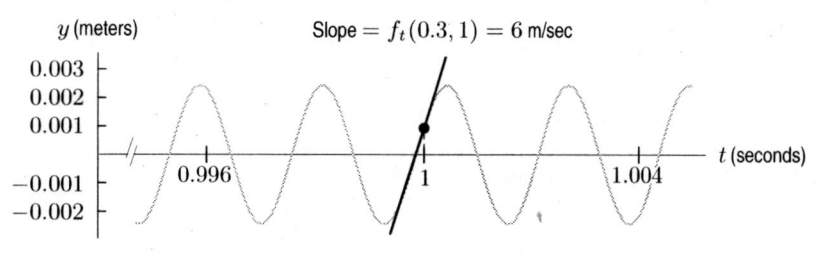

Figure 14.17: Graph of $f(0.3, t)$: Position versus time graph of the point $x = 0.3$ m from the end of the guitar string

Exercises and Problems for Section 14.2

Exercises

Find the partial derivatives in Exercises 1–35. Assume the variables are restricted to a domain on which the function is defined.

1. f_x and f_y if $f(x, y) = 5x^2y^3 + 8xy^2 - 3x^2$
2. $f_x(1, 2)$ and $f_y(1, 2)$ if $f(x, y) = x^3 + 3x^2y - 2y^2$
3. $\dfrac{\partial}{\partial y}(3x^5y^7 - 32x^4y^3 + 5xy)$
4. $\dfrac{\partial z}{\partial x}$ and $\dfrac{\partial z}{\partial y}$ if $z = (x^2 + x - y)^7$
5. z_x if $z = x^2y + 2x^5y$
6. V_r if $V = \frac{1}{3}\pi r^2 h$
7. z_y if $z = \dfrac{3x^2y^7 - y^2}{15xy - 8}$
8. $\dfrac{\partial}{\partial T}\left(\dfrac{2\pi r}{T}\right)$
9. $\dfrac{\partial}{\partial x}(a\sqrt{x})$
10. $\dfrac{\partial}{\partial x}(xe^{\sqrt{xy}})$
11. F_m if $F = mg$
12. $z_x(2, 3)$ if $z = (\cos x) + y$
13. $\dfrac{\partial A}{\partial h}$ if $A = \frac{1}{2}(a + b)h$
14. $\dfrac{\partial}{\partial m}\left(\dfrac{1}{2}mv^2\right)$
15. $\dfrac{\partial}{\partial B}\left(\dfrac{1}{u_0}B^2\right)$
16. $\dfrac{\partial}{\partial r}\left(\dfrac{2\pi r}{v}\right)$
17. F_v if $F = \dfrac{mv^2}{r}$
18. $\dfrac{\partial}{\partial v_0}(v_0 + at)$
19. $\dfrac{\partial F}{\partial m_2}$ if $F = \dfrac{Gm_1m_2}{r^2}$
20. a_v if $a = \dfrac{v^2}{r}$
21. $\dfrac{\partial}{\partial t}\left(v_0 t + \frac{1}{2}at^2\right)$
22. $\dfrac{\partial f_0}{\partial L}$ if $f_0 = \dfrac{1}{2\pi\sqrt{LC}}$
23. $\dfrac{\partial T}{\partial l}$ if $T = 2\pi\sqrt{\dfrac{l}{g}}$
24. $\dfrac{\partial}{\partial M}\left(\dfrac{2\pi r^{3/2}}{\sqrt{GM}}\right)$
25. f_a if $f(a, b) = e^a \sin(a + b)$
26. z_x if $z = \sin(5x^3y - 3xy^2)$
27. g_x if $g(x, y) = \ln(ye^{xy})$
28. $\dfrac{\partial V}{\partial r}$ and $\dfrac{\partial V}{\partial h}$ if $V = \frac{4}{3}\pi r^2 h$
29. u_E if $u = \dfrac{1}{2}\epsilon_0 E^2 + \dfrac{1}{2\mu_0}B^2$
30. $\dfrac{\partial}{\partial x}\left(\dfrac{1}{\sqrt{2\pi}\sigma}e^{-(x-\mu)^2/(2\sigma^2)}\right)$
31. $\dfrac{\partial Q}{\partial K}$ if $Q = c(a_1 K^{b_1} + a_2 L^{b_2})^\gamma$
32. z_x and z_y for $z = x^7 + 2^y + x^y$
33. $\dfrac{\partial m}{\partial v}$ if $m = \dfrac{m_0}{\sqrt{1 - v^2/c^2}}$
34. $\dfrac{\partial z}{\partial y}\bigg|_{(1, 0.5)}$ if $z = e^{x+2y}\sin y$
35. $\dfrac{\partial f}{\partial x}\bigg|_{(\pi/3, 1)}$ if $f(x, y) = x\ln(y\cos x)$

Problems

36. (a) Let $f(u,v) = u(u^2+v^2)^{3/2}$. Use a difference quotient to approximate $f_u(1,3)$ with $h = 0.001$.
(b) Now evaluate $f_u(1,3)$ exactly. Was the approximation in part (a) reasonable?

37. (a) Let $f(x,y) = x^2 + y^2$. Estimate $f_x(2,1)$ and $f_y(2,1)$ using the contour diagram for f in Figure 14.18.
(b) Estimate $f_x(2,1)$ and $f_y(2,1)$ from a table of values for f with $x = 1.9, 2, 2.1$ and $y = 0.9, 1, 1.1$.
(c) Compare your estimates in parts (a) and (b) with the exact values of $f_x(2,1)$ and $f_y(2,1)$ found algebraically.

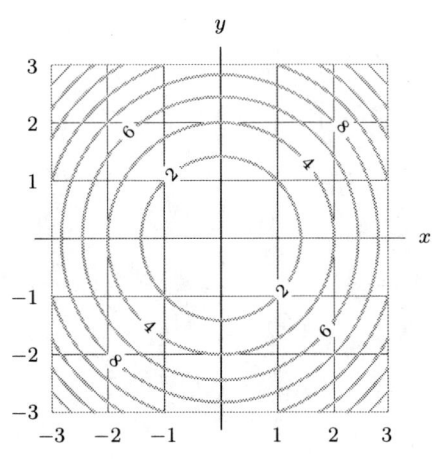

Figure 14.18

38. The energy, E, of a body of mass m moving with speed v is given by the formula

$$E = mc^2\left(\frac{1}{\sqrt{1-v^2/c^2}} - 1\right).$$

The speed, v, is nonnegative and less than the speed of light, c, which is a constant.

(a) Find $\partial E/\partial m$. What would you expect the sign of $\partial E/\partial m$ to be? Explain.
(b) Find $\partial E/\partial v$. Explain what you would expect the sign of $\partial E/\partial v$ to be and why.

39. A one-meter long bar is heated unevenly, with temperature in °C at a distance x meters from one end at time t given by

$$H(x,t) = 100e^{-0.1t}\sin(\pi x) \qquad 0 \le x \le 1.$$

(a) Sketch a graph of H against x for $t=0$ and $t=1$.
(b) Calculate $H_x(0.2, t)$ and $H_x(0.8, t)$. What is the practical interpretation (in terms of temperature) of these two partial derivatives? Explain why each one has the sign it does.

(c) Calculate $H_t(x,t)$. What is its sign? What is its interpretation in terms of temperature?

40. Money in a bank account earns interest at a continuous rate, r. The amount of money, $\$B$, in the account depends on the amount deposited, $\$P$, and the time, t, it has been in the bank according to the formula

$$B = Pe^{rt}.$$

Find $\partial B/\partial t$ and $\partial B/\partial P$ and interpret each in financial terms.

41. The gravitational force, F newtons, exerted on a mass of m kg at a distance of r meters from the center of the earth is given by

$$F = \frac{GMm}{r^2}$$

where the mass of the earth $M = 6 \cdot 10^{24}$ kilograms, and $G = 6.67 \cdot 10^{-11}$. Find the gravitational force on a person with mass 70 kg at the surface of the earth ($r = 6.4 \cdot 10^6$). Calculate $\partial F/\partial m$ and $\partial F/\partial r$ for these values of m and r. Interpret these partial derivatives in terms of gravitational force.

42. Let $h(x,t) = 5 + \cos(0.5x - t)$ describe a wave. The value of $h(x,t)$ gives the depth of the water in cm at a distance x meters from a fixed point and at time t seconds. Evaluate $h_x(2,5)$ and $h_t(2,5)$ and interpret each in terms of the wave.

43. The Dubois formula relates a person's surface area, s, in m², to weight, w, in kg, and height, h, in cm, by

$$s = f(w,h) = 0.01w^{0.25}h^{0.75}.$$

Find $f(65,160)$, $f_w(65,160)$, and $f_h(65,160)$. Interpret your answers in terms of surface area, height, and weight.

44. Show that the Cobb-Douglas function

$$Q = bK^\alpha L^{1-\alpha} \quad \text{where} \quad 0 < \alpha < 1$$

satisfies the equation

$$K\frac{\partial Q}{\partial K} + L\frac{\partial Q}{\partial L} = Q.$$

45. Is there a function f which has the following partial derivatives? If so what is it? Are there any others?

$$f_x(x,y) = 4x^3y^2 - 3y^4,$$
$$f_y(x,y) = 2x^4y - 12xy^3.$$

14.3 LOCAL LINEARITY AND THE DIFFERENTIAL

In Sections 14.1 and 14.2 we studied a function of two variables by allowing one variable at a time to change. We now let both variables change at once to develop a linear approximation for functions of two variables.

Zooming In to See Local Linearity

For a function of one variable, local linearity means that as we zoom in on the graph, it looks like a straight line. As we zoom in on the graph of a two-variable function, the graph usually looks like a plane, which is the graph of a linear function of two variables. (See Figure 14.19.)

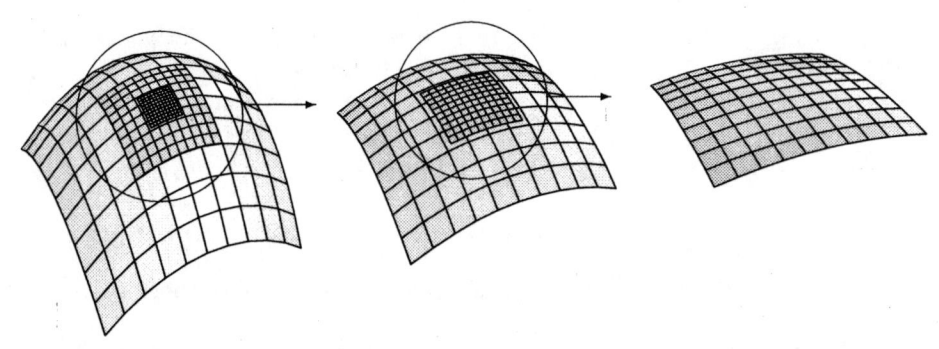

Figure 14.19: Zooming in on the graph of a function of two variables until the graph looks like a plane

Similarly, Figure 14.20 shows three successive views of the contours near a point. As we zoom in, the contours look more like equally spaced parallel lines, which are the contours of a linear function. (As we zoom in, we have to add more contours.)

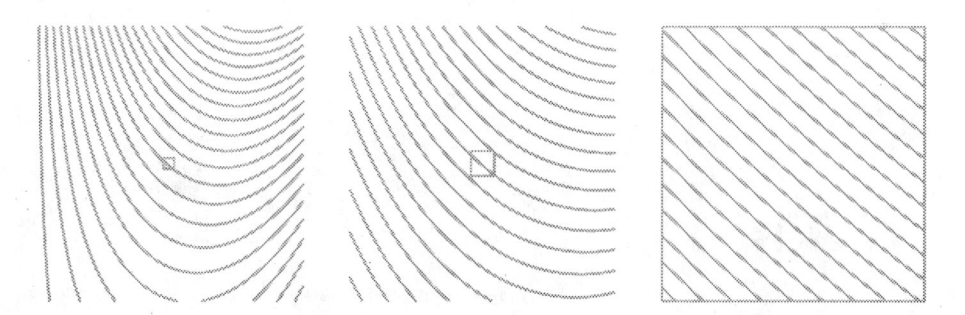

Figure 14.20: Zooming in on a contour diagram until the lines look parallel and equally spaced

This effect can also be seen numerically by zooming in with tables of values. Table 14.4 shows three tables of values for $f(x, y) = x^2 + y^3$ near $x = 2$, $y = 1$, each one a closer view than the previous one. Notice how each table looks more like the table of a linear function.

Table 14.4 *Zooming in on values of $f(x,y) = x^2 + y^3$ near $(2, 1)$ until the table looks linear*

	y=0	y=1	y=2
x=1	1	2	9
x=2	4	5	12
x=3	9	10	17

	y=0.9	y=1.0	y=1.1
x=1.9	4.34	4.61	4.94
x=2.0	4.73	5.00	5.33
x=2.1	5.14	5.41	5.74

	y=0.99	y=1.00	y=1.01
x=1.99	4.93	4.96	4.99
x=2.00	4.97	5.00	5.03
x=2.01	5.01	5.04	5.07

Zooming in Algebraically: Differentiability

Seeing a plane when we zoom in at a point tells us (provided the plane is not vertical) that $f(x, y)$ is closely approximated near that point by a linear function, $L(x, y)$:

$$f(x, y) \approx L(x, y).$$

The graph of the function $z = L(x, y)$ is the tangent plane at that point. Provided the approximation is sufficiently good, we say that $f(x, y)$ is *differentiable* at the point. Section 14.8 on page 734 defines precisely what is meant by the approximation being sufficiently good. The functions we encounter are differentiable at most points in their domain. We usually only consider the question of differentiability at points that are in the interior of the domain of f, not on the boundary[1].

The Tangent Plane

The plane that we see when we zoom in on a surface is called the *tangent plane* to the surface at the point. Figure 14.21 shows the graph of a function with a tangent plane.

What is the equation of the tangent plane? At the point (a, b), the x-slope of the graph of f is the partial derivative $f_x(a, b)$ and the y-slope is $f_y(a, b)$. Thus, using the equation for a plane on page 629 of Chapter 12, we have the following result:

Tangent Plane to the Surface $z = f(x, y)$ at the Point (a, b)

Assuming f is differentiable at (a, b), the equation of the tangent plane is

$$z = f(a, b) + f_x(a, b)(x - a) + f_y(a, b)(y - b).$$

Here we are thinking of a and b as fixed, so $f(a, b)$, and $f_x(a, b)$, and $f_y(a, b)$ are constants. Thus, the right side of the equation is a linear function of x and y.

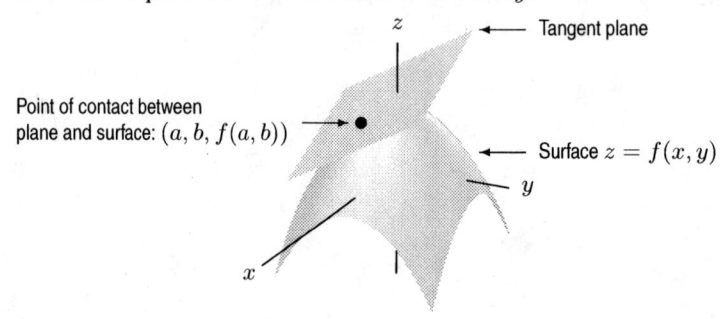

Figure 14.21: The tangent plane to the surface $z = f(x, y)$ at the point (a, b)

Example 1 Find the equation for the tangent plane to the surface $z = x^2 + y^2$ at the point $(3, 4)$.

Solution We have $f_x(x, y) = 2x$, so $f_x(3, 4) = 6$, and $f_y(x, y) = 2y$, so $f_y(3, 4) = 8$. Also, $f(3, 4) = 3^2 + 4^2 = 25$. Thus, the equation for the tangent plane at $(3, 4)$ is

$$z = 25 + 6(x - 3) + 8(y - 4) = -25 + 6x + 8y.$$

Local Linearization

Since the tangent plane lies close to the surface near the point at which they meet, z-values on the tangent plane are close to values of $f(x, y)$ for points near (a, b). Thus, replacing z by $f(x, y)$ in the equation of the tangent plane, we get the following approximation:

---

[1] Interior points and boundary points are defined precisely on page 761

Tangent Plane Approximation to $f(x, y)$ for (x, y) Near the Point (a, b)

Provided f is differentiable at (a, b), we can approximate $f(x, y)$:

$$f(x, y) \approx f(a, b) + f_x(a, b)(x - a) + f_y(a, b)(y - b).$$

We are thinking of a and b as fixed, so the expression on the right side is linear in x and y. The right side of this approximation is called the **local linearization** of f near $x = a$, $y = b$.

Figure 14.22 shows the tangent plane approximation graphically.

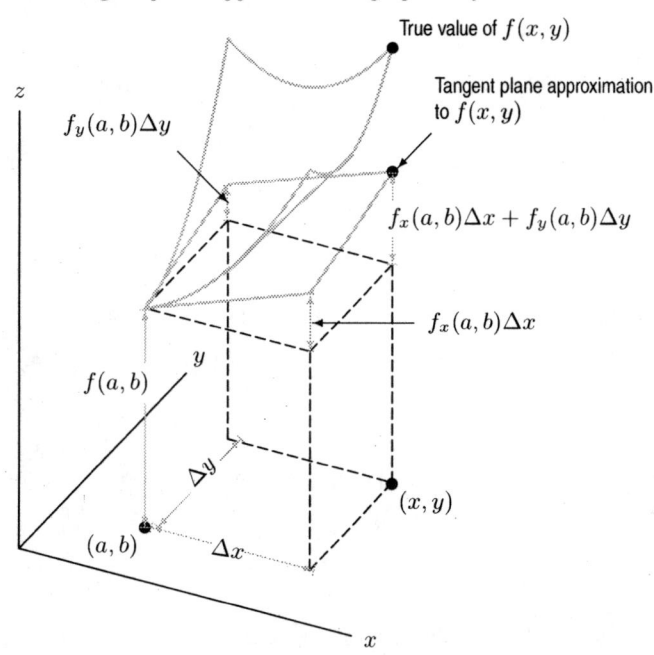

Figure 14.22: Local linearization: Approximating $f(x, y)$ by the z-value from the tangent plane

Example 2 Find the local linearization of $f(x, y) = x^2 + y^2$ at the point $(3, 4)$. Estimate $f(2.9, 4.2)$ and $f(2, 2)$ using the linearization and compare your answers to the true values.

Solution Let $z = f(x, y) = x^2 + y^2$. In Example 1 on page 698, we found the equation of the tangent plane at $(3, 4)$ to be

$$z = 25 + 6(x - 3) + 8(y - 4).$$

Therefore, for (x, y) near $(3, 4)$, we have the local linearization

$$f(x, y) \approx 25 + 6(x - 3) + 8(y - 4).$$

Substituting $x = 2.9, y = 4.2$ gives

$$f(2.9, 4.2) \approx 25 + 6(-0.1) + 8(0.2) = 26.$$

This compares favorably with the true value $f(2.9, 4.2) = (2.9)^2 + (4.2)^2 = 26.05$.

However, the local linearization does not give a good approximation at points far away from $(3, 4)$. For example, if $x = 2, y = 2$, the local linearization gives

$$f(2, 2) \approx 25 + 6(-1) + 8(-2) = 3,$$

whereas the true value of the function is $f(2, 2) = 2^2 + 2^2 = 8$.

Chapter Fourteen   DIFFERENTIATING FUNCTIONS OF SEVERAL VARIABLES

Example 3   Designing safe boilers depends on knowing how steam behaves under changes in temperature and pressure. Steam tables, such as Table 14.5, are published giving values of the function $V = f(T, P)$ where V is the volume (in ft^3) of one pound of steam at a temperature T (in °F) and pressure P (in lb/in^2).

(a) Give a linear function approximating $V = f(T, P)$ for T near 500°F and P near 24 lb/in^2.
(b) Estimate the volume of a pound of steam at a temperature of 505°F and a pressure of 24.3 lb/in^2.

Table 14.5   *Volume (in cubic feet) of one pound of steam at various temperatures and pressures*

		Pressure P (lb/in^2)			
		20	22	24	26
Temperature T (°F)	480	27.85	25.31	23.19	21.39
	500	28.46	25.86	23.69	21.86
	520	29.06	26.41	24.20	22.33
	540	29.66	26.95	24.70	22.79

Solution   (a) We want the local linearization around the point $T = 500$, $P = 24$, which is

$$f(T, P) \approx f(500, 24) + f_T(500, 24)(T - 500) + f_P(500, 24)(P - 24).$$

We read the value $f(500, 24) = 23.69$ from the table.

Next we approximate $f_T(500, 24)$ by a difference quotient. From the $P = 24$ column, we compute the average rate of change between $T = 500$ and $T = 520$:

$$f_T(500, 24) \approx \frac{f(520, 24) - f(500, 24)}{520 - 500} = \frac{24.20 - 23.69}{20} = 0.0255.$$

Note that $f_T(500, 24)$ is positive, because steam expands when heated.

Next we approximate $f_P(500, 24)$ by looking at the $T = 500$ row and computing the average rate of change between $P = 24$ and $P = 26$:

$$f_P(500, 24) \approx \frac{f(500, 26) - f(500, 24)}{26 - 24} = \frac{21.86 - 23.69}{2} = -0.915.$$

Note that $f_P(500, 24)$ is negative, because increasing the pressure on steam decreases its volume. Using these approximations for the partial derivatives, we obtain the local linearization:

$$V = f(T, P) \approx 23.69 + 0.0255(T - 500) - 0.915(P - 24) \text{ ft}^3 \quad \begin{array}{l} \text{for } T \text{ near } 500 \text{ °F} \\ \text{and } P \text{ near } 24 \text{ lb/in}^2. \end{array}$$

(b) We are interested in the volume at $T = 505$°F and $P = 24.3$ lb/in^2. Since these values are close to $T = 500$°F and $P = 24$ lb/in^2, we use the linear relation obtained in part (a).

$$V \approx 23.69 + 0.0255(505 - 500) - 0.915(24.3 - 24) = 23.54 \text{ ft}^3.$$

Local Linearity With Three or More Variables

Local linear approximations for functions of three or more variables follow the same pattern as for functions of two variables. The local linearization of $f(x, y, z)$ at (a, b, c) is given by

$$f(x, y, z) \approx f(a, b, c) + f_x(a, b, c)(x - a) + f_y(a, b, c)(y - b) + f_z(a, b, c)(z - c).$$

The Differential

We are often interested in the change in the value of the function as we move from the point (a, b) to a nearby point (x, y). Then we use the notation

$$\Delta f = f(x,y) - f(a,b) \quad \text{and} \quad \Delta x = x - a \quad \text{and} \quad \Delta y = y - b$$

to rewrite the tangent plane approximation

$$f(x,y) \approx f(a,b) + f_x(a,b)(x-a) + f_y(a,b)(y-b)$$

in the form

$$\Delta f \approx f_x(a,b)\Delta x + f_y(a,b)\Delta y.$$

For fixed a and b, the right side of this is a linear function of Δx and Δy that can be used to estimate Δf. We call this linear function the *differential*. To define the differential in general, we introduce new variables dx and dy to represent changes in x and y.

> **The Differential of a Function $z = f(x, y)$**
>
> The **differential**, df (or dz), at a point (a, b) is the linear function of dx and dy given by the formula
>
> $$df = f_x(a,b)\,dx + f_y(a,b)\,dy.$$
>
> The differential at a general point is often written $df = f_x\,dx + f_y\,dy$.

Note that the differential, df, is a function of four variables a, b, and dx, dy.

Example 4 Compute the differentials of the following functions.
(a) $f(x,y) = x^2 e^{5y}$ (b) $z = x\sin(xy)$ (c) $f(x,y) = x\cos(2x)$

Solution (a) Since $f_x(x,y) = 2xe^{5y}$ and $f_y(x,y) = 5x^2 e^{5y}$, we have

$$df = 2xe^{5y}\,dx + 5x^2 e^{5y}\,dy.$$

(b) Since $\partial z/\partial x = \sin(xy) + xy\cos(xy)$ and $\partial z/\partial y = x^2\cos(xy)$, we have

$$dz = (\sin(xy) + xy\cos(xy))\,dx + x^2\cos(xy)\,dy.$$

(c) Since $f_x(x,y) = \cos(2x) - 2x\sin(2x)$ and $f_y(x,y) = 0$, we have

$$df = (\cos(2x) - 2x\sin(2x))\,dx + 0\,dy = (\cos(2x) - 2x\sin(2x))\,dx.$$

Example 5 The density ρ (in g/cm^3) of carbon dioxide gas CO_2 depends upon its temperature T (in °C) and pressure P (in atmospheres). The ideal gas model for CO_2 gives what is called the state equation

$$\rho = \frac{0.5363P}{T + 273.15}.$$

Compute the differential $d\rho$. Explain the signs of the coefficients of dT and dP.

Solution The differential for $\rho = f(T, P)$ is

$$d\rho = f_T(T,P)\,dT + f_P(T,P)\,dP = \frac{-0.5363P}{(T+273.15)^2}\,dT + \frac{0.5363}{T+273.15}\,dP.$$

The coefficient of dT is negative because increasing the temperature expands the gas (if the pressure is kept constant) and therefore decreases its density. The coefficient of dP is positive because increasing the pressure compresses the gas (if the temperature is kept constant) and therefore increases its density.

Where Does the Notation for the Differential Come From?

We write the differential as a linear function of the new variables dx and dy. You may wonder why we chose these names for our variables. The reason is historical: The people who invented calculus thought of dx and dy as "infinitesimal" changes in x and y. The equation

$$df = f_x dx + f_y dy$$

was regarded as an infinitesimal version of the local linear approximation

$$\Delta f \approx f_x \Delta x + f_y \Delta y.$$

In spite of the problems with defining exactly what "infinitesimal" means, some mathematicians, scientists, and engineers think of the differential in terms of infinitesimals.

Figure 14.23 illustrates a way of thinking about differentials that combines the definition with this informal point of view. It shows the graph of f along with a view of the graph around the point $(a, b, f(a, b))$ under a microscope. Since f is locally linear at the point, the magnified view looks like the tangent plane. Under the microscope, we use a magnified coordinate system with its origin at the point $(a, b, f(a, b))$ and with coordinates dx, dy, and dz along the three axes. The graph of the differential df is the tangent plane, which has equation $df = f_x(a, b)\, dx + f_y(a, b)\, dy$ in the magnified coordinates.

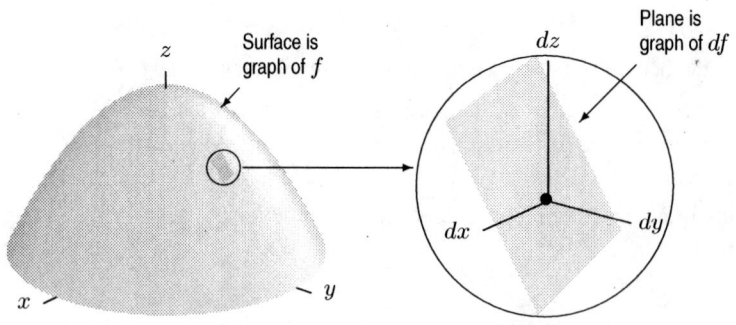

Figure 14.23: The graph of f, with a view through a microscope showing the tangent plane in the magnified coordinate system

Exercises and Problems for Section 14.3

Exercises

For the functions in Exercises 1–4, find the equation of the tangent plane at the given point.

1. $z = \frac{1}{2}(x^2 + 4y^2)$ at the point $(2, 1, 4)$.
2. $z = ye^{x/y}$ at the point $(1, 1, e)$.
3. $z = e^y + x + x^2 + 6$ at the point $(1, 0, 9)$.
4. $z = \ln(x^2 + 1) + y^2$ at the point $(0, 3, 9)$

Find the differentials of the functions in Exercises 5–8.

5. $g(u, v) = u^2 + uv$
6. $f(x, y) = \sin(xy)$
7. $z = e^{-x} \cos y$
8. $h(x, t) = e^{-3t} \sin(x + 5t)$

Find differentials of the functions in Exercises 9–12 at the given point.

9. $f(x, y) = xe^{-y}$ at $(1, 0)$
10. $F(m, r) = Gm/r^2$ at $(100, 10)$
11. $g(x, t) = x^2 \sin(2t)$ at $(2, \pi/4)$
12. $P(L, K) = 1.01L^{0.25}K^{0.75}$ at $(100, 1)$

Problems

13. Find the local linearization of the function $f(x, y) = x^2 y$ at the point $(3, 1)$.
14. A student was asked to find the equation of the tangent plane to the surface $z = x^3 - y^2$ at the point $(x, y) = (2, 3)$. The student's answer was

$$z = 3x^2(x - 2) - 2y(y - 3) - 1.$$

(a) At a glance, how do you know this is wrong?
(b) What mistake did the student make?
(c) Answer the question correctly.

15. **(a)** Check the local linearity of $f(x,y) = e^{-x} \sin y$ near $x = 1$, $y = 2$ by making a table of values of f for $x = 0.9$, 1.0, 1.1 and $y = 1.9$, 2.0, 2.1. Express values of f with 4 digits after the decimal point. Then make a table of values for $x = 0.99$, 1.00, 1.01 and $y = 1.99$, 2.00, 2.01, again showing 4 digits after the decimal point. Do both tables look nearly linear? Does the second table look more linear than the first?

    (b) Give the local linearization of $f(x,y) = e^{-x} \sin y$ at $(1,2)$, first using your tables, and second using the fact that $f_x(x,y) = -e^{-x} \sin y$ and $f_y(x,y) = e^{-x} \cos y$.

16. Find the differential of $f(x,y) = \sqrt{x^2 + y^3}$ at the point $(1,2)$. Use it to estimate $f(1.04, 1.98)$.

17. Give the local linearization for the monthly car-loan payment function at each of the points investigated in Problem 15 on page 691.

18. An unevenly heated plate has temperature $T(x,y)$ in °C at the point (x,y). If $T(2,1) = 135$, and $T_x(2,1) = 16$, and $T_y(2,1) = -15$, estimate the temperature at the point $(2.04, 0.97)$.

19. In Example 3 on page 700 we found a linear approximation for $V = f(T, P)$ near $(500, 24)$. Now find a linear approximation near $(480, 20)$.

20. In Example 3 on page 700 we found a linear approximation for $V = f(T, p)$ near $(500, 24)$.

    (a) Test the accuracy of this approximation by comparing its predicted value with the four neighboring values in the table. What do you notice? Which predicted values are accurate? Which are not? Explain your answer.

    (b) Suggest a linear approximation for $f(T,p)$ near $(500, 24)$ that does not have the property you noticed in part (a). [Hint: Estimate the partial derivatives in a different way.]

21. A company uses x hours of unskilled labor and y hours of skilled labor to produce $F(x,y) = 60x^{2/3}y^{1/3}$ units of output. It currently employs 400 hours of unskilled labor and 50 hours of skilled labor. The company is planning to hire an additional 5 hours of skilled labor.

    (a) Use a linear approximation to decide by about how much the company can reduce its use of unskilled labor and keep its output at current level.

    (b) Calculate the exact value of the reduction.

22. The gas equation for one mole of oxygen relates its pressure, P (in atmospheres), its temperature, T (in K), and its volume, V (in cubic decimeters, dm^3):

    $$T = 16.574\frac{1}{V} - 0.52754\frac{1}{V^2} - 0.3879P + 12.187VP.$$

    (a) Find the temperature T and differential dT if the volume is 25 dm^3 and the pressure is 1 atmosphere.

    (b) Use your answer to part (a) to estimate how much the volume would have to change if the pressure increased by 0.1 atmosphere and the temperature remained constant.

23. The coefficient, β, of thermal expansion of a liquid relates the change in the volume V (in m^3) of a fixed quantity of a liquid to an increase in its temperature T (in °C):

    $$dV = \beta V \, dT.$$

    (a) Let ρ be the density (in kg/m^3) of water as a function of temperature. Write an expression for $d\rho$ in terms of ρ and dT.

    (b) The graph in Figure 14.24 shows density of water as a function of temperature. Use it to estimate β when $T = 20°C$ and when $T = 80°C$.

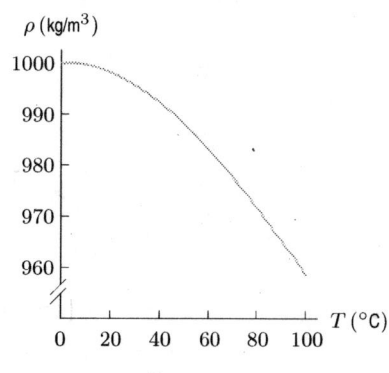

    Figure 14.24

24. One mole of ammonia gas is contained in a vessel which is capable of changing its volume (a compartment sealed by a piston, for example). The total energy U (in joules) of the ammonia, is a function of the volume V (in m^3) of the container, and the temperature T (in K) of the gas. The differential dU is given by

    $$dU = 840 \, dV + 27.32 \, dT.$$

    (a) How does the energy change if the volume is held constant and the temperature is increased slightly?

    (b) How does the energy change if the temperature is held constant and the volume is increased slightly?

    (c) Find the approximate change in energy if the gas is compressed by 100 cm^3 and heated by 2 K.

25. A fluid moves through a tube of radius $r = 0.005 \pm 0.00025$ m, and length 1 m under a pressure $p = 10^5 \pm 1000$ pascals, at a rate $v = 0.625 \cdot 10^{-9}$ m^3 per unit time. Find the maximum error in the viscosity η given by

    $$\eta = \frac{\pi}{8}\frac{pr^4}{v}.$$

26. The period, T, of oscillation in seconds of a pendulum clock is given by $T = 2\pi\sqrt{l/g}$, where g is the acceleration due to gravity. The length of the pendulum, l, depends on the temperature, t, according to the formula $l = l_0(1 + \alpha t)$ where l_0 is the length of the pendulum at temperature t_0 and α is a constant which characterizes the clock. The clock is set to the correct period at the temperature t_0. How many seconds a day does the clock lose or gain when the temperature is $t_0 + \Delta t$? Show that this loss or gain is independent of l_0.

27. (a) Write a formula for the number π using only the perimeter L and the area A of a circle.
(b) Suppose that L and A are determined experimentally. Show that if the relative, or percent, errors in the measured values of L and A are λ and μ, respectively, then the resulting relative, or percent, error in π is $2\lambda - \mu$.

14.4 GRADIENTS AND DIRECTIONAL DERIVATIVES IN THE PLANE

The Rate of Change in an Arbitrary Direction: The Directional Derivative

The partial derivatives of a function f tell us the rate of change of f in the directions parallel to the coordinate axes. In this section we see how to compute the rate of change of f in an arbitrary direction.

Example 1 Figure 14.25 shows the temperature, in °C, at the point (x, y). Estimate the average rate of change of temperature as we walk from point A to point B.

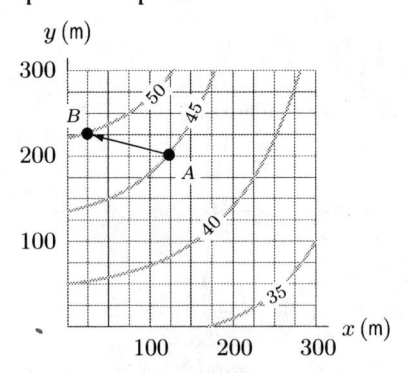

Figure 14.25: Estimating rate of change on a temperature map

Solution At the point A we are on the $H = 45°C$ contour. At B we are on the $H = 50°C$ contour. The displacement vector from A to B has x component approximately $-100\vec{i}$ and y component approximately $25\vec{j}$, so its length is $\sqrt{(-100)^2 + 25^2} \approx 103$. Thus the temperature rises by $5°C$ as we move 103 meters, so the average rate of change of the temperature in that direction is about $5/103 \approx 0.05°C/m$.

Suppose we want to compute the rate of change of a function $f(x, y)$ at the point $P = (a, b)$ in the direction of the unit vector $\vec{u} = u_1\vec{i} + u_2\vec{j}$. For $h > 0$, consider the point $Q = (a+hu_1, b+hu_2)$ whose displacement from P is $h\vec{u}$. (See Figure 14.26.) Since $\|\vec{u}\| = 1$, the distance from P to Q is h. Thus,

$$\frac{\text{Average rate of change}}{\text{in } f \text{ from } P \text{ to } Q} = \frac{\text{Change in } f}{\text{Distance from } P \text{ to } Q} = \frac{f(a + hu_1, b + hu_2) - f(a, b)}{h}.$$

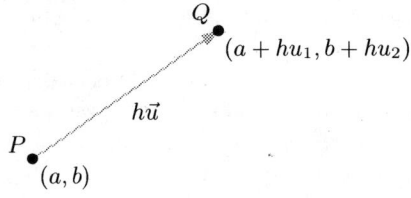

Figure 14.26: Displacement of $h\vec{u}$ from the point (a, b)

14.4 GRADIENTS AND DIRECTIONAL DERIVATIVES IN THE PLANE

Taking the limit as $h \to 0$ gives the instantaneous rate of change and the following definition:

Directional Derivative of f at (a, b) in the Direction of a Unit Vector $\vec{u}$

If $\vec{u} = u_1\vec{i} + u_2\vec{j}$ is a unit vector, we define the directional derivative, $f_{\vec{u}}$, by

$$f_{\vec{u}}(a, b) = \begin{matrix} \text{Rate of change} \\ \text{of } f \text{ in direction} \\ \text{of } \vec{u} \text{ at } (a, b) \end{matrix} = \lim_{h \to 0} \frac{f(a + hu_1, b + hu_2) - f(a, b)}{h},$$

provided the limit exists.

Notice that if $\vec{u} = \vec{i}$, so $u_1 = 1, u_2 = 0$, then the directional derivative is f_x, since

$$f_{\vec{i}}(a, b) = \lim_{h \to 0} \frac{f(a + h, b) - f(a, b)}{h} = f_x(a, b).$$

Similarly, if $\vec{u} = \vec{j}$ then the directional derivative $f_{\vec{j}} = f_y$.

What If We Do Not Have a Unit Vector?

We defined $f_{\vec{u}}$ for $\vec{u}$ a unit vector. If $\vec{v}$ is not a unit vector, $\vec{v} \neq \vec{0}$, we construct a unit vector $\vec{u} = \vec{v}/\|\vec{v}\|$ in the same direction as $\vec{v}$ and define the rate of change of f in the direction of $\vec{v}$ as $f_{\vec{u}}$.

Example 2 For each of the functions f, g, and h in Figure 14.27, decide whether the directional derivative at the indicated point is positive, negative, or zero, in the direction of the vector $\vec{v} = \vec{i} + 2\vec{j}$, and in the direction of the vector $\vec{w} = 2\vec{i} + \vec{j}$.

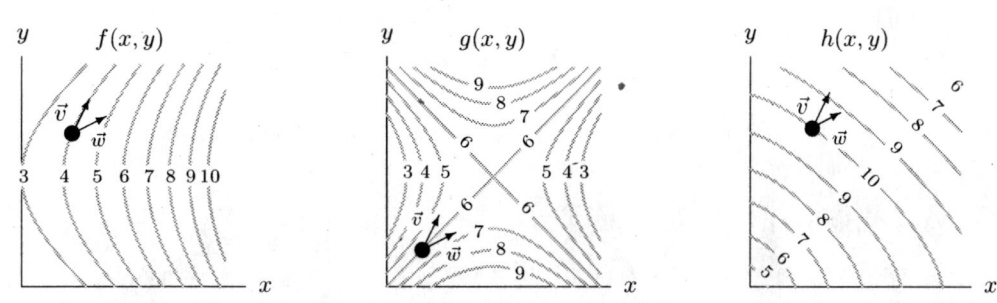

Figure 14.27: Contour diagrams of three functions with direction vectors $\vec{v} = \vec{i} + 2\vec{j}$ and $\vec{w} = 2\vec{i} + \vec{j}$ marked on each

Solution On the contour diagram for f, the vector $\vec{v} = \vec{i} + 2\vec{j}$ appears to be tangent to the contour. Thus, in this direction, the value of the function is not changing, so the directional derivative in the direction of $\vec{v}$ is zero. The vector $\vec{w} = 2\vec{i} + \vec{j}$ points from the contour marked 4 toward the contour marked 5. Thus, the values of the function are increasing and the directional derivative in the direction of $\vec{w}$ is positive.

On the contour diagram for g, the vector $\vec{v} = \vec{i} + 2\vec{j}$ points from the contour marked 6 toward the contour marked 5, so the function is decreasing in that direction. Thus, the rate of change is negative. On the other hand, the vector $\vec{w} = 2\vec{i} + \vec{j}$ points from the contour marked 6 toward the contour marked 7, and hence the directional derivative in the direction of $\vec{w}$ is positive.

Finally, on the contour diagram for h, both vectors point from the $h = 10$ contour to the $h = 9$ contour, so both directional derivatives are negative.

Example 3 Calculate the directional derivative of $f(x,y) = x^2 + y^2$ at $(1,0)$ in the direction of the vector $\vec{i} + \vec{j}$.

Solution First we have to find the unit vector in the same direction as the vector $\vec{i} + \vec{j}$. Since this vector has magnitude $\sqrt{2}$, the unit vector is

$$\vec{u} = \frac{1}{\sqrt{2}}(\vec{i} + \vec{j}) = \frac{1}{\sqrt{2}}\vec{i} + \frac{1}{\sqrt{2}}\vec{j}.$$

Thus,

$$f_{\vec{u}}(1,0) = \lim_{h \to 0} \frac{f(1 + h/\sqrt{2}, h/\sqrt{2}) - f(1,0)}{h} = \lim_{h \to 0} \frac{(1 + h/\sqrt{2})^2 + (h/\sqrt{2})^2 - 1}{h}$$

$$= \lim_{h \to 0} \frac{\sqrt{2}h + h^2}{h} = \lim_{h \to 0} (\sqrt{2} + h) = \sqrt{2}.$$

Computing Directional Derivatives From Partial Derivatives

If f is differentiable, we will now see how to use local linearity to find a formula for the directional derivative which does not involve a limit. If $\vec{u}$ is a unit vector, the definition of $f_{\vec{u}}$ says

$$f_{\vec{u}}(a,b) = \lim_{h \to 0} \frac{f(a + hu_1, b + hu_2) - f(a,b)}{h} = \lim_{h \to 0} \frac{\Delta f}{h},$$

where $\Delta f = f(a + hu_1, b + hu_2) - f(a,b)$ is the change in f. We write Δx for the change in x, so $\Delta x = (a + hu_1) - a = hu_1$; similarly $\Delta y = hu_2$. Using local linearity, we have

$$\Delta f \approx f_x(a,b)\Delta x + f_y(a,b)\Delta y = f_x(a,b)hu_1 + f_y(a,b)hu_2.$$

Thus, dividing by h gives

$$\frac{\Delta f}{h} \approx \frac{f_x(a,b)hu_1 + f_y(a,b)hu_2}{h} = f_x(a,b)u_1 + f_y(a,b)u_2.$$

This approximation becomes exact as $h \to 0$, so we have the following formula:

$$f_{\vec{u}}(a,b) = f_x(a,b)u_1 + f_y(a,b)u_2.$$

Example 4 Use the preceding formula to compute the directional derivative in Example 3. Check that we get the same answer as before.

Solution We calculate $f_{\vec{u}}(1,0)$, where $f(x,y) = x^2 + y^2$ and $\vec{u} = \frac{1}{\sqrt{2}}\vec{i} + \frac{1}{\sqrt{2}}\vec{j}$.

The partial derivatives are $f_x(x,y) = 2x$ and $f_y(x,y) = 2y$. So, as before

$$f_{\vec{u}}(1,0) = f_x(1,0)u_1 + f_y(1,0)u_2 = (2)\left(\frac{1}{\sqrt{2}}\right) + (0)\left(\frac{1}{\sqrt{2}}\right) = \sqrt{2}.$$

The Gradient Vector

Notice that the expression for $f_{\vec{u}}(a,b)$ can be written as a dot product of $\vec{u}$ and a new vector:

$$f_{\vec{u}}(a,b) = f_x(a,b)u_1 + f_y(a,b)u_2 = (f_x(a,b)\vec{i} + f_y(a,b)\vec{j}) \cdot (u_1\vec{i} + u_2\vec{j}).$$

The new vector, $f_x(a,b)\vec{i} + f_y(a,b)\vec{j}$, turns out to be important. Thus, we make the following definition:

> **The Gradient Vector** of a differentiable function f at the point (a,b) is
>
> $$\operatorname{grad} f(a,b) = f_x(a,b)\vec{i} + f_y(a,b)\vec{j}$$

14.4 GRADIENTS AND DIRECTIONAL DERIVATIVES IN THE PLANE

The formula for the directional derivative can be written in terms of the gradient as follows:

The Directional Derivative and the Gradient

If f is differentiable at (a, b) and $\vec{u} = u_1 \vec{i} + u_2 \vec{j}$ is a unit vector, then

$$f_{\vec{u}}(a, b) = f_x(a, b)u_1 + f_y(a, b)u_2 = \text{grad } f(a, b) \cdot \vec{u}.$$

Example 5 Find the gradient vector of $f(x, y) = x + e^y$ at the point $(1, 1)$.

Solution Using the definition we have

$$\text{grad } f = f_x \vec{i} + f_y \vec{j} = \vec{i} + e^y \vec{j},$$

so at the point $(1, 1)$

$$\text{grad } f(1, 1) = \vec{i} + e\vec{j}.$$

Alternative Notation for the Gradient

You can think of $\dfrac{\partial f}{\partial x}\vec{i} + \dfrac{\partial f}{\partial y}\vec{j}$ as the result of applying the vector operator (pronounced "del")

$$\nabla = \frac{\partial}{\partial x}\vec{i} + \frac{\partial}{\partial y}\vec{j}$$

to the function f. Thus, we get the alternative notation

$$\text{grad } f = \nabla f.$$

If $z = f(x, y)$, we can write grad z or ∇z for grad f or for ∇f.

What Does the Gradient Tell Us?

The fact that $f_{\vec{u}} = \text{grad } f \cdot \vec{u}$ enables us to see what the gradient vector represents. Suppose θ is the angle between the vectors grad f and $\vec{u}$. At the point (a, b), we have

$$f_{\vec{u}} = \text{grad } f \cdot \vec{u} = \|\text{grad } f\| \underbrace{\|\vec{u}\|}_{1} \cos \theta = \|\text{grad } f\| \cos \theta.$$

Imagine that grad f is fixed and that $\vec{u}$ can rotate. (See Figure 14.28.) The maximum value of $f_{\vec{u}}$ occurs when $\cos \theta = 1$, so $\theta = 0$ and $\vec{u}$ is pointing in the direction of grad f. Then

$$\text{Maximum } f_{\vec{u}} = \|\text{grad } f\| \cos 0 = \|\text{grad } f\|.$$

The minimum value of $f_{\vec{u}}$ occurs when $\cos \theta = -1$, so $\theta = \pi$ and $\vec{u}$ is pointing in the direction opposite to grad f. Then

$$\text{Minimum } f_{\vec{u}} = \|\text{grad } f\| \cos \pi = -\|\text{grad } f\|.$$

When $\theta = \pi/2$ or $3\pi/2$, so $\cos \theta = 0$, the directional derivative is zero.

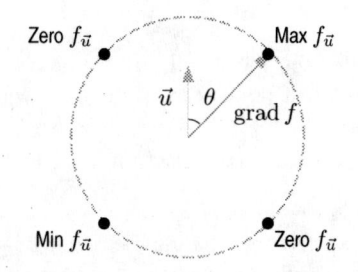

Figure 14.28: Values of the directional derivative at different angles to the gradient

Properties of The Gradient Vector

We have seen that the gradient vector points in the direction of the greatest rate of change at a point and the magnitude of the gradient vector is that rate of change.

Figure 14.29 shows that the gradient vector at any point is perpendicular to the contour through that point. Assuming f is differentiable at the point (a, b), local linearity tells us that the contours of f around the point (a, b) appear straight, parallel, and equally spaced. The greatest rate of change is obtained by moving in the direction that takes us to the next contour in the shortest possible distance, which is the direction perpendicular to the contour. Thus, we have the following:

> **Geometric Properties of the Gradient Vector in the Plane**
> If f is a differentiable function at the point (a, b) and $\operatorname{grad} f(a, b) \neq \vec{0}$, then:
> - The direction of $\operatorname{grad} f(a, b)$ is
>   - Perpendicular to the contour of f through (a, b)
>   - In the direction of increasing f
> - The magnitude of the gradient vector, $\|\operatorname{grad} f\|$, is
>   - The maximum rate of change of f at that point
>   - Large when the contours are close together and small when they are far apart.

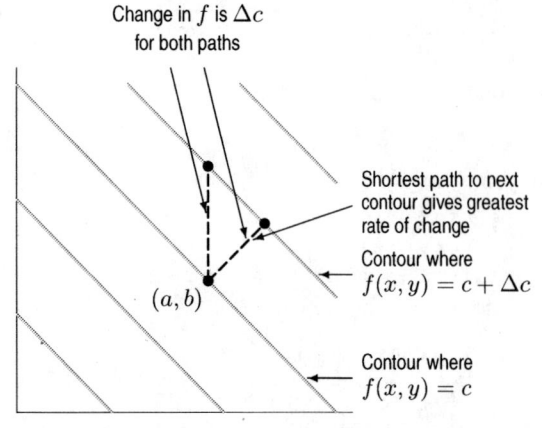

Figure 14.29: Close-up view of the contours around (a, b), showing the gradient is perpendicular to the contours

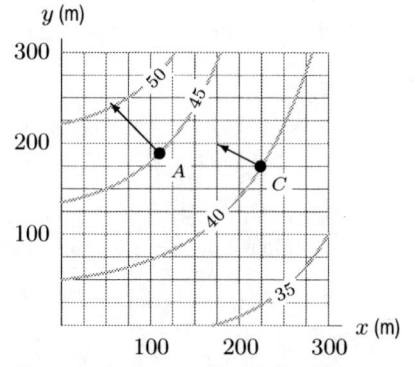

Figure 14.30: A temperature map showing directions and relative magnitudes of two gradient vectors

Examples of Directional Derivatives and Gradient Vectors

Example 6 Explain why the gradient vectors at points A and C in Figure 14.30 have the direction and the relative magnitudes they do.

Solution The gradient vector points in the direction of greatest increase of the function. This means that in Figure 14.30, the gradient points directly toward warmer temperatures. The magnitude of the gradient vector measures the rate of change. The gradient vector at A is longer than the gradient vector at C because the contours are closer together at A, so the rate of change is larger.

Example 2 on page 705 shows how the contour diagram can tell us the sign of the directional derivative. In the next example we compute the directional derivative in three directions, two close to that of the gradient vector and one that is not.

14.4 GRADIENTS AND DIRECTIONAL DERIVATIVES IN THE PLANE

Example 7 Use the gradient to find the directional derivative of $f(x,y) = x + e^y$ at the point $(1,1)$ in the direction of the vectors $\vec{i} - \vec{j}, \vec{i} + 2\vec{j}, \vec{i} + 3\vec{j}$.

Solution In Example 5 we found
$$\text{grad } f(1,1) = \vec{i} + e\vec{j}.$$
A unit vector in the direction of $\vec{i} - \vec{j}$ is $\vec{s} = (\vec{i} - \vec{j})/\sqrt{2}$, so
$$f_{\vec{s}}(1,1) = \text{grad } f(1,1) \cdot \vec{s} = (\vec{i} + e\vec{j}) \cdot \left(\frac{\vec{i} - \vec{j}}{\sqrt{2}}\right) = \frac{1-e}{\sqrt{2}} \approx -1.215.$$
A unit vector in the direction of $\vec{i} + 2\vec{j}$ is $\vec{v} = (\vec{i} + 2\vec{j})/\sqrt{5}$, so
$$f_{\vec{v}}(1,1) = \text{grad } f(1,1) \cdot \vec{v} = (\vec{i} + e\vec{j}) \cdot \left(\frac{\vec{i} + 2\vec{j}}{\sqrt{5}}\right) = \frac{1+2e}{\sqrt{5}} \approx 2.879.$$
A unit vector in the direction of $\vec{i} + 3\vec{j}$ is $\vec{w} = (\vec{i} + 3\vec{j})/\sqrt{10}$, so
$$f_{\vec{w}}(1,1) = \text{grad } f(1,1) \cdot \vec{w} = (\vec{i} + e\vec{j}) \cdot \left(\frac{\vec{i} + 3\vec{j}}{\sqrt{10}}\right) = \frac{1+3e}{\sqrt{10}} \approx 2.895.$$

Now look back at the answers and compare with the value of $\|\text{grad } f\| = \sqrt{1+e^2} \approx 2.896$. One answer is not close to this value; the other two, $f_{\vec{v}} = 2.879$ and $f_{\vec{w}} = 2.895$, are close but slightly smaller than $\|\text{grad } f\|$. Since $\|\text{grad } f\|$ is the maximum rate of change of f at the point, we would expect for *any* unit vector $\vec{u}$:
$$f_{\vec{u}}(1,1) \leq \|\text{grad } f\|.$$
with equality when $\vec{u}$ is in the direction of grad f. Since $e \approx 2.718$, the vectors $\vec{i} + 2\vec{j}$ and $\vec{i} + 3\vec{j}$ both point roughly, but not exactly, in the direction of the gradient vector grad $f(1,1) = \vec{i} + e\vec{j}$. Thus, the values of $f_{\vec{v}}$ and $f_{\vec{w}}$ are both close to the value of $\|\text{grad } f\|$. The direction of the vector $\vec{i} - \vec{j}$ is not close to the direction of grad f and the value of $f_{\vec{s}}$ is not close to the value of $\|\text{grad } f\|$.

Exercises and Problems for Section 14.4

Exercises

In Exercises 1–6, use the contour diagram of f in Figure 14.31 to decide if the specified directional derivative is positive, negative, or approximately zero.

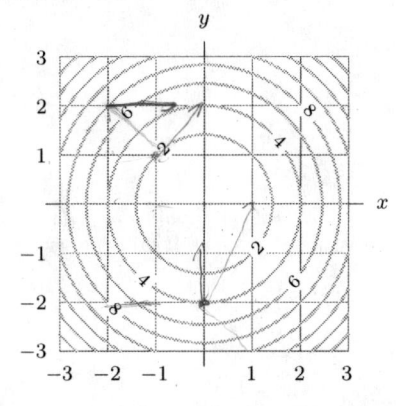

Figure 14.31

1. At point $(-2, 2)$, in direction $\vec{i}$.
2. At point $(0, -2)$, in direction $\vec{j}$.
3. At point $(-1, 1)$, in direction $\vec{i} + \vec{j}$.
4. At point $(-1, 1)$, in direction $-\vec{i} + \vec{j}$.
5. At point $(0, -2)$, in direction $\vec{i} + 2\vec{j}$.
6. At point $(0, -2)$, in direction $\vec{i} - 2\vec{j}$.

In Exercises 7–14, use the contour diagram of f in Figure 14.31 to find the approximate direction of the gradient vector at the given point.

7. $(2, 0)$ 8. $(0, 2)$ 9. $(-2, 0)$ 10. $(0, -2)$
11. $(2, 2)$ 12. $(2, -2)$ 13. $(-2, 2)$ 14. $(-2, -2)$

In Exercises 15–28 find the gradient of the given function. Assume the variables are restricted to a domain on which the function is defined.

15. $f(x,y) = \frac{3}{2}x^5 - \frac{4}{7}y^6$ 16. $f(m,n) = m^2 + n^2$
17. $Q = 50K + 100L$ 18. $z = xe^y$
19. $z = (x+y)e^y$ 20. $f(K, L) = K^{0.3}L^{0.7}$
21. $f(x,y) = \sqrt{x^2 + y^2}$ 22. $f(r, h) = \pi r^2 h$

23. $f(r, \theta) = r \sin \theta$
24. $z = \sin(x^2 + y^2)$
25. $z = \sin(x/y)$
26. $z = \tan^{-1}(x/y)$
27. $f(\alpha, \beta) = \dfrac{2\alpha + 3\beta}{2\alpha - 3\beta}$
28. $z = x\dfrac{e^y}{x+y}$

32. $f(x, y) = \sin(x^2) + \cos y$, at $(\tfrac{\sqrt{\pi}}{2}, 0)$
33. $f(x, y) = 1/(x^2 + y^2)$, at $(-1, 3)$
34. $f(x, y) = \sqrt{\tan x + y}$, at $(0, 1)$

In Exercises 35–38, find the directional derivative $f_{\vec{u}}(1, 2)$ for the function f with $\vec{u} = (3\vec{i} - 4\vec{j})/5$.

In Exercises 29–34, find the gradient at the given point.

29. $f(x, y) = x^2 y + 7xy^3$, at $(1, 2)$
30. $f(m, n) = 5m^2 + 3n^4$, at $(5, 2)$
31. $f(r, h) = 2\pi rh + \pi r^2$, at $(2, 3)$

35. $f(x, y) = 3x - 4y$
36. $f(x, y) = x^2 - y^2$
37. $f(x, y) = xy + y^3$
38. $f(x, y) = \sin(2x - y)$

Problems

In Problems 39–44, use the contour diagram for $f(x, y)$ in Figure 14.32 to estimate the directional derivative of $f(x, y)$ in the direction $\vec{v}$ at the point given.

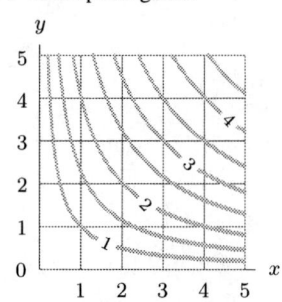

Figure 14.32

39. $\vec{v} = \vec{i}$ at $(1, 1)$
40. $\vec{v} = \vec{j}$ at $(1, 1)$
41. $\vec{v} = \vec{i} + \vec{j}$ at $(1, 1)$
42. $\vec{v} = \vec{i} + \vec{j}$ at $(4, 1)$
43. $\vec{v} = -2\vec{i} + \vec{j}$ at $(3, 3)$
44. $\vec{v} = -2\vec{i} + \vec{j}$ at $(4, 1)$

45. Figure 14.32 shows the level curves of $f(x, y)$. At the points $(1, 1)$ and $(1, 4)$, draw a vector representing grad f. Explain how you know the direction and length of each vector.

46. Figure 14.33 shows grad $f(x, y)$. In each of the following cases, list the marked points (if any) at which
   (a) $f_x > 0$ (b) $f_y < 0$
   (c) $f_{xx} > 0$ (d) $f_{yy} < 0$

47. Figure 14.33 shows grad $f(x, y)$; the scales along the x and y-axes are equal. In each of the following cases, list the marked points at which
   (a) grad $f \cdot (\vec{i} + \vec{j})$ is maximum.
   (b) The rate of change of $f(x, y)$ is maximum in the $\vec{i} - \vec{j}$ direction.
   (c) The rate of change of $f(x, y)$ is zero in the $\vec{i} + \vec{j}$ direction.
   (d) The rate of change of $f(x, y)$ is negative in the $\vec{i} + \vec{j}$ direction.

48. The surface $z = g(x, y)$ is in Figure 14.34. What is the sign of each of the following directional derivatives?
   (a) $g_{\vec{u}}(2, 5)$ where $\vec{u} = (\vec{i} - \vec{j})/\sqrt{2}$.
   (b) $g_{\vec{u}}(2, 5)$ where $\vec{u} = (\vec{i} + \vec{j})/\sqrt{2}$.

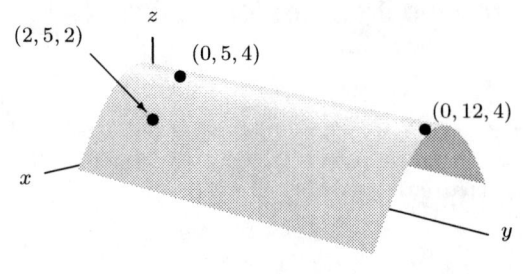

Figure 14.34

49. The table gives values of a differentiable function $f(x, y)$. At the point $(1.2, 0)$, into which quadrant does the gradient vector of f point? Justify your answer.

	$y = -1$	$y = 0$	$y = 1$
$x = 1.0$	0.7	0.1	−0.5
$x = 1.2$	4.8	4.2	3.6
$x = 1.4$	8.9	8.3	7.7

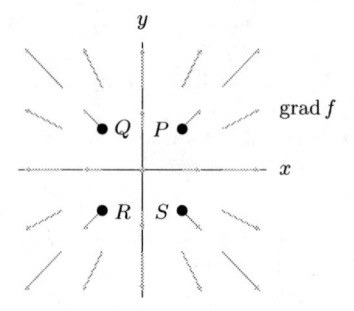

Figure 14.33

50. Figure 14.35 represents the level curves $f(x, y) = c$; the values of f on each curve are marked. In each of the following parts, decide whether the given quantity is positive, negative or zero. Explain your answer.

   (a) The value of $\nabla f \cdot \vec{i}$ at P.
   (b) The value of $\nabla f \cdot \vec{j}$ at P.
   (c) $\partial f/\partial x$ at Q.
   (d) $\partial f/\partial y$ at Q.

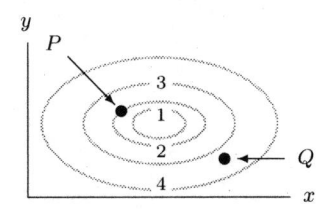

Figure 14.35

51. In Figure 14.35, which is larger: $\|\nabla f\|$ at P or $\|\nabla f\|$ at Q? Explain how you know.

52. (a) Let $f(x, y) = (x+y)/(1+x^2)$. Find the directional derivative of f at $P = (1, -2)$ in the direction of:
      (i) $\vec{v} = 3\vec{i} - 2\vec{j}$ (ii) $\vec{v} = -\vec{i} + 4\vec{j}$
   (b) What is the direction of greatest increase of f at P?

53. (a) Let $f(x, y) = x^2 + \ln y$. Find the average rate of change of f as you go from $(3, 1)$ to $(1, 2)$.
   (b) Find the instantaneous rate of change of f as you leave the point $(3, 1)$ heading toward $(1, 2)$.

54. (a) What is the rate of change of $f(x, y) = 3xy + y^2$ at the point $(2, 3)$ in the direction $\vec{v} = 3\vec{i} - \vec{j}$?
   (b) What is the direction of maximum rate of change of f at $(2, 3)$?
   (c) What is the maximum rate of change?

55. Let $f(x, y) = x^2 y^3$. At the point $(-1, 2)$, find a vector
   (a) In the direction of maximum rate of change.
   (b) In the direction of minimum rate of change.
   (c) In a direction in which the rate of change is zero.

56. Let P be a fixed point in the plane and let $f(x, y)$ be the distance from P to (x, y). Answer the following questions using geometric interpretations, not formulas.
   (a) What are the level curves of f?
   (b) What direction does $\text{grad } f(x, y)$ point?
   (c) What is the magnitude $\|\text{grad } f(x, y)\|$?

57. The directional derivative of $z = f(x, y)$ at $(2, 1)$ in the direction toward the point $(1, 3)$ is $-2/\sqrt{5}$, and the directional derivative in the direction toward the point $(5, 5)$ is 1. Compute $\partial z/\partial x$ and $\partial z/\partial y$ at $(2, 1)$.

58. Consider the function $f(x, y)$. If you start at the point $(4, 5)$ and move toward the point $(5, 6)$, the directional derivative is 2. Starting at the point $(4, 5)$ and moving toward the point $(6, 6)$ gives a directional derivative of 3. Find ∇f at the point $(4, 5)$.

59. Find the directional derivative of $z = x^2 y$ at $(1, 2)$ in the direction making an angle of $5\pi/4$ with the x-axis. In which direction is the directional derivative the largest?

In Problems 60–63, check that the point $(2, 3)$ lies on the curve. Then, viewing the curve as a contour of $f(x, y)$, use grad $f(2, 3)$ to find a vector normal to the curve at $(2, 3)$ and an equation for the tangent line to the curve at $(2, 3)$.

60. $x^2 + y^2 = 13$ 61. $xy = 6$
62. $y = x^2 - 1$ 63. $(y - x)^2 + 2 = xy - 3$

64. An ant is crawling across a heated plate with velocity $\vec{v}$ cm/sec, and the temperature of the plate at position (x, y) is $H(x, y)$ degrees, where x and y are in centimeters. Which of the following (if any) is correct? The rate of change in deg/sec of the temperature felt by the ant is:

   (a) $\|\text{grad } H\| \|\vec{v}\|$, because it is the product of the ant's speed and the rate of change of H with respect to distance.
   (b) $\text{grad } H \cdot \vec{v}$, because it is the product of the ant's speed and the directional derivative of H in the direction of $\vec{v}$.
   (c) $H_{\vec{u}}$, where $\vec{u} = \vec{v}/\|\vec{v}\|$, because it is the rate of change of H in the direction of $\vec{v}$.

65. The depth, in feet, of a lake at a point x miles east and y miles north of a buoy is given by

$$h(x, y) = 150 - 30x^2 - 20y^2.$$

   (a) A rowboat is 1 mile east and 2 miles south of the buoy. At what rate is the depth changing with respect to distance in the direction of the buoy?
   (b) The boat starts moving toward the buoy at a rate of 3 mph. At what rate is the depth of the lake beneath the boat changing with respect to time?

66. The temperature at any point in the plane is given by the function

$$T(x, y) = \frac{100}{x^2 + y^2 + 1}.$$

   (a) What shape are the level curves of T?
   (b) Where on the plane is it hottest? What is the temperature at that point?
   (c) Find the direction of the greatest increase in temperature at the point $(3, 2)$. What is the magnitude of that greatest increase?
   (d) Find the direction of the greatest decrease in temperature at the point $(3, 2)$.
   (e) Find a direction at the point $(3, 2)$ in which the temperature does not increase or decrease.

67. At a certain point on a heated plate, the greatest rate of temperature increase, 5° C per meter, is toward the northeast. If an object at this point moves directly north, at what rate is the temperature increasing?

68. You are climbing a mountain by the steepest route at a slope of 20° when you come upon a trail branching off at a 30° angle from yours. What is the angle of ascent of the branch trail?

69. Figure 14.37 is a graph of the directional derivative, $f_{\vec{u}}$, at the point (a, b) versus θ, the angle in Figure 14.36.

(a) Which points on the graph in Figure 14.37 correspond to the greatest rate of increase of f? The greatest rate of decrease?

(b) Mark points on the circle in Figure 14.36 corresponding to the points P, Q, R, S.

(c) What is the amplitude of the function graphed in Figure 14.37? What is its formula?

h in the direction of $\vec{u}$, then the directional derivative is approximated by the difference quotient

$$\frac{\text{Change in } f \text{ between } P \text{ and } Q}{h}.$$

(a) Use the gradient to show that

$$\text{Change in } f \approx \|\text{grad } f\|(h \cos \theta).$$

(b) Use part (a) to obtain $f_{\vec{u}}(a, b) = \text{grad } f(a, b) \cdot \vec{u}$.

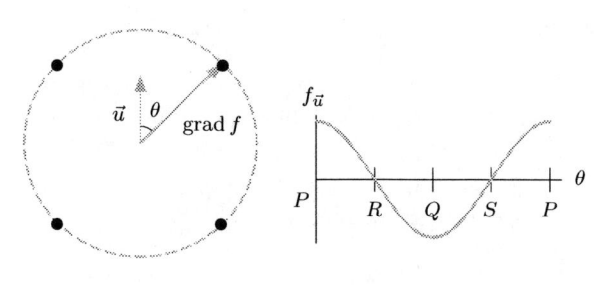

Figure 14.36

Figure 14.37

70. In this problem we see another way of obtaining the formula $f_{\vec{u}}(a, b) = \text{grad } f(a, b) \cdot \vec{u}$. Imagine zooming in on a function $f(x, y)$ at a point (a, b). By local linearity, the contours around (a, b) look like the contours of a linear function. See Figure 14.38. Suppose you want to find the directional derivative $f_{\vec{u}}(a, b)$ in the direction of a unit vector $\vec{u}$. If you move from P to Q, a small distance

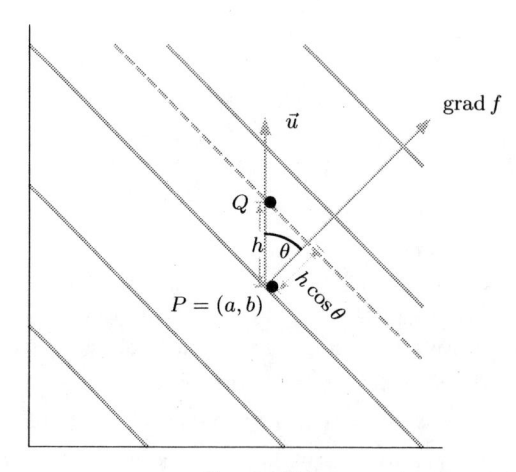

Figure 14.38

14.5 GRADIENTS AND DIRECTIONAL DERIVATIVES IN SPACE

Directional Derivatives of Functions of Three Variables

We calculate directional derivatives of a function of three variables in the same way as for a function of two variables. If the function f is differentiable at the point (a, b, c), then the rate of change of $f(x, y, z)$ at the point (a, b, c) in the direction of a unit vector $\vec{u} = u_1\vec{i} + u_2\vec{j} + u_3\vec{k}$ is

$$f_{\vec{u}}(a, b, c) = f_x(a, b, c)u_1 + f_y(a, b, c)u_2 + f_z(a, b, c)u_3.$$

This can be justified using local linearity in the same way as for functions of two variables.

Example 1 Find the directional derivative of $f(x, y, z) = xy + z$ at the point $(-1, 0, 1)$ in the direction of the vector $\vec{v} = 2\vec{i} + \vec{k}$.

Solution The magnitude of $\vec{v}$ is $\|\vec{v}\| = \sqrt{2^2 + 1} = \sqrt{5}$, so a unit vector in the same direction as $\vec{v}$ is

$$\vec{u} = \frac{\vec{v}}{\|\vec{v}\|} = \frac{2}{\sqrt{5}}\vec{i} + 0\vec{j} + \frac{1}{\sqrt{5}}\vec{k}.$$

The partial derivatives of f are $f_x(x, y, z) = y$ and $f_y(x, y, z) = x$ and $f_z(x, y, z) = 1$. Thus,

$$f_{\vec{u}}(-1, 0, 1) = f_x(-1, 0, 1)u_1 + f_y(-1, 0, 1)u_2 + f_z(-1, 0, 1)u_3$$

$$= (0)\left(\frac{2}{\sqrt{5}}\right) + (-1)(0) + (1)\left(\frac{1}{\sqrt{5}}\right) = \frac{1}{\sqrt{5}}.$$

The Gradient Vector of a Function of Three Variables

The gradient of a function of three variables is defined in the same way as for two variables:

$$\operatorname{grad} f(a,b,c) = f_x(a,b,c)\vec{i} + f_y(a,b,c)\vec{j} + f_z(a,b,c)\vec{k}.$$

Directional derivatives are related to gradients in the same way as for functions of two variables:

$$f_{\vec{u}}(a,b,c) = f_x(a,b,c)u_1 + f_y(a,b,c)u_2 + f_z(a,b,c)u_3 = \operatorname{grad} f(a,b,c) \cdot \vec{u}.$$

Since $\operatorname{grad} f(a,b,c) \cdot \vec{u} = \|\operatorname{grad} f(a,b,c)\| \cos\theta$, where θ is the angle between $\operatorname{grad} f(a,b,c)$ and $\vec{u}$, the value of $f_{\vec{u}}(a,b,c)$ is largest when $\theta = 0$, that is, when $\vec{u}$ is in the same direction as $\operatorname{grad} f(a,b,c)$. In addition, $f_{\vec{u}}(a,b,c) = 0$ when $\theta = \pi/2$, so $\operatorname{grad} f(a,b,c)$ is perpendicular to the level surface of f. The properties of gradients in space are similar to those in the plane:

Properties of the Gradient Vector in Space

If f is differentiable at (a,b,c) and $\vec{u}$ is a unit vector, then

$$f_{\vec{u}}(a,b,c) = \operatorname{grad} f(a,b,c) \cdot \vec{u}.$$

If, in addition, $\operatorname{grad} f(a,b,c) \neq \vec{0}$, then
- $\operatorname{grad} f(a,b,c)$ is in the direction of the greatest rate of increase of f
- $\operatorname{grad} f(a,b,c)$ is perpendicular to the level surface of f at (a,b,c)
- $\|\operatorname{grad} f(a,b,c)\|$ is the maximum rate of change of f at (a,b,c).

Example 2 Let $f(x,y,z) = x^2 + y^2$ and $g(x,y,z) = -x^2 - y^2 - z^2$. What can we say about the direction of the following vectors?
(a) $\operatorname{grad} f(0,1,1)$ (b) $\operatorname{grad} f(1,0,1)$ (c) $\operatorname{grad} g(0,1,1)$ (d) $\operatorname{grad} g(1,0,1)$.

Solution The cylinder $x^2 + y^2 = 1$ in Figure 14.39 is a level surface of f and contains both the points $(0,1,1)$ and $(1,0,1)$. Since the value of f does not change at all in the z-direction, all the gradient vectors are horizontal. They are perpendicular to the cylinder and point outward because the value of f increases as we move out.

Similarly, the points $(0,1,1)$ and $(1,0,1)$ also lie on the same level surface of g, namely $g(x,y,z) = -x^2 - y^2 - z^2 = -2$, which is the sphere $x^2 + y^2 + z^2 = 2$. Part of this level surface is shown in Figure 14.40. This time the gradient vectors point inward, since the negative signs mean that the function increases (from large negative values to small negative values) as we move inward.

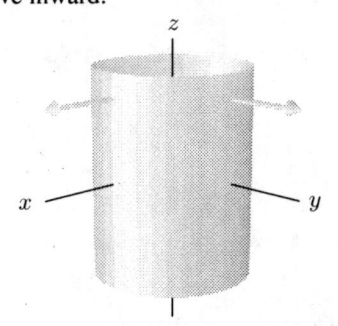

Figure 14.39: The level surface $f(x,y,z) = x^2 + y^2 = 1$ with two gradient vectors

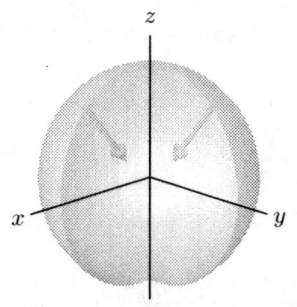

Figure 14.40: The level surface $g(x,y,z) = -x^2 - y^2 - z^2 = -2$ with two gradient vectors

Example 3 Consider the functions $f(x, y) = 4-x^2-2y^2$ and $g(x, y) = 4-x^2$. Calculate a vector perpendicular to each of the following:

(a) The level curve of f at the point $(1, 1)$
(b) The surface $z = f(x, y)$ at the point $(1, 1, 1)$
(c) The level curve of g at the point $(1, 1)$
(d) The surface $z = g(x, y)$ at the point $(1, 1, 3)$

Solution (a) The vector we want is a 2-vector in the plane. Since $\operatorname{grad} f = -2x\vec{i} - 4y\vec{j}$, we have

$$\operatorname{grad} f(1, 1) = -2\vec{i} - 4\vec{j}.$$

Any nonzero multiple of this vector is perpendicular to the level curve at the point $(1, 1)$.

(b) In this case we want a 3-vector in space. To find it we rewrite $z = 4 - x^2 - 2y^2$ as the level surface of the function F, where

$$F(x, y, z) = 4 - x^2 - 2y^2 - z = 0.$$

Then

$$\operatorname{grad} F = -2x\vec{i} - 4y\vec{j} - \vec{k},$$

so

$$\operatorname{grad} F(1, 1, 1) = -2\vec{i} - 4\vec{j} - \vec{k},$$

and $\operatorname{grad} F(1, 1, 1)$ is perpendicular to the surface $z = 4-x^2-2y^2$ at the point $(1, 1, 1)$. Notice that $-2\vec{i} - 4\vec{j} - \vec{k}$ is not the only possible answer: any multiple of this vector will do.

(c) We are looking for a 2-vector. Since $\operatorname{grad} g = -2x\vec{i} + 0\vec{j}$, we have

$$\operatorname{grad} g(1, 1) = -2\vec{i}.$$

Any multiple of this vector is perpendicular to the level curve also.

(d) We are looking for a 3-vector. We rewrite $z = 4 - x^2$ as the level surface of the function G, where

$$G(x, y, z) = 4 - x^2 - z = 0.$$

Then

$$\operatorname{grad} G = -2x\vec{i} - \vec{k}$$

So

$$\operatorname{grad} G(1, 1, 3) = -2\vec{i} - \vec{k},$$

and any multiple of $\operatorname{grad} G(1, 1, 3)$ is perpendicular to the surface $z = 4 - x^2$ at this point.

Example 4 (a) A hiker on the surface $f(x, y) = 4 - x^2 - 2y^2$ at the point $(1, -1, 1)$ starts to climb along the path of steepest ascent. What is the relation between the vector $\operatorname{grad} f(1, -1)$ and a vector tangent to the path at the point $(1, -1, 1)$ and pointing uphill?

(b) Consider the surface $g(x, y) = 4 - x^2$. What is the relation between $\operatorname{grad} g(-1, -1)$ and a vector tangent to the path of steepest ascent at $(-1, -1, 3)$?

(c) At the point $(1, -1, 1)$ on the surface $f(x, y) = 4 - x^2 - 2y^2$, calculate a vector perpendicular to the surface and a vector, $\vec{T}$, tangent to the curve of steepest ascent.

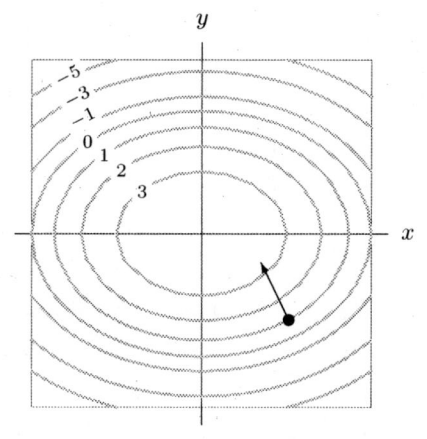

Figure 14.41: Contour diagram for $z = f(x,y) = 4 - x^2 - 2y^2$ showing direction of grad $f(1,-1)$

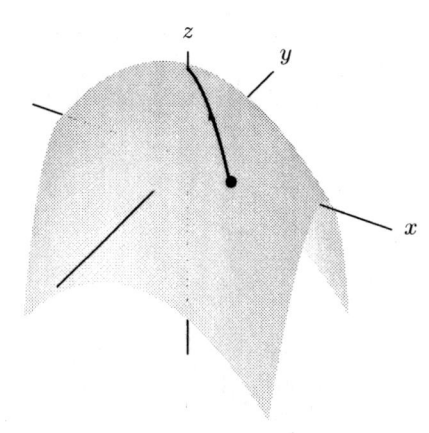

Figure 14.42: Graph of $f(x,y) = 4 - x^2 - 2y^2$ showing path of steepest ascent from the point $(1,-1,1)$

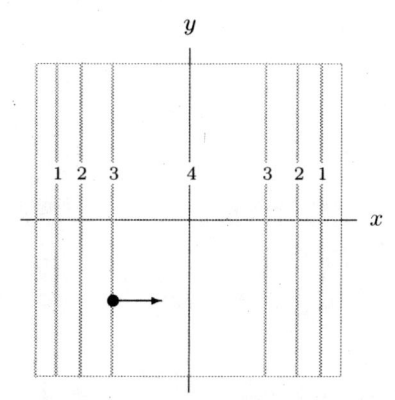

Figure 14.43: Contour diagram for $z = g(x,y) = 4 - x^2$ showing direction of grad $g(-1,-1)$

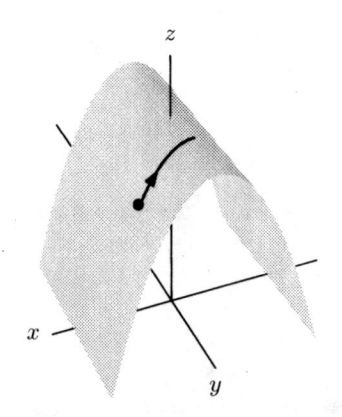

Figure 14.44: Graph of $g(x,y) = 4 - x^2$ showing path of steepest ascent from the point $(-1,-1,3)$

Solution

(a) The hiker at the point $(1, -1, 1)$ lies directly above the point $(1, -1)$ in the xy-plane. The vector grad $f(1, -1)$ lies in 2-space, pointing like a compass in the direction in which f increases most rapidly. Therefore, grad $f(1, -1)$ lies directly under a vector tangent to the hiker's path at $(1, -1, 1)$ and pointing uphill. (See Figures 14.41 and 14.42.)

(b) The point $(-1, -1, 3)$ lies above the point $(-1, -1)$. The vector grad $g(-1, -1)$ points in the direction in which g increases most rapidly and lies directly under the path of steepest ascent. (See Figures 14.43 and 14.44.)

(c) The surface is represented by $F(x, y, z) = 4 - x^2 - 2y^2 - z = 0$. Since grad $F = -2x\vec{i} - 4y\vec{j} - \vec{k}$, the normal, $\vec{N}$, to the surface is given by

$$\vec{N} = \operatorname{grad} F(1, -1, 1) = -2(1)\vec{i} - 4(-1)\vec{j} - \vec{k} = -2\vec{i} + 4\vec{j} - \vec{k}.$$

We take the $\vec{i}$ and $\vec{j}$ components of $\vec{T}$ to be the vector grad $f(1, -1) = -2\vec{i} + 4\vec{j}$. Thus we have that, for some $a > 0$,

$$\vec{T} = -2\vec{i} + 4\vec{j} + a\vec{k}$$

We want $\vec{N} \cdot \vec{T} = 0$, so

$$\vec{N} \cdot \vec{T} = (-2\vec{i} + 4\vec{j} - \vec{k}) \cdot (-2\vec{i} + 4\vec{j} + a\vec{k}) = 4 + 16 - a = 0$$

So $a = 20$ and hence

$$\vec{T} = -2\vec{i} + 4\vec{j} + 20\vec{k}.$$

Example 5 Find the equation of the tangent plane to the sphere $x^2 + y^2 + z^2 = 14$ at the point $(1, 2, 3)$.

Solution We write the sphere as a level surface as follows:
$$f(x, y, z) = x^2 + y^2 + z^2 = 14.$$
We have
$$\text{grad } f = 2x\vec{i} + 2y\vec{j} + 2z\vec{k},$$
so the vector
$$\text{grad } f(1, 2, 3) = 2\vec{i} + 4\vec{j} + 6\vec{k}$$
is perpendicular to the sphere at the point $(1, 2, 3)$. Since the vector grad $f(1, 2, 3)$ is normal to the tangent plane, the equation of the plane is
$$2x + 4y + 6z = 2(1) + 4(2) + 6(3) = 28 \quad \text{or} \quad x + 2y + 3z = 14.$$

The equation of a tangent plane to a level surface can always be found by the method of Example 5. If the level surface $f(x, y, z) = k$ cannot be solved for z in terms of x and y, this is the only method.

Tangent Plane to a Level Surface

If $f(x, y, z)$ is differentiable at (a, b, c), then an equation for the tangent plane to the level surface of f at the point (a, b, c) is

$$f_x(a, b, c)(x - a) + f_y(a, b, c)(y - b) + f_z(a, b, c)(z - c) = 0.$$

Caution: Units and the Geometric Interpretation of the Gradient

When we interpreted the gradient of a function geometrically (page 708), we tacitly assumed that the scales along the x and y axes were the same. If they are not, the gradient vector may not look perpendicular to the contours. Consider the function $f(x, y) = x^2 + y$ with gradient vector given by grad $f = 2x\vec{i} + \vec{j}$. Figure 14.45 shows the gradient vector at $(1, 1)$ using the same scales in the x and y directions. As expected, the gradient vector is perpendicular to the contour line. Figure 14.46 shows contours of the same function with unequal scales on the two axes. Notice that the gradient vector no longer appears perpendicular to the contour lines. Thus we see that the geometric interpretation of the gradient vector requires that the same scale be used on the two axes.

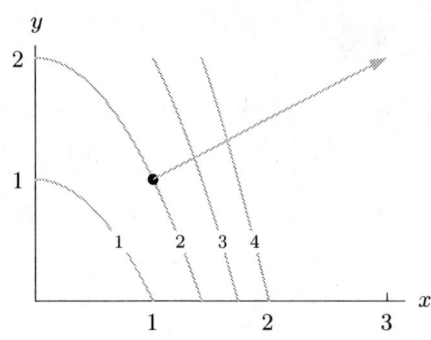

Figure 14.45: The gradient vector with x and y scales equal

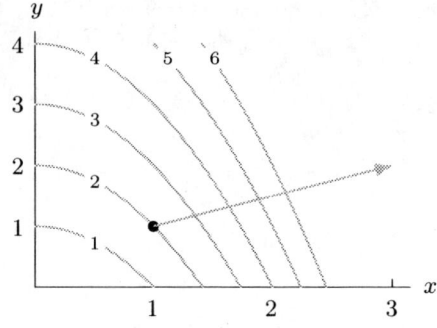

Figure 14.46: The gradient vector with x and y scales unequal

Exercises and Problems for Section 14.5

Exercises

In Exercises 1–10, find the gradient of the given function.

1. $f(x,y,z) = x^2 + y^3 - z^4$
2. $f(x,y,z) = \cos(x+y) + \sin(y+z)$
3. $f(x,y,z) = 1/(x^2 + y^2 + z^2)$
4. $f(x,y,z) = \sqrt{x^2 + y^2 + z^2}$
5. $f(x,y,z) = xe^y + \ln(xz)$
6. $f(x,y,z) = xe^y \sin z$
7. $f(x_1, x_2, x_3) = x_1^2 x_2^3 x_3^4$
8. $f(p,q,r) = e^p + \ln q + e^{r^2}$
9. $f(x,y,z) = xyz$ at $(1,2,3)$
10. $f(x,y,z) = e^{xy} \cos z$ at $(1,0,0)$

In Exercises 11–16, find the directional derivative using $f(x,y,z) = xy + z^2$.

11. At $(1,2,3)$ in the direction of $\vec{i} + \vec{j} + \vec{k}$.
12. At $(1,1,1)$ in the direction of $\vec{i} + 2\vec{j} + 3\vec{k}$.
13. As you leave the point $(1,1,0)$ heading in the direction of the point $(0,1,1)$.
14. As you arrive at $(0,1,1)$ from the direction of $(1,1,0)$.
15. At the point $(2,3,4)$ in the direction of a vector making an angle of $3\pi/4$ with grad $f(2,3,4)$.
16. At the point $(2,3,4)$ in the direction of the maximum rate of change of f.

In Exercises 17–22, check that the point $(-1, 1, 2)$ lies on the given surface. Then, viewing the surface as a level surface for a function $f(x,y,z)$, find a vector normal to the surface and an equation for the tangent plane to the surface at $(-1,1,2)$.

17. $x^2 - y^2 + z^2 = 4$
18. $x^2 - xyz = 3$
19. $z = x^2 + y^2$
20. $y^2 = z^2 - 3$
21. $\cos(x+y) = e^{xz+2}$
22. $y = 4/(2x + 3z)$

Problems

23. Find the directional derivative of $f(x,y,z) = 3x^2 y^2 + 2yz$ at the point $(-1, 0, 4)$ in the following directions
    (a) $\vec{i} - \vec{k}$ (b) $-\vec{i} + 3\vec{j} + 3\vec{k}$.

24. Let $f(x,y,z) = e^x \sin y + z$.
    (a) Find a vector at the point $(0, \pi/2, 2)$ pointing in the direction in which f
       (i) Increases fastest (ii) Decreases fastest
    (b) Is there a direction at the point $(0, \pi/2, 2)$ in which the function does not change initially? If so, find a vector pointing in that direction.

25. You are standing above the point $(1, 3)$ on the surface $z = 20 - (2x^2 + y^2)$.
    (a) In which direction should you walk to descend fastest? (Give your answer as a 2-vector.)
    (b) If you start to move in this direction, what is the slope of your path?

26. (a) Find the tangent plane to the surface $x^2 + y^2 + 3z^2 = 4$ at the point $(0.6, 0.8, 1)$.
    (b) Is there a point on the surface $x^2 + y^2 + 3z^2 = 4$ at which the tangent plane is parallel to the plane $8x + 6y + 30z = 1$? If so, find it. If not, explain why not.

27. The surface S is represented by the equation $F = 0$ where $F(x,y,z) = x^2 - (y/z^2)$.
    (a) Find the unit vectors $\vec{u}_1$ and $\vec{u}_2$ pointing in the direction of maximum increase of F at the points $(0,0,1)$ and $(1,1,1)$ respectively.
    (b) Find the tangent plane to S at the points $(0,0,1)$ and $(1,1,1)$.
    (c) Find all points on S where a normal vector is parallel to the xy-plane.

28. Find the equation of the tangent plane at $(2, 3, 1)$ to the surface $x^2 + y^2 - xyz = 7$. Do this in two ways:
    (a) Viewing the surface as the level set of a function of three variables, $F(x,y,z)$.
    (b) Viewing the surface as the graph of a function of two variables $z = f(x,y)$.

29. Consider the function $f(x,y) = (e^x - x)\cos y$. Suppose S is the surface $z = f(x,y)$.
    (a) Find a vector which is perpendicular to the level curve of f through the point $(2,3)$ in the direction in which f decreases most rapidly.
    (b) Suppose $\vec{v} = 5\vec{i} + 4\vec{j} + a\vec{k}$ is a vector in 3-space which is tangent to the surface S at the point P lying on the surface above $(2,3)$. What is a?

30. A differentiable function $f(x,y)$ has the property that $f(1,3) = 7$ and grad $f(1,3) = 2\vec{i} - 5\vec{j}$.
    (a) Find the equation of the tangent line to the level curve of f through the point $(1,3)$.
    (b) Find the equation of the tangent plane to the surface $z = f(x,y)$ at the point $(1,3,7)$.

31. Let $g_x(2,1,7) = 3$, $g_y(2,1,7) = 10$, $g_z(2,1,7) = -5$. Find the equation of the tangent plane to $g(x,y,z) = 0$ at the point $(2,1,7)$.

32. The temperature of a gas at the point (x,y,z) is given by $G(x,y,z) = x^2 - 5xy + y^2 z$.

(a) What is the rate of change in the temperature at the point $(1,2,3)$ in the direction $\vec{v} = 2\vec{i} + \vec{j} - 4\vec{k}$?

(b) What is the direction of maximum rate of change of temperature at the point $(1,2,3)$?

(c) What is the maximum rate of change at the point $(1,2,3)$?

33. The temperature at the point (x,y,z) in 3-space is given, in degrees Celsius, by $T(x,y,z) = e^{-(x^2+y^2+z^2)}$.

(a) Describe in words the shape of surfaces on which the temperature is constant.

(b) Find grad T.

(c) You travel from the point $(1,0,0)$ to the point $(2,1,0)$ at a speed of 3 units per second. Find the instantaneous rate of change of the temperature as you leave the point $(1,0,0)$. Give units.

34. At an altitude of h feet above sea level, the air pressure, P, in inches of mercury (in Hg), is given by

$$P = 30e^{-3.23 \times 10^{-5} h}.$$

An unpressurized seaplane takes off at an angle of $30°$ to the horizontal and a speed of 100 mph. What is the rate of change of pressure in the plane with respect to time at take-off, in inches of mercury per second?

35. A spaceship is plunging into the atmosphere of a planet. The pressure of the atmosphere at a position (x,y,z), with respect to the center of the planet, where the coordinates are in miles, is given by

$$P = 5e^{-0.1\sqrt{x^2+y^2+z^2}} \text{ atmospheres.}$$

The velocity vector, in miles/sec, of the spaceship at $(0,0,1)$ is $\vec{v} = \vec{i} - 2.5\vec{k}$. At $(0,0,1)$. What is the rate of change with respect to time of the pressure experienced by the spacecraft?

36. The earth has mass M and is located at the origin in 3-space, while the moon has mass m. Newton's Law of Gravitation states that if the moon is located at the point (x,y,z) then the attractive force exerted by the earth on the moon is given by the vector

$$\vec{F} = -GMm \frac{\vec{r}}{\|\vec{r}\|^3},$$

where $\vec{r} = x\vec{i} + y\vec{j} + z\vec{k}$. Show that $\vec{F} = \text{grad }\varphi$, where φ is the function given by

$$\varphi(x,y,z) = \frac{GMm}{\|\vec{r}\|}.$$

37. Two surfaces are said to be *tangential* at a point P if they have the same tangent plane at P. Show that the surfaces $z = \sqrt{2x^2 + 2y^2 - 25}$ and $z = \frac{1}{5}(x^2+y^2)$ are tangential at the point $(4,3,5)$.

38. Two surfaces are said to be *orthogonal* to each other at a point P if the normals to their tangent planes are perpendicular at P. Show that the surfaces $z = \frac{1}{2}(x^2+y^2-1)$ and $z = \frac{1}{2}(1-x^2-y^2)$ are orthogonal at all points of intersection.

39. Let $\vec{r}$ be the position vector of the point (x,y,z). If $\vec{\mu} = \mu_1 \vec{i} + \mu_2 \vec{j} + \mu_3 \vec{k}$ is a constant vector, show that

$$\text{grad}(\vec{\mu} \cdot \vec{r}) = \vec{\mu}.$$

40. Let $\vec{r}$ be the position vector of the point (x,y,z). Show that, if a is a constant,

$$\text{grad}(\|\vec{r}\|^a) = a\|\vec{r}\|^{a-2}\vec{r}, \qquad \vec{r} \neq \vec{0}.$$

41. Let f and g be functions on 3-space. Suppose f is differentiable and that

$$\text{grad } f(x,y,z) = (x\vec{i} + y\vec{j} + z\vec{k})g(x,y,z).$$

Explain why f must be constant on any sphere centered at the origin.

14.6 THE CHAIN RULE

Composition of Functions of Many Variables and Rates of Change

The chain rule enables us to differentiate *composite functions*. If we have a function of two variables $z = f(x,y)$ and we substitute $x = g(t), y = h(t)$ into $z = f(x,y)$, then we have a composite function in which z is a function of t:

$$z = f(g(t), h(t)).$$

If, on the other hand, we substitute $x = g(u,v), y = h(u,v)$, then we have a different composite function in which z is a function of u and v:

$$z = f(g(u,v), h(u,v)).$$

The next example shows how to calculate the rate of change of a composite function.

14.6 THE CHAIN RULE

Example 1 Corn production, C, depends on annual rainfall, R, and average temperature, T, so $C = f(R, T)$. Global warming predicts that both rainfall and temperature depend on time. Suppose that according to a particular model of global warming, rainfall is decreasing at 0.2 cm per year and temperature is increasing at 0.1°C per year. Use the fact that at current levels of production, $f_R = 3.3$ and $f_T = -5$ to estimate the current rate of change, dC/dt.

Solution By local linearity, we know that changes ΔR and ΔT generate a change, ΔC, in C given approximately by
$$\Delta C \approx f_R \Delta R + f_T \Delta T = 3.3 \Delta R - 5 \Delta T.$$
We want to know how ΔC depends on the time increment, Δt. A change Δt causes changes ΔR and ΔT, which in turn cause a change ΔC. The model of global warming tells us that
$$\frac{dR}{dt} = -0.2 \quad \text{and} \quad \frac{dT}{dt} = 0.1.$$
Thus, a time increment, Δt, generates changes of ΔR and ΔT given by
$$\Delta R \approx -0.2 \Delta t \quad \text{and} \quad \Delta T \approx 0.1 \Delta t.$$
Substituting for ΔR and ΔT in the expression for ΔC gives us
$$\Delta C \approx 3.3(-0.2\Delta t) - 5(0.1\Delta t) = -1.16\Delta t.$$
Thus,
$$\frac{\Delta C}{\Delta t} \approx -1.16 \quad \text{and, therefore,} \quad \frac{dC}{dt} \approx -1.16.$$

The relationship between ΔC and Δt, which gives the value of dC/dt, is an example of the *chain rule*. The argument in Example 1 leads to more general versions of the chain rule.

The Chain Rule for $z = f(x, y)$, $x = g(t)$, $y = h(t)$

Since $z = f(g(t), h(t))$ is a function of t, we can consider the derivative dz/dt. The chain rule gives dz/dt in terms of the derivatives of $f, g,$ and h. Since dz/dt represents the rate of change of z with t, we look at the change Δz generated by a small change, Δt.

We substitute the local linearizations
$$\Delta x \approx \frac{dx}{dt} \Delta t \quad \text{and} \quad \Delta y \approx \frac{dy}{dt} \Delta t$$
into the local linearization
$$\Delta z \approx \frac{\partial z}{\partial x} \Delta x + \frac{\partial z}{\partial y} \Delta y,$$
yielding
$$\Delta z \approx \frac{\partial z}{\partial x} \frac{dx}{dt} \Delta t + \frac{\partial z}{\partial y} \frac{dy}{dt} \Delta t$$
$$= \left(\frac{\partial z}{\partial x} \frac{dx}{dt} + \frac{\partial z}{\partial y} \frac{dy}{dt} \right) \Delta t.$$
Thus,
$$\frac{\Delta z}{\Delta t} \approx \frac{\partial z}{\partial x} \frac{dx}{dt} + \frac{\partial z}{\partial y} \frac{dy}{dt}.$$
Taking the limit as $\Delta t \to 0$, we get the following result.

> If $f, g,$ and h are differentiable and if $z = f(x, y)$, and $x = g(t)$, and $y = h(t)$, then
> $$\frac{dz}{dt} = \frac{\partial z}{\partial x} \frac{dx}{dt} + \frac{\partial z}{\partial y} \frac{dy}{dt}.$$

Visualizing the Chain Rule with a Diagram

The diagram in Figure 14.47 provides a way of remembering the chain rule. It shows the chain of dependence: z depends on x and y, which in turn depend on t. Each line in the diagram is labeled with a derivative relating the variables at its ends.

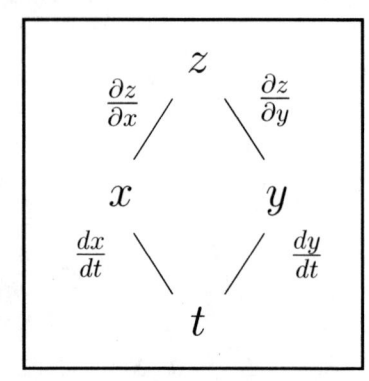

Figure 14.47: Diagram for $z = f(x, y)$, $x = g(t)$, $y = h(t)$. Lines represent dependence of z on x and y, and of x and y on t

The diagram keeps track of how a change in t propagates through the chain of composed functions. There are two paths from t to z, one through x and one through y. For each path, we multiply together the derivatives along the path. Then, to calculate dz/dt, we add the contributions from the two paths.

Example 2 Suppose that $z = f(x, y) = x \sin y$, where $x = t^2$ and $y = 2t + 1$. Let $z = g(t)$. Compute $g'(t)$ directly and using the chain rule.

Solution Since $z = g(t) = f(t^2, 2t + 1) = t^2 \sin(2t + 1)$, it is possible to compute $g'(t)$ directly by one-variable methods:

$$g'(t) = t^2 \frac{d}{dt}(\sin(2t+1)) + \left(\frac{d}{dt}(t^2)\right)\sin(2t+1) = 2t^2\cos(2t+1) + 2t\sin(2t+1).$$

The chain rule provides an alternative route to the same answer. We have

$$\frac{dz}{dt} = \frac{\partial z}{\partial x}\frac{dx}{dt} + \frac{\partial z}{\partial y}\frac{dy}{dt} = (\sin y)(2t) + (x\cos y)(2) = 2t\sin(2t+1) + 2t^2\cos(2t+1).$$

Example 3 The capacity, C, of a communication channel, such as a telephone line, to carry information depends on the ratio of the signal strength, S, to the noise, N. For some positive constant k,

$$C = k\ln\left(1 + \frac{S}{N}\right).$$

Suppose that the signal and noise are given as a function of time, t in seconds, by

$$S(t) = 4 + \cos(4\pi t) \qquad N(t) = 2 + \sin(2\pi t).$$

What is dC/dt one second after transmission started? Is the capacity increasing or decreasing at that instant?

Solution By the chain rule

$$\frac{dC}{dt} = \frac{\partial C}{\partial S}\frac{dS}{dt} + \frac{\partial C}{\partial N}\frac{dN}{dt}$$

$$= \frac{k}{1+S/N} \cdot \frac{1}{N}(-4\pi \sin 4\pi t) + \frac{k}{1+S/N}\left(-\frac{S}{N^2}\right)(2\pi\cos 2\pi t).$$

When $t = 1$, the first term is zero, $S(1) = 5$, and $N(1) = 2$, so

$$\frac{dC}{dt} = \frac{k}{1 + S(1)/N(1)} \left(-\frac{S(1)}{(N(1))^2}\right) \cdot 2\pi = \frac{k}{1 + \frac{5}{2}} \left(-\frac{5}{4}\right) \cdot 2\pi.$$

Since dC/dt is negative, the capacity is decreasing at time $t = 1$ second.

How to Formulate a General Chain Rule

A diagram can be used to write the chain rule for general compositions.

> To find the rate of change of one variable with respect to another in a chain of composed differentiable functions:
> - Draw a diagram expressing the relationship between the variables, and label each link in the diagram with the derivative relating the variables at its ends.
> - For each path between the two variables, multiply together the derivatives from each step along the path.
> - Add the contributions from each path.

The diagram keeps track of all the ways in which a change in one variable can cause a change in another; the diagram generates all the terms we would get from the appropriate substitutions into the local linearizations.

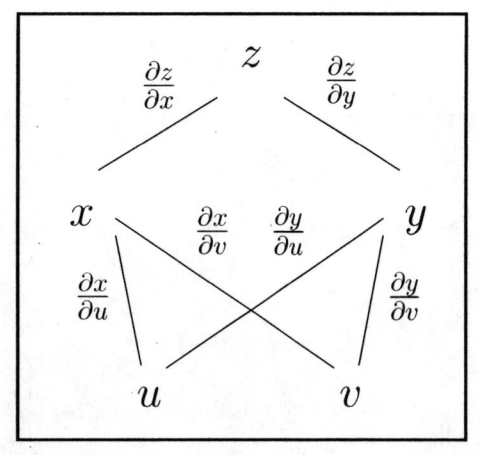

Figure 14.48: Diagram for $z = f(x, y)$, $x = g(u, v)$, $y = h(u, v)$. Lines represent dependence of z on x and y, and of x and y on u and v

For example, we can use Figure 14.48 to find formulas for $\partial z/\partial u$ and $\partial z/\partial v$. Adding the contributions for the two paths from z to u, we get the following results:

> If f, g, h are differentiable and if $z = f(x, y)$, with $x = g(u, v)$ and $y = h(u, v)$, then
> $$\frac{\partial z}{\partial u} = \frac{\partial z}{\partial x}\frac{\partial x}{\partial u} + \frac{\partial z}{\partial y}\frac{\partial y}{\partial u},$$
> $$\frac{\partial z}{\partial v} = \frac{\partial z}{\partial x}\frac{\partial x}{\partial v} + \frac{\partial z}{\partial y}\frac{\partial y}{\partial v}.$$

Chapter Fourteen DIFFERENTIATING FUNCTIONS OF SEVERAL VARIABLES

Example 4 Let $w = x^2 e^y$, $x = 4u$, and $y = 3u^2 - 2v$. Compute $\partial w/\partial u$ and $\partial w/\partial v$ using the chain rule.

Solution Using the previous result, we have

$$\frac{\partial w}{\partial u} = \frac{\partial w}{\partial x}\frac{\partial x}{\partial u} + \frac{\partial w}{\partial y}\frac{\partial y}{\partial u} = 2xe^y(4) + x^2 e^y(6u) = (8x + 6x^2 u)e^y$$

$$= (32u + 96u^3)e^{3u^2 - 2v}.$$

Similarly,

$$\frac{\partial w}{\partial v} = \frac{\partial w}{\partial x}\frac{\partial x}{\partial v} + \frac{\partial w}{\partial y}\frac{\partial y}{\partial v} = 2xe^y(0) + x^2 e^y(-2) = -2x^2 e^y$$

$$= -32u^2 e^{3u^2 - 2v}.$$

Example 5 A quantity z can be expressed either as a function of x and y, so that $z = f(x, y)$, or as a function of u and v, so that $z = g(u, v)$. The two coordinate systems are related by

$$x = u + v, \quad y = u - v.$$

(a) Use the chain rule to express $\partial z/\partial u$ and $\partial z/\partial v$ in terms of $\partial z/\partial x$ and $\partial z/\partial y$.
(b) Solve the equations in part (a) for $\partial z/\partial x$ and $\partial z/\partial y$.
(c) Show that the expressions we get in part (b) are the same as we get by expressing u and v in terms of x and y and using the chain rule.

Solution (a) We have $\partial x/\partial u = 1$ and $\partial x/\partial v = 1$, and also $\partial y/\partial u = 1$ and $\partial y/\partial v = -1$. Thus,

$$\frac{\partial z}{\partial u} = \frac{\partial z}{\partial x}(1) + \frac{\partial z}{\partial y}(1) = \frac{\partial z}{\partial x} + \frac{\partial z}{\partial y}$$

and

$$\frac{\partial z}{\partial v} = \frac{\partial z}{\partial x}(1) + \frac{\partial z}{\partial y}(-1) = \frac{\partial z}{\partial x} - \frac{\partial z}{\partial y}.$$

(b) Adding together the equations for $\partial z/\partial u$ and $\partial z/\partial v$ we get

$$\frac{\partial z}{\partial u} + \frac{\partial z}{\partial v} = 2\frac{\partial z}{\partial x}, \quad \text{so} \quad \frac{\partial z}{\partial x} = \frac{1}{2}\frac{\partial z}{\partial u} + \frac{1}{2}\frac{\partial z}{\partial v}.$$

Similarly, subtracting the equations for $\partial z/\partial u$ and $\partial z/\partial v$ yields

$$\frac{\partial z}{\partial y} = \frac{1}{2}\frac{\partial z}{\partial u} - \frac{1}{2}\frac{\partial z}{\partial v}.$$

(c) Alternatively, we can solve the equations

$$x = u + v, \quad y = u - v$$

for u and v, which yields

$$u = \frac{1}{2}x + \frac{1}{2}y, \quad v = \frac{1}{2}x - \frac{1}{2}y.$$

Now we can think of z as a function of u and v, and u and v as functions of x and y, and apply the chain rule again. This gives us

$$\frac{\partial z}{\partial x} = \frac{\partial z}{\partial u}\frac{\partial u}{\partial x} + \frac{\partial z}{\partial v}\frac{\partial v}{\partial x} = \frac{1}{2}\frac{\partial z}{\partial u} + \frac{1}{2}\frac{\partial z}{\partial v}$$

and

$$\frac{\partial z}{\partial y} = \frac{\partial z}{\partial u}\frac{\partial u}{\partial y} + \frac{\partial z}{\partial v}\frac{\partial v}{\partial y} = \frac{1}{2}\frac{\partial z}{\partial u} - \frac{1}{2}\frac{\partial z}{\partial v}.$$

These are the same expressions we got in part (b).

An Application to Physical Chemistry

A chemist investigating the properties of a gas such as carbon dioxide may want to know how the internal energy U of a given quantity of the gas depends on its temperature, T, pressure, P, and volume, V. The three quantities T, P, and V are not independent, however. For instance, according to the ideal gas law, they satisfy the equation

$$PV = kT$$

where k is a constant which depends only upon the quantity of the gas. The internal energy can then be thought of as a function of any two of the three quantities T, P, and V:

$$U = U_1(T, P) = U_2(T, V) = U_3(P, V).$$

The chemist writes, for example, $\left(\frac{\partial U}{\partial T}\right)_P$ to indicate the partial derivative of U with respect to T *holding P constant*, signifying that for this computation U is viewed as a function of T and P. Thus, we interpret $\left(\frac{\partial U}{\partial T}\right)_P$ as

$$\left(\frac{\partial U}{\partial T}\right)_P = \frac{\partial U_1(T, P)}{\partial T}.$$

If U is to be viewed as a function of T and V, the chemist writes $\left(\frac{\partial U}{\partial T}\right)_V$ for the partial derivative of U with respect to T holding V constant: thus, $\left(\frac{\partial U}{\partial T}\right)_V = \frac{\partial U_2(T,V)}{\partial T}$.

Each of the functions U_1, U_2, U_3 gives rise to one of the following formulas for the differential dU:

$$dU = \left(\frac{\partial U}{\partial T}\right)_P dT + \left(\frac{\partial U}{\partial P}\right)_T dP \quad \text{corresponds to } U_1$$

$$dU = \left(\frac{\partial U}{\partial T}\right)_V dT + \left(\frac{\partial U}{\partial V}\right)_T dV \quad \text{corresponds to } U_2,$$

$$dU = \left(\frac{\partial U}{\partial P}\right)_V dP + \left(\frac{\partial U}{\partial V}\right)_P dV \quad \text{corresponds to } U_3.$$

All the six partial derivatives appearing in formulas for dU have physical meaning, but they are not all equally easy to measure experimentally. A relationship among the partial derivatives, usually derived from the chain rule, may make it possible to evaluate one of the partials in terms of others that are more easily measured.

Example 6 Suppose a gas satisfies the equation $PV = 2T$ and $P = 3$ when $V = 4$. If $\left(\frac{\partial U}{\partial P}\right)_V = 7$ and $\left(\frac{\partial U}{\partial V}\right)_P = 8$, find the values of $\left(\frac{\partial U}{\partial P}\right)_T$ and $\left(\frac{\partial U}{\partial T}\right)_P$.

Solution Since we know the values of $\left(\frac{\partial U}{\partial P}\right)_V$ and $\left(\frac{\partial U}{\partial V}\right)_P$, we think of U as a function of P and V and use the function U_3 to write

$$dU = \left(\frac{\partial U}{\partial P}\right)_V dP + \left(\frac{\partial U}{\partial V}\right)_P dV$$
$$dU = 7dP + 8dV.$$

To calculate $\left(\frac{\partial U}{\partial P}\right)_T$ and $\left(\frac{\partial U}{\partial T}\right)_P$, we think of U as a function of T and P. Thus we want to substitute for dV in terms of dT and dP. Since $PV = 2T$, we have

$$P\,dV + V\,dP = 2dT,$$
$$3dV + 4dP = 2dT.$$

Solving gives $dV = (2dT - 4dP)/3$ so

$$dU = 7dP + 8\left(\frac{2dT - 4dP}{3}\right)$$
$$dU = -\frac{11}{3}dP + \frac{16}{3}dT.$$

Comparing with the formula for dU obtained from U_1:

$$dU = \left(\frac{\partial U}{\partial T}\right)_P dT + \left(\frac{\partial U}{\partial P}\right)_T dP,$$

we have

$$\left(\frac{\partial U}{\partial T}\right)_P = \frac{16}{3} \quad \text{and} \quad \left(\frac{\partial U}{\partial P}\right)_T = -\frac{11}{3}.$$

In Example 6, we could have substituted for dP instead of dV, leading to values of $\left(\frac{\partial U}{\partial T}\right)_V$ and $\left(\frac{\partial U}{\partial V}\right)_T$. See Problem 31.

In general, if for some particular P, V, and T, we can measure two of the six quantities $\left(\frac{\partial U}{\partial P}\right)_V$, $\left(\frac{\partial U}{\partial V}\right)_P$, $\left(\frac{\partial U}{\partial P}\right)_T$, $\left(\frac{\partial U}{\partial T}\right)_P$, $\left(\frac{\partial U}{\partial V}\right)_T$, $\left(\frac{\partial U}{\partial T}\right)_V$, then we can compute the other four using the relationship between dP, dV, and dT given by the gas law.

General formulas for each partial derivative in terms of others can be obtained in the same way. See the following example and Problem 31.

Example 7 Express $\left(\frac{\partial U}{\partial T}\right)_P$ in terms of $\left(\frac{\partial U}{\partial T}\right)_V$ and $\left(\frac{\partial U}{\partial V}\right)_T$ and $\left(\frac{\partial V}{\partial T}\right)_P$

Solution Since we are interested in the derivatives $\left(\frac{\partial U}{\partial T}\right)_V$ and $\left(\frac{\partial U}{\partial V}\right)_T$, we think of U as a function of T and V and use the formula

$$dU = \left(\frac{\partial U}{\partial T}\right)_V dT + \left(\frac{\partial U}{\partial V}\right)_T dV \quad \text{corresponding to } U_2.$$

We want to find a formula for $\left(\frac{\partial U}{\partial T}\right)_P$, which means thinking of U as a function of T and P. Thus, we want to substitute for dV. Since V is a function of T and P, we have

$$dV = \left(\frac{\partial V}{\partial T}\right)_P dT + \left(\frac{\partial V}{\partial P}\right)_T dP.$$

Substituting for dV into the formula for dU corresponding to U_2 gives

$$dU = \left(\frac{\partial U}{\partial T}\right)_V dT + \left(\frac{\partial U}{\partial V}\right)_T \left(\left(\frac{\partial V}{\partial T}\right)_P dT + \left(\frac{\partial V}{\partial P}\right)_T dP\right).$$

Collecting the terms containing dT and the terms containing dP gives

$$dU = \left(\left(\frac{\partial U}{\partial T}\right)_V + \left(\frac{\partial U}{\partial V}\right)_T \left(\frac{\partial V}{\partial T}\right)_P\right) dT + \left(\frac{\partial U}{\partial V}\right)_T \left(\frac{\partial V}{\partial P}\right)_T dP.$$

But we also have the formula

$$dU = \left(\frac{\partial U}{\partial T}\right)_P dT + \left(\frac{\partial U}{\partial P}\right)_T dP \quad \text{corresponding to } U_1.$$

We now have two formulas for dU in terms of dT and dP. The coefficients of dT must be identical, so we conclude

$$\left(\frac{\partial U}{\partial T}\right)_P = \left(\frac{\partial U}{\partial T}\right)_V + \left(\frac{\partial U}{\partial V}\right)_T \left(\frac{\partial V}{\partial T}\right)_P.$$

Example 7 expresses $\left(\frac{\partial U}{\partial T}\right)_P$ in terms of three other partial derivatives. Two of them, namely $\left(\frac{\partial U}{\partial T}\right)_V$, the constant-volume heat capacity, and $\left(\frac{\partial V}{\partial T}\right)_P$, the expansion coefficient, can be easily measured experimentally. The third, the internal pressure, $\left(\frac{\partial U}{\partial V}\right)_T$, cannot be measured directly but can be related to $\left(\frac{\partial P}{\partial T}\right)_V$, which is measurable. Thus, $\left(\frac{\partial U}{\partial T}\right)_P$ can be determined indirectly using this identity.

Exercises and Problems for Section 14.6

Exercises

For Exercises 1–6, find dz/dt using the chain rule. Assume the variables are restricted to domains on which the functions are defined.

1. $z = xy^2$, $x = e^{-t}$, $y = \sin t$
2. $z = \ln(x^2 + y^2)$, $x = 1/t$, $y = \sqrt{t}$
3. $z = xe^y$, $x = 2t$, $y = 1 - t^2$
4. $z = (x+y)e^y$, $x = 2t$, $y = 1 - t^2$
5. $z = x \sin y + y \sin x$, $x = t^2$, $y = \ln t$
6. $z = \sin(x/y)$, $x = 2t$, $y = 1 - t^2$

For Exercises 7–14, find $\partial z / \partial u$ and $\partial z / \partial v$. The variables are restricted to domains on which the functions are defined.

7. $z = xe^y$, $x = \ln u$, $y = v$
8. $z = (x+y)e^y$, $x = \ln u$, $y = v$
9. $z = xe^y$, $x = u^2 + v^2$, $y = u^2 - v^2$
10. $z = (x+y)e^y$, $x = u^2 + v^2$, $y = u^2 - v^2$
11. $z = \sin(x/y)$, $x = \ln u$, $y = v$
12. $z = \tan^{-1}(x/y)$, $x = u^2 + v^2$, $y = u^2 - v^2$
13. $z = xe^{-y} + ye^{-x}$, $x = u \sin v$, $y = v \cos u$
14. $z = \cos(x^2 + y^2)$, $x = u \cos v$, $y = u \sin v$

Problems

15. Suppose $w = f(x, y, z)$ and that x, y, z are functions of u and v. Use a tree diagram to write down the chain rule formula for $\partial w / \partial u$ and $\partial w / \partial v$.

16. Suppose $w = f(x, y, z)$ and that x, y, z are all functions of t. Use a tree diagram to write down the chain rule for dw/dt.

17. Corn production, C, is a function of rainfall, R, and temperature, T. Figures 14.49 and 14.50 show how rainfall and temperature are predicted to vary with time because of global warming. Suppose we know that $\Delta C \approx 3.3 \Delta R - 5 \Delta T$. Use this to estimate the change in corn production between the year 2020 and the year 2021. Hence, estimate dC/dt when $t = 2020$.

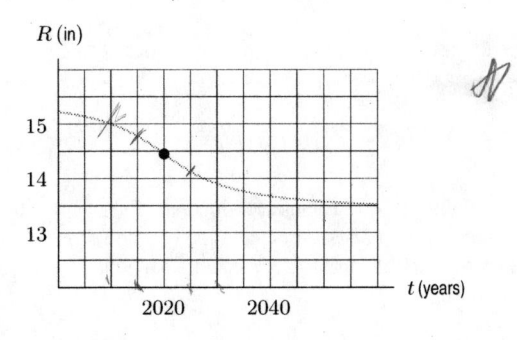

Figure 14.49: Rainfall as a function of time

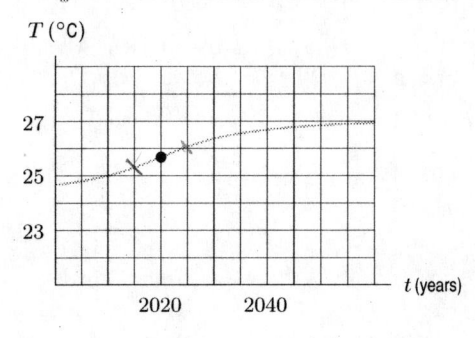

Figure 14.50: Temperature as a function of time

18. Let $z = g(u, v, w)$ and $u = u(s, t)$, $v = v(s, t)$, $w = w(s, t)$. How many terms are there in the expression for $\partial z / \partial t$?

19. The voltage, V, (in volts) across a circuit is given by Ohm's law: $V = IR$, where I is the current (in amps) flowing through the circuit and R is the resistance (in ohms). If we place two circuits, with resistance R_1 and R_2, in parallel, then their combined resistance, R, is given by
$$\frac{1}{R} = \frac{1}{R_1} + \frac{1}{R_2}.$$
Suppose the current is 2 amps and increasing at 10^{-2} amp/sec and R_1 is 3 ohms and increasing at 0.5 ohm/sec, while R_2 is 5 ohms and decreasing at 0.1 ohm/sec. Calculate the rate at which the voltage is changing.

20. Air pressure decreases at a rate of 2 pascals per kilometer in the eastward direction. In addition, the air pressure is dropping at a constant rate with respect to time everywhere. A ship sailing eastward at 10 km/hour past an island takes barometer readings and records a pressure drop of 50 pascals in 2 hours. Estimate the time rate of change of air pressure on the island. (A pascal is a unit of air pressure.)

21. Given $z = f(x, y)$, $x = x(u, v)$, $y = y(u, v)$ and $x(1, 2) = 5$, $y(1, 2) = 3$, calculate $z_u(1, 2)$ in terms of some of the numbers a, b, c, d, e, k, p, q, where

$f_x(1, 2) = a \quad f_y(1, 2) = c \quad x_u(1, 2) = e \quad y_u(1, 2) = p$
$f_x(5, 3) = b \quad f_y(5, 3) = d \quad x_v(1, 2) = k \quad y_v(1, 2) = q$

22. Let $F(u, v)$ be a function of two variables. Find $f'(x)$ if
   (a) $f(x) = F(x, 3)$
   (b) $f(x) = F(3, x)$
   (c) $f(x) = F(x, x)$
   (d) $f(x) = F(5x, x^2)$

23. Let $F(u, v, w)$ be a function of three variables. Find $G_x(x, y)$ if
   (a) $G(x, y) = F(x, y, 3)$
   (b) $G(x, y) = F(3, y, x)$
   (c) $G(x, y) = F(x, y, x)$
   (d) $G(x, y) = F(x, y, xy)$

24. The equation $f(x,y) = f(a,b)$ defines a level curve through a point (a,b) where grad $f(a,b) \neq \vec{0}$. Use implicit differentiation and the chain rule to show that the slope of the line tangent to this curve at the point (a,b) is $-f_x(a,b)/f_y(a,b)$ if $f_y(a,b) \neq 0$.

25. A function $f(x,y)$ is *homogeneous of degree p* if $f(tx, ty) = t^p f(x,y)$ for all t. Show that any differentiable, homogeneous function of degree p satisfies Euler's Theorem:
$$x f_x(x,y) + y f_y(x,y) = p f(x,y).$$
[Hint: Define $g(t) = f(tx, ty)$ and compute $g'(1)$.]

26. Let $F(x,y,z)$ be a function and define a function $z = f(x,y)$ implicitly by letting $F(x,y,f(x,y)) = 0$. Use the chain rule to show that
$$\frac{\partial z}{\partial x} = -\frac{\partial F/\partial x}{\partial F/\partial z} \quad \text{and} \quad \frac{\partial z}{\partial y} = -\frac{\partial F/\partial y}{\partial F/\partial z}.$$

For Problems 27–28, suppose the quantity z can be expressed either as a function of Cartesian coordinates (x,y) or as a function of polar coordinates (r, θ), so that $z = f(x,y) = g(r, \theta)$. [Recall that $x = r\cos\theta, y = r\sin\theta$ and $r = \sqrt{x^2 + y^2}, \theta = \arctan(y/x)$]

27. (a) Use the chain rule to find $\partial z/\partial r$ and $\partial z/\partial \theta$ in terms of $\partial z/\partial x$ and $\partial z/\partial y$.
(b) Solve the equations you have just written down for $\partial z/\partial x$ and $\partial z/\partial y$ in terms of $\partial z/\partial r$ and $\partial z/\partial \theta$.
(c) Show that the expressions you get in part (b) are the same as you would get by using the chain rule to find $\partial z/\partial x$ and $\partial z/\partial y$ in terms of $\partial z/\partial r$ and $\partial z/\partial \theta$.

28. Show that
$$\left(\frac{\partial z}{\partial x}\right)^2 + \left(\frac{\partial z}{\partial y}\right)^2 = \left(\frac{\partial z}{\partial r}\right)^2 + \frac{1}{r^2}\left(\frac{\partial z}{\partial \theta}\right)^2.$$

Problems 29–35 are continuations of the physical chemistry example on page 724.

29. Write $\left(\frac{\partial U}{\partial P}\right)_V$ as a partial derivative of one of the functions $U_1, U_2,$ or U_3.

30. Write $\left(\frac{\partial U}{\partial P}\right)_T$ as a partial derivative of one of the functions U_1, U_2, U_3.

31. For the gas in Example 6, find $\left(\frac{\partial U}{\partial T}\right)_V$ and $\left(\frac{\partial U}{\partial V}\right)_T$. [Hint: Use the same method as the example, but substitute for dP instead of dV.]

32. Show that $\left(\frac{\partial T}{\partial V}\right)_P = 1 \Big/ \left(\frac{\partial V}{\partial T}\right)_P$.

33. Express $\left(\frac{\partial U}{\partial P}\right)_T$ in terms of $\left(\frac{\partial U}{\partial V}\right)_T$ and $\left(\frac{\partial V}{\partial P}\right)_T$.

34. Use Example 7 and Problem 32 to show that
$$\left(\frac{\partial U}{\partial V}\right)_P = \left(\frac{\partial U}{\partial V}\right)_T + \frac{\left(\frac{\partial U}{\partial T}\right)_V}{\left(\frac{\partial V}{\partial T}\right)_P}.$$

35. In Example 6, we calculated values of $(\partial U/\partial T)_P$ and $(\partial U/\partial P)_T$ using the relationship $PV = 2T$ for a specific gas. In this problem, you will derive general relationships for these two partial derivatives.

(a) Think of V as a function of P and T and write an expression for dV.
(b) Substitute for dV into the following formula for dU (thinking of U as a function of P and V):
$$dU = \left(\frac{\partial U}{\partial P}\right)_V dP + \left(\frac{\partial U}{\partial V}\right)_P dV.$$
(c) Thinking of U as a function of P and T, write an expression for dU.
(d) By comparing coefficients of dP and dT in your answers to parts (b) and (c), show that
$$\left(\frac{\partial U}{\partial T}\right)_P = \left(\frac{\partial U}{\partial V}\right)_P \cdot \left(\frac{\partial V}{\partial T}\right)_P$$
$$\left(\frac{\partial U}{\partial P}\right)_T = \left(\frac{\partial U}{\partial P}\right)_V + \left(\frac{\partial U}{\partial V}\right)_P \cdot \left(\frac{\partial V}{\partial P}\right)_T.$$

Problems 36–38 concern differentiating an integral in one variable, y, which also involves another variable x, either in the integrand, or in the limits, or both:
$$\int_0^5 (x^2 y + 4)\, dy \quad \text{or} \quad \int_0^x (y+4)\, dy \quad \text{or} \quad \int_0^x (xy+4)\, dy.$$

To differentiate the first integral with respect to x, it can be shown that in most cases we can differentiate with respect to x inside the integral:
$$\frac{d}{dx}\left(\int_0^5 (x^2 y + 4)\, dy\right) = \int_0^5 2xy\, dy.$$

Differentiating the second integral with respect to x uses the Fundamental Theorem of Calculus:
$$\frac{d}{dx}\int_0^x (y+4)\, dy = x + 4.$$

Differentiating the third integral involves the chain rule, as shown in Problem 38. Assume that the function F is continuously differentiable and b is constant throughout.

36. Let $f(x) = \int_0^b F(x,y)\, dy$. Find $f'(x)$.

37. Let $f(x) = \int_0^x F(b,y)\, dy$. Find $f'(x)$.

38. Let $f(x) = \int_0^x F(x,y)\, dy$. Use Problem 36 and Problem 37 to find $f'(x)$ by the following steps:
(a) Let $G(u,w) = \int_0^w F(u,y)\, dy$. Find $G_u(u,w)$ and $G_w(u,w)$.
(b) Use part (a) and the chain rule applied to $G(x,x) = f(x)$ to show:
$$f'(x) = \int_0^x F_u(x,y)\, dy + F(x,x).$$

14.7 SECOND-ORDER PARTIAL DERIVATIVES

Since the partial derivatives of a function are themselves functions, we can differentiate them, giving *second-order partial derivatives*. A function $z = f(x, y)$ has two first-order partial derivatives, f_x and f_y, and four second-order partial derivatives.

The Second-Order Partial Derivatives of $z = f(x, y)$

$$\frac{\partial^2 z}{\partial x^2} = f_{xx} = (f_x)_x, \qquad \frac{\partial^2 z}{\partial x \partial y} = f_{yx} = (f_y)_x,$$

$$\frac{\partial^2 z}{\partial y \partial x} = f_{xy} = (f_x)_y, \qquad \frac{\partial^2 z}{\partial y^2} = f_{yy} = (f_y)_y.$$

It is usual to omit the parentheses, writing f_{xy} instead of $(f_x)_y$ and $\dfrac{\partial^2 z}{\partial y \partial x}$ instead of $\dfrac{\partial}{\partial y}\left(\dfrac{\partial z}{\partial x}\right)$.

Example 1 Compute the four second-order partial derivatives of $f(x, y) = xy^2 + 3x^2 e^y$.

Solution From $f_x(x, y) = y^2 + 6xe^y$ we get

$$f_{xx}(x, y) = \frac{\partial}{\partial x}(y^2 + 6xe^y) = 6e^y \quad \text{and} \quad f_{xy}(x, y) = \frac{\partial}{\partial y}(y^2 + 6xe^y) = 2y + 6xe^y.$$

From $f_y(x, y) = 2xy + 3x^2 e^y$ we get

$$f_{yx}(x, y) = \frac{\partial}{\partial x}(2xy + 3x^2 e^y) = 2y + 6xe^y \quad \text{and} \quad f_{yy}(x, y) = \frac{\partial}{\partial y}(2xy + 3x^2 e^y) = 2x + 3x^2 e^y.$$

Observe that $f_{xy} = f_{yx}$ in this example.

Example 2 Use the values of the function $f(x, y)$ in Table 14.6 to estimate $f_{xy}(1, 2)$ and $f_{yx}(1, 2)$.

Table 14.6 *Values of $f(x, y)$*

$y \backslash x$	0.9	1.0	1.1
1.8	4.72	5.83	7.06
2.0	6.48	8.00	9.60
2.2	8.62	10.65	12.88

Solution Since $f_{xy} = (f_x)_y$, we first estimate f_x

$$f_x(1, 2) \approx \frac{f(1.1, 2) - f(1, 2)}{0.1} = \frac{9.60 - 8.00}{0.1} = 16.0,$$

$$f_x(1, 2.2) \approx \frac{f(1.1, 2.2) - f(1, 2.2)}{0.1} = \frac{12.88 - 10.65}{0.1} = 22.3.$$

Thus,

$$f_{xy}(1, 2) \approx \frac{f_x(1, 2.2) - f_x(1, 2)}{0.2} = \frac{22.3 - 16.0}{0.2} = 31.5.$$

Similarly,

$$f_{yx}(1,2) \approx \frac{f_y(1.1,2) - f_y(1,2)}{0.1} \approx \frac{1}{0.1}\left(\frac{f(1.1,2.2) - f(1.1,2)}{0.2} - \frac{f(1,2.2) - f(1,2)}{0.2}\right)$$

$$= \frac{1}{0.1}\left(\frac{12.88 - 9.60}{0.2} - \frac{10.65 - 8.00}{0.2}\right) = 31.5.$$

Observe that in this example also, $f_{xy} = f_{yx}$.

The Mixed Partial Derivatives Are Equal

It is not an accident that the estimates for $f_{xy}(1,2)$ and $f_{yx}(1,2)$ are equal in Example 2, because the same values of the function are used to calculate each one. The fact that $f_{xy} = f_{yx}$ in Examples 1 and 2 corroborates the following general result; Problem 36 suggests why you might expect it to be true.[2]

Theorem 14.1: Equality of Mixed Partial Derivatives

If f_{xy} and f_{yx} are continuous at (a,b), an interior point of their domain, then

$$f_{xy}(a,b) = f_{yx}(a,b).$$

For most functions f we encounter and most points (a,b) in their domains, not only are f_{xy} and f_{yx} continuous at (a,b), but all their higher order partial derivatives (such as f_{xxy} or f_{xyyy}) exist and are continuous at (a,b). In that case we say f is *smooth* at (a,b). We say f is smooth on a region R if it is smooth at every point of R.

What Do the Second-Order Partial Derivatives Tell Us?

Example 3 Let us return to the guitar string of Example 4, page 694. The string is 1 meter long and at time t seconds, the point x meters from one end is displaced $f(x,t)$ meters from its rest position, where

$$f(x,t) = 0.003 \sin(\pi x) \sin(2765t).$$

Compute the four second-order partial derivatives of f at the point $(x,t) = (0.3, 1)$ and describe the meaning of their signs in practical terms.

Solution First we compute $f_x(x,t) = 0.003\pi \cos(\pi x) \sin(2765t)$, from which we get

$$f_{xx}(x,t) = \frac{\partial}{\partial x}(f_x(x,t)) = -0.003\pi^2 \sin(\pi x) \sin(2765t), \quad \text{so} \quad f_{xx}(0.3, 1) \approx -0.01;$$

and

$$f_{xt}(x,t) = \frac{\partial}{\partial t}(f_x(x,t)) = (0.003)(2765)\pi \cos(\pi x) \cos(2765t), \quad \text{so} \quad f_{xt}(0.3, 1) \approx 14.$$

On page 694 we saw that $f_x(x,t)$ gives the slope of the string at any point and time. Therefore, $f_{xx}(x,t)$ measures the concavity of the string. The fact that $f_{xx}(0.3, 1) < 0$ means the string is concave down at the point $x = 0.3$ when $t = 1$. (See Figure 14.51.)

On the other hand, $f_{xt}(x,t)$ is the rate of change of the slope of the string with respect to time. Thus $f_{xt}(0.3, 1) > 0$ means that at time $t = 1$ the slope at the point $x = 0.3$ is increasing. (See Figure 14.52.)

---

[2] For a proof, see M. Spivak, *Calculus on Manifolds*, p. 26 (New York: Benjamin, 1965).

14.7 SECOND-ORDER PARTIAL DERIVATIVES

The slope at B is less than the slope at A

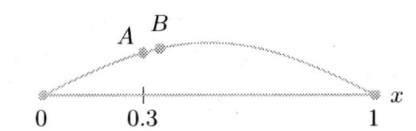

Figure 14.51: Interpretation of $f_{xx}(0.3, 1) < 0$: The concavity of the string at $t = 1$

The slope at B is greater than the slope at A

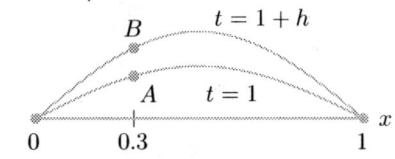

Figure 14.52: Interpretation of $f_{xt}(0.3, 1) > 0$: The slope of one point on the string at two different times

Now we compute $f_t(x, t) = (0.003)(2765) \sin(\pi x) \cos(2765t)$, from which we get

$$f_{tx}(x, t) = \frac{\partial}{\partial x}(f_t(x, t)) = (0.003)(2765)\pi \cos(\pi x) \cos(2765t), \quad \text{so} \quad f_{tx}(0.3, 1) \approx 14$$

and

$$f_{tt}(x, t) = \frac{\partial}{\partial t}(f_t(x, t)) = -(0.003)(2765)^2 \sin(\pi x) \sin(2765t), \quad \text{so} \quad f_{tt}(0.3, 1) \approx -7200.$$

On page 694 we saw that $f_t(x, t)$ gives the velocity of the string at any point and time. Therefore, $f_{tx}(x, t)$ and $f_{tt}(x, t)$ will both be rates of change of velocity. That $f_{tx}(0.3, 1) > 0$ means that at time $t = 1$ the velocities of points just to the right of $x = 0.3$ are greater than the velocity at $x = 0.3$. (See Figure 14.53.) That $f_{tt}(0.3, 1) < 0$ means that the velocity of the point $x = 0.3$ is decreasing at time $t = 1$. Thus, $f_{tt}(0.3, 1) = -7200$ m/sec^2 is the acceleration of this point. (See Figure 14.54.)

The velocity at B is greater than the velocity at A

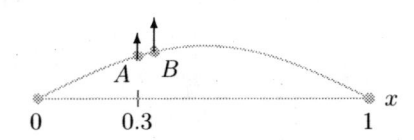

Figure 14.53: Interpretation of $f_{tx}(0.3, 1) > 0$: The velocity of different points on the string at $t = 1$

The velocity at B is less than the velocity at A

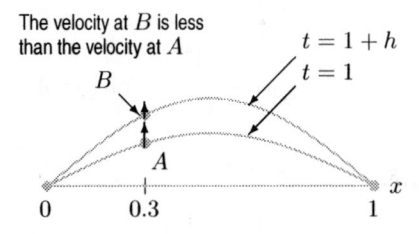

Figure 14.54: Interpretation of $f_{tt}(0.3, 1) < 0$: Negative acceleration. The velocity of one point on the string at two different times

Taylor Approximations

We use second derivatives to construct quadratic Taylor approximations. In Section 14.3, we saw how to approximate $f(x, y)$ by a linear function (its local linearization). We now see how to improve this approximation of $f(x, y)$ using a quadratic function.

Linear and Quadratic Approximations Near (0,0)

For a function of one variable, local linearity tells us that the best *linear* approximation is the degree 1 Taylor polynomial

$$f(x) \approx f(a) + f'(a)(x - a) \quad \text{for } x \text{ near } a.$$

A better approximation to $f(x)$ is given by the degree 2 Taylor polynomial:

$$f(x) \approx f(a) + f'(a)(x - a) + \frac{f''(a)}{2}(x - a)^2 \quad \text{for } x \text{ near } a.$$

For a function of two variables the local linearization for (x, y) near (a, b) is

$$f(x, y) \approx L(x, y) = f(a, b) + f_x(a, b)(x - a) + f_y(a, b)(y - b).$$

In the case $(a, b) = (0, 0)$, we have:

Taylor Polynomial of Degree 1 Approximating $f(x, y)$ for (x, y) near $(0,0)$
If f has continuous first-order partial derivatives, then

$$f(x,y) \approx L(x,y) = f(0,0) + f_x(0,0)x + f_y(0,0)y.$$

We get a better approximation to f by using a quadratic polynomial. We choose a quadratic polynomial $Q(x, y)$, with the same partial derivatives as the original function f. You can check that the following Taylor polynomial of degree 2 has this property.

Taylor Polynomial of Degree 2 Approximating $f(x, y)$ for (x, y) near $(0,0)$ If f has continuous second-order partial derivatives, then

$$f(x,y) \approx Q(x,y)$$
$$= f(0,0) + f_x(0,0)x + f_y(0,0)y + \frac{f_{xx}(0,0)}{2}x^2 + f_{xy}(0,0)xy + \frac{f_{yy}(0,0)}{2}y^2.$$

Example 4 Let $f(x,y) = \cos(2x+y) + 3\sin(x+y)$
(a) Compute the linear and quadratic Taylor polynomials, L and Q, approximating f near $(0,0)$.
(b) Explain why the contour plots of L and Q for $-1 \le x \le 1$, $-1 \le y \le 1$ look the way they do.

Solution (a) We have $f(0,0) = 1$. The derivatives we need are as follows:

$$\begin{aligned}
f_x(x,y) &= -2\sin(2x+y) + 3\cos(x+y) & \text{so} && f_x(0,0) &= 3, \\
f_y(x,y) &= -\sin(2x+y) + 3\cos(x+y) & \text{so} && f_y(0,0) &= 3, \\
f_{xx}(x,y) &= -4\cos(2x+y) - 3\sin(x+y) & \text{so} && f_{xx}(0,0) &= -4, \\
f_{xy}(x,y) &= -2\cos(2x+y) - 3\sin(x+y) & \text{so} && f_{xy}(0,0) &= -2, \\
f_{yy}(x,y) &= -\cos(2x+y) - 3\sin(x+y) & \text{so} && f_{yy}(0,0) &= -1.
\end{aligned}$$

Thus, the linear approximation, $L(x, y)$, to $f(x, y)$ at $(0, 0)$ is given by

$$f(x,y) \approx L(x,y) = f(0,0) + f_x(0,0)x + f_y(0,0)y = 1 + 3x + 3y.$$

The quadratic approximation, $Q(x, y)$, to $f(x, y)$ near $(0, 0)$ is given by

$$f(x,y) \approx Q(x,y)$$
$$= f(0,0) + f_x(0,0)x + f_y(0,0)y + \frac{f_{xx}(0,0)}{2}x^2 + f_{xy}(0,0)xy + \frac{f_{yy}(0,0)}{2}y^2$$
$$= 1 + 3x + 3y - 2x^2 - 2xy - \frac{1}{2}y^2.$$

Notice that the linear terms in $Q(x, y)$ are the same as the linear terms in $L(x, y)$. The quadratic terms in $Q(x, y)$ can be thought of as "correction terms" to the linear approximation.
(b) The contour plots of $f(x, y)$, $L(x, y)$, and $Q(x, y)$ are in Figures 14.55–14.57.

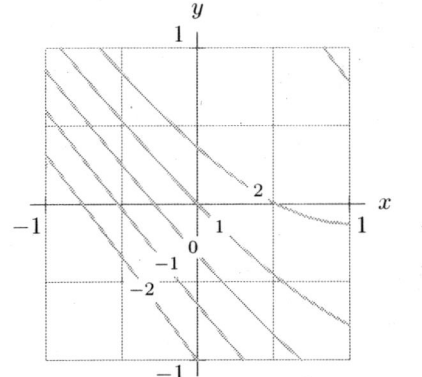

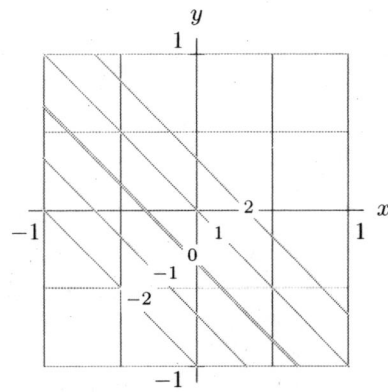

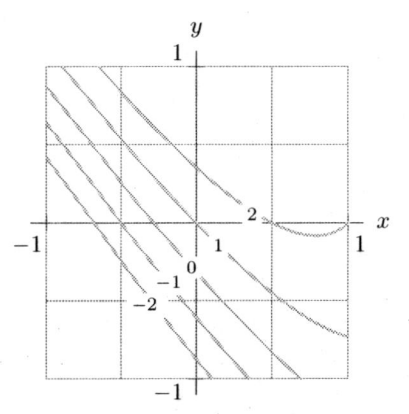

Figure 14.55: Original function, $f(x, y)$

Figure 14.56: Linear approximation, $L(x, y)$

Figure 14.57: Quadratic approximation, $Q(x, y)$

Notice that the contour plot of Q is more similar to the contour plot of f than is the contour plot of L. Since L is linear, the contour plot of L consists of parallel, equally spaced lines.

An alternative, and much quicker, way to find the Taylor polynomial in the previous example is to use the single-variable approximations. For example, since

$$\cos u = 1 - \frac{u^2}{2!} + \frac{u^4}{4!} + \cdots \quad \text{and} \quad \sin v = v - \frac{v^3}{3!} + \cdots,$$

we can substitute $u = 2x + y$ and $v = x + y$ and expand. We discard terms beyond the second (since we want the quadratic polynomial) getting

$$\cos(2x+y) = 1 - \frac{(2x+y)^2}{2!} + \frac{(2x+y)^4}{4!} + \cdots \approx 1 - \frac{1}{2}(4x^2 + 4xy + y^2) = 1 - 2x^2 - 2xy - \frac{1}{2}y^2$$

and

$$\sin(x+y) = (x+y) - \frac{(x+y)^3}{3!} + \cdots \approx x + y.$$

Combining these results, we get

$$\cos(2x+y) + 3\sin(x+y) \approx 1 - 2x^2 - 2xy - \frac{1}{2}y^2 + 3(x+y) = 1 + 3x + 3y - 2x^2 - 2xy - \frac{1}{2}y^2.$$

Linear and Quadratic Approximations Near (a, b)

The local linearization for a function $f(x, y)$ at a point (a, b) is

> **Taylor Polynomial of Degree 1 Approximating $f(x, y)$ for (x, y) near (a, b)**
> If f has continuous first-order partial derivatives, then
>
> $$f(x, y) \approx L(x, y) = f(a, b) + f_x(a, b)(x - a) + f_y(a, b)(y - b).$$

This suggests that a quadratic polynomial approximation $Q(x, y)$ for $f(x, y)$ near a point (a, b) should be written in terms of $(x - a)$ and $(y - b)$ instead of x and y. If we require that $Q(a, b) = f(a, b)$ and that the first- and second-order partial derivatives of Q and f at (a, b) be equal, then we get the following polynomial:

Taylor Polynomial of Degree 2 Approximating $f(x,y)$ for (x,y) near (a,b)
If f has continuous second-order partial derivatives, then

$$f(x,y) \approx Q(x,y)$$
$$= f(a,b) + f_x(a,b)(x-a) + f_y(a,b)(y-b)$$
$$+ \frac{f_{xx}(a,b)}{2}(x-a)^2 + f_{xy}(a,b)(x-a)(y-b) + \frac{f_{yy}(a,b)}{2}(y-b)^2.$$

These coefficients are derived in exactly the same way as for $(a,b) = (0,0)$.

Example 5 Find the Taylor polynomial of degree 2 at the point $(1,2)$ for the function $f(x,y) = \dfrac{1}{xy}$.

Solution Table 14.7 contains the partial derivatives and their values at the point $(1,2)$.

Table 14.7 *Partial derivatives of $f(x,y) = 1/(xy)$*

Derivative	Formula	Value at (1,2)	Derivative	Formula	Value at (1,2)
$f(x,y)$	$1/(xy)$	$1/2$	$f_{xx}(x,y)$	$2/(x^3 y)$	1
$f_x(x,y)$	$-1/(x^2 y)$	$-1/2$	$f_{xy}(x,y)$	$1/(x^2 y^2)$	$1/4$
$f_y(x,y)$	$-1/(xy^2)$	$-1/4$	$f_{yy}(x,y)$	$2/(xy^3)$	$1/4$

So, the quadratic Taylor polynomial for f near $(1,2)$ is

$$\frac{1}{xy} \approx Q(x,y)$$
$$= \frac{1}{2} - \frac{1}{2}(x-1) - \frac{1}{4}(y-2) + \frac{1}{2}(1)(x-1)^2 + \frac{1}{4}(x-1)(y-2) + \left(\frac{1}{2}\right)\left(\frac{1}{4}\right)(y-2)^2$$
$$= \frac{1}{2} - \frac{x-1}{2} - \frac{y-2}{4} + \frac{(x-1)^2}{2} + \frac{(x-1)(y-2)}{4} + \frac{(y-2)^2}{8}.$$

Exercises and Problems for Section 14.7

Exercises

In Exercises 1–10, calculate all four second-order partial derivatives and check that $f_{xy} = f_{yx}$. Assume the variables are restricted to a domain on which the function is defined.

1. $f(x,y) = 3x^2 y + 5xy^3$
2. $f(x,y) = (x+y)^2$
3. $f(x,y) = (x+y)^3$
4. $f(x,y) = xe^y$
5. $f(x,y) = e^{2xy}$
6. $f(x,y) = (x+y)e^y$
7. $f(x,y) = \sqrt{x^2 + y^2}$
8. $f(x,y) = \sin(x/y)$
9. $f(x,y) = \sin(x^2 + y^2)$
10. $f(x,y) = 3\sin 2x \cos 5y$

Find the quadratic Taylor polynomials about $(0,0)$ for the functions in Exercises 11–18.

11. $(x - y + 1)^2$
12. $(y-1)(x+1)^2$
13. $e^{-2x^2 - y^2}$
14. $1/(1 + 2x - y)$
15. $e^x \cos y$
16. $\cos(x + 3y)$
17. $\ln(1 + x^2 - y)$
18. $\sin 2x + \cos y$

In Exercises 19–28, use the level curves of the function $z = f(x, y)$ to decide the sign (positive, negative, or zero) of each of the following partial derivatives at the point P. Assume the x- and y-axes are in the usual positions.

(a) $f_x(P)$ (b) $f_y(P)$ (c) $f_{xx}(P)$
(d) $f_{yy}(P)$ (e) $f_{xy}(P)$

19.

20.

21.

22.

23.

24.

25.

26.

27.

28.

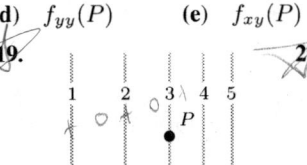

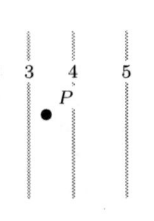

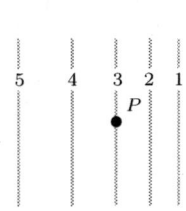

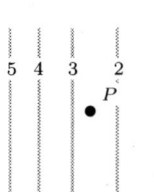

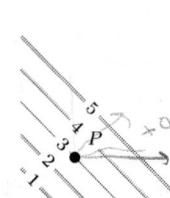

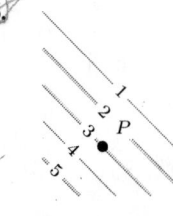

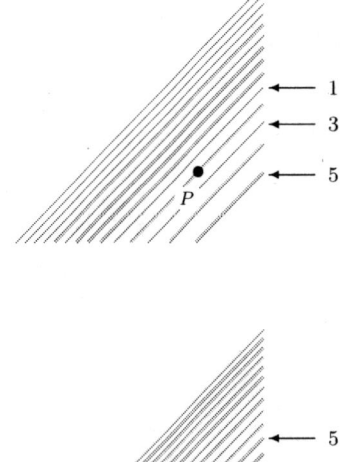

In Exercises 29–32, find the linear, $L(x, y)$, and quadratic, $Q(x, y)$, Taylor polynomials valid near $(1, 0)$. Compare the values of the approximations $L(0.9, 0.2)$ and $Q(0.9, 0.2)$ with the exact value of the function $f(0.9, 0.2)$.

29. $f(x, y) = x^2 y$ **30.** $f(x, y) = \sqrt{x + 2y}$

31. $f(x, y) = xe^{-y}$

32. $f(x, y) = \sin(x - 1) \cos y$

Problems

33. Figure 14.58 shows a graph of $z = f(x, y)$. Is $f_{xx}(0, 0)$ positive or negative? Is $f_{yy}(0, 0)$ positive or negative? Give reasons for your answers.

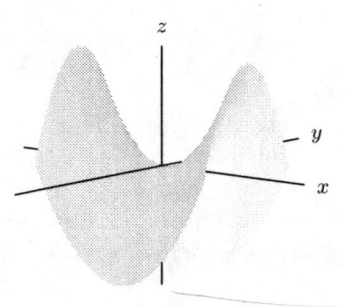

Figure 14.58

34. If $z = f(x) + yg(x)$, what can you say about z_{yy}? Explain your answer.

35. If $z_{xy} = 4y$, what can you say about the value of
(a) z_{yx}?   (b) z_{xyx}?   (c) z_{xyy}?

36. Give an explanation of why you might expect $f_{xy}(a, b) = f_{yx}(a, b)$ using the following steps.

(a) Write the definition of $f_x(a, b)$.
(b) Write a definition of $f_{xy}(a, b)$ as $(f_x)_y$.
(c) Substitute for f_x in the definition of f_{xy}.
(d) Write an expression for f_{yx} similar to the one for f_{xy} you obtained in part (c).
(e) Compare your answers to parts (c) and (d). What do you have to assume to conclude that f_{xy} and f_{yx} are equal?

37. A contour diagram for the smooth function $z = f(x, y)$ is in Figure 14.59.

  (a) Is z an increasing or decreasing function of x? Of y?
  (b) Is f_x positive or negative? How about f_y?
  (c) Is f_{xx} positive or negative? How about f_{yy}?
  (d) Sketch the direction of grad f at points P and Q.
  (e) Is grad f longer at P or at Q? How do you know?

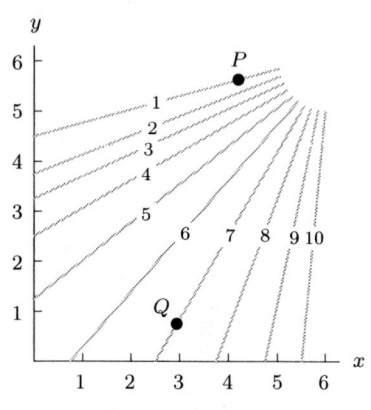

Figure 14.59

38. Figure 14.60 shows the level curves of a function $f(x, y)$ around a maximum or minimum, M. One of the points P and Q has coordinates (x_1, y_1) and the other has coordinates (x_2, y_2). Suppose $b > 0$ and $c > 0$. Consider the two linear approximations to f given by

$$f(x, y) \approx a + b(x - x_1) + c(y - y_1)$$
$$f(x, y) \approx k + m(x - x_2) + n(y - y_2).$$

  (a) What is the relationship between the values of a and k?
  (b) What are the coordinates of P?
  (c) Is M a maximum or a minimum?
  (d) What can you say about the sign of the constants m and n?

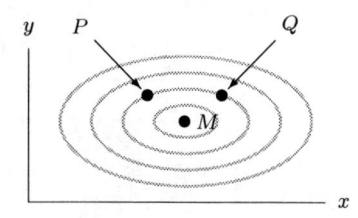

Figure 14.60

39. Consider the function $f(x, y) = (\sin x)(\sin y)$.

  (a) Find the Taylor polynomials of degree 2 for f about the points $(0, 0)$ and $(\pi/2, \pi/2)$.
  (b) Use the Taylor polynomials to sketch the contours of f close to each of the points $(0, 0)$ and $(\pi/2, \pi/2)$.

40. Let $f(x, y) = \sqrt{x + 2y + 1}$.

  (a) Compute the local linearization of f at $(0, 0)$.
  (b) Compute the quadratic Taylor polynomial for f at $(0, 0)$.
  (c) Compare the values of the linear and quadratic approximations in part (a) and part (b) with the true values for $f(x, y)$ at the points $(0.1, 0.1)$, $(-0.1, 0.1)$, $(0.1, -0.1)$, $(-0.1, -0.1)$. Which approximation gives the closest values?

41. Using a computer and your answer to Problem 40, draw the six contour diagrams of $f(x, y) = \sqrt{x + 2y + 1}$ and its linear and quadratic approximations, $L(x, y)$ and $Q(x, y)$, in the two windows $[-0.6, 0.6] \times [-0.6, 0.6]$ and $[-2, 2] \times [-2, 2]$. Explain the shape of the contours, their spacing, and the relationship between the contours of f, L, and Q.

42. Give contour diagrams of two different functions $f(x, y)$ and $g(x, y)$ that have the same quadratic approximations near $(0, 0)$.

43. Suppose that $f(x, y)$ has continuous partial derivatives f_x and f_y. Using the Fundamental Theorem of Calculus to evaluate the integrals, show that

$$f(a, b) = f(0, 0) + \int_{t=0}^{a} f_x(t, 0) dt + \int_{t=0}^{b} f_y(a, t) dt.$$

44. Suppose that $f(x, y)$ has continuous partial derivatives and that $f(0, 0) = 0$ and $|f_x(x, y)| \leq A$ and $|f_y(x, y)| \leq B$ for all points (x, y) in the plane. Use Problem 43 to show that $|f(x, y)| \leq A|x| + B|y|$.

This inequality shows how bounds on the partial derivatives of f limit the growth of f.

14.8 DIFFERENTIABILITY

In Section 14.3 we gave an informal introduction to the concept of differentiability. We called a function $f(x, y)$ *differentiable* at a point (a, b) if it is well-approximated by a linear function near (a, b). This section focuses on the precise meaning of the phrase "well-approximated." By looking at examples, we shall see that local linearity requires the existence of partial derivatives, but they do not tell the whole story. In particular, existence of partial derivatives at a point is not sufficient to guarantee local linearity at that point.

We begin by discussing the relation between continuity and differentiability. As an illustration,

14.8 DIFFERENTIABILITY

take a sheet of paper, crumple it into a ball and smooth it out again. Wherever there is a crease it would be difficult to approximate the surface by a plane—these are points of nondifferentiability of the function giving the height of the paper above the floor. Yet the sheet of paper models a graph which is continuous—there are no breaks. As in the case of one-variable calculus, continuity does not imply differentiability. But differentiability does *require* continuity: there cannot be linear approximations to a surface at points where there are abrupt changes in height.

Differentiability For Functions Of Two Variables

For a function of two variables, as for a function of one variable, we define differentiability at a point in terms of the error and the distance from the point. If the point is (a, b) and a nearby point is $(a + h, b + k)$, the distance between them is $\sqrt{h^2 + k^2}$. (See Figure 14.61.)

A function $f(x, y)$ is **differentiable at the point** (a, b) if there is a linear function $L(x, y) = f(a, b) + m(x - a) + n(y - b)$ such that if the *error* $E(x, y)$ is defined by

$$f(x, y) = L(x, y) + E(x, y),$$

and if $h = x - a, k = y - b$, then the *relative error* $E(a + h, b + k)/\sqrt{h^2 + k^2}$ satisfies

$$\lim_{\substack{h \to 0 \\ k \to 0}} \frac{E(a + h, b + k)}{\sqrt{h^2 + k^2}} = 0.$$

The function f is **differentiable on a region** R if it is differentiable at each point of R. The function $L(x, y)$ is called the *local linearization* of $f(x, y)$ near (a, b).

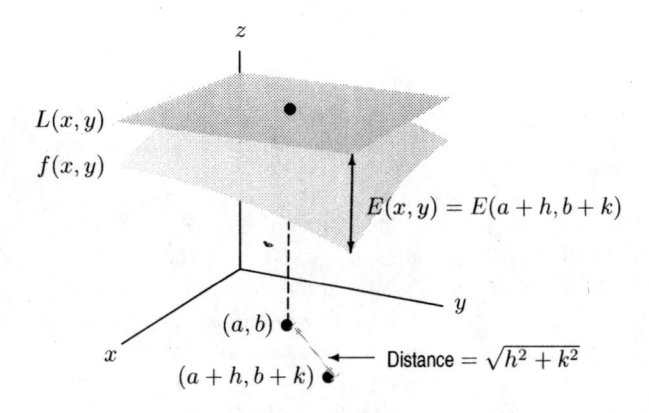

Figure 14.61: Graph of function $z = f(x, y)$ and its local linearization $z = L(x, y)$ near the point (a, b)

Partial Derivatives and Differentiability

In the next example, we show that this definition of differentiability is consistent with our previous notion — that is, that $m = f_x$ and $n = f_y$ and that the graph of $L(x, y)$ is the tangent plane.

Example 1 Show that if f is a differentiable function with local linearization $L(x, y) = f(a, b) + m(x - a) + n(y - b)$, then $m = f_x(a, b)$ and $n = f_y(a, b)$.

Solution Since f is differentiable, we know that the relative error in $L(x, y)$ tends to 0 as we get close to (a, b). Suppose $h > 0$ and $k = 0$. Then we know that

$$0 = \lim_{h \to 0} \frac{E(a + h, b)}{\sqrt{h^2 + k^2}} = \lim_{h \to 0} \frac{E(a + h, b)}{h} = \lim_{h \to 0} \frac{f(a + h, b) - L(a + h, b)}{h}$$
$$= \lim_{h \to 0} \frac{f(a + h, b) - f(a, b) - mh}{h}$$
$$= \lim_{h \to 0} \left(\frac{f(a + h, b) - f(a, b)}{h} \right) - m = f_x(a, b) - m.$$

A similar result holds if $h < 0$, so we have $m = f_x(a, b)$. The result $n = f_y(a, b)$ is found in a similar manner.

The previous example shows that if a function is differentiable at a point, it has partial derivatives there. Therefore, if any of the partial derivatives fail to exist, then the function cannot be differentiable. This is what happens in the following example of a cone.

Example 2 Consider the function $f(x, y) = \sqrt{x^2 + y^2}$. Is f differentiable at the origin?

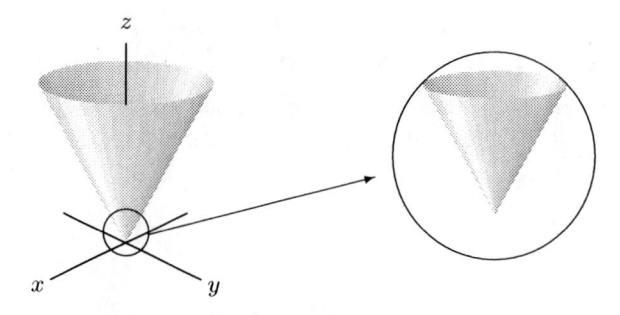

Figure 14.62: The function $f(x, y) = \sqrt{x^2 + y^2}$ is not locally linear at $(0, 0)$: Zooming in around $(0, 0)$ does not make the graph look like a plane

Solution If we zoom in on the graph of the function $f(x, y) = \sqrt{x^2 + y^2}$ at the origin, as shown in Figure 14.62, the sharp point remains; the graph never flattens out to look like a plane. Near its vertex, the graph does not look like it is well approximated (in any reasonable sense) by any plane.

Judging from the graph of f, we would not expect f to be differentiable at $(0, 0)$. Let us check this by trying to compute the partial derivatives of f at $(0, 0)$:

$$f_x(0, 0) = \lim_{h \to 0} \frac{f(h, 0) - f(0, 0)}{h} = \lim_{h \to 0} \frac{\sqrt{h^2 + 0} - 0}{h} = \lim_{h \to 0} \frac{|h|}{h}.$$

Since $|h|/h = \pm 1$, depending on whether h approaches 0 from the left or right, this limit does not exist and so neither does the partial derivative $f_x(0, 0)$. Thus, f cannot be differentiable at the origin. If it were, both of the partial derivatives, $f_x(0, 0)$ and $f_y(0, 0)$, would exist.

Alternatively, we could show directly that there is no linear approximation near $(0, 0)$ that satisfies the small relative error criterion for differentiability. Any plane passing through the point $(0, 0, 0)$ has the form $L(x, y) = mx + ny$ for some constants m and n. If $E(x, y) = f(x, y) - L(x, y)$, then

$$E(x, y) = \sqrt{x^2 + y^2} - mx - ny.$$

Then for f to be differentiable at the origin, we would need to show that

$$\lim_{\substack{h \to 0 \\ k \to 0}} \frac{\sqrt{h^2 + k^2} - mh - nk}{\sqrt{h^2 + k^2}} = 0.$$

Taking $k = 0$ gives

$$\lim_{h \to 0} \frac{|h| - mh}{|h|} = 1 - m \lim_{h \to 0} \frac{h}{|h|}.$$

This limit exists only if $m = 0$ for the same reason as before. But then the value of the limit is 1 and not 0 as required. Thus, we again conclude f is not differentiable.

In Example 2 the partial derivatives f_x and f_y did not exist at the origin and this was sufficient to establish nondifferentiability there. We might expect that if both partial derivatives do exist, then *f is* differentiable. But the next example shows that this not necessarily true: the existence of both partial derivatives at a point is *not* sufficient to guarantee differentiability.

Example 3 Consider the function $f(x, y) = x^{1/3} y^{1/3}$. Show that the partial derivatives $f_x(0, 0)$ and $f_y(0, 0)$ exist, but that f is not differentiable at $(0, 0)$.

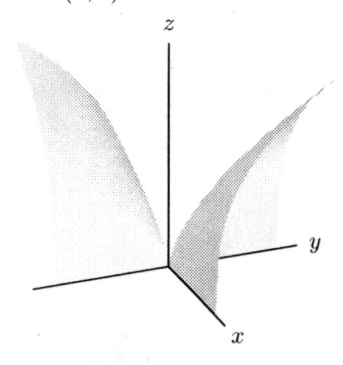

Figure 14.63: Graph of $z = x^{1/3} y^{1/3}$ for $z \geq 0$

Solution See Figure 14.63 for the part of the graph of $z = x^{1/3} y^{1/3}$ when $z \geq 0$. We have $f(0, 0) = 0$ and we compute the partial derivatives using the definition:

$$f_x(0, 0) = \lim_{h \to 0} \frac{f(h, 0) - f(0, 0)}{h} = \lim_{h \to 0} \frac{0 - 0}{h} = 0,$$

and similarly

$$f_y(0, 0) = 0.$$

So, if there did exist a linear approximation near the origin, it would have to be $L(x, y) = 0$. But we can show that this choice of $L(x, y)$ does not result in the small relative error that is required for differentiability. In fact, since $E(x, y) = f(x, y) - L(x, y) = f(x, y)$, we need to look at the limit

$$\lim_{\substack{h \to 0 \\ k \to 0}} \frac{h^{1/3} k^{1/3}}{\sqrt{h^2 + k^2}}.$$

If this limit exists, we get the same value no matter how h and k approach 0. Suppose we take $k = h > 0$. Then the limit becomes

$$\lim_{h \to 0} \frac{h^{1/3} h^{1/3}}{\sqrt{h^2 + h^2}} = \lim_{h \to 0} \frac{h^{2/3}}{h \sqrt{2}} = \lim_{h \to 0} \frac{1}{h^{1/3} \sqrt{2}}.$$

But this limit does not exist, since small values for h will make the fraction arbitrarily large. So the only possible candidate for a linear approximation at the origin does not have a sufficiently small relative error. Thus, this function is *not* differentiable at the origin, even though the partial derivatives $f_x(0, 0)$ and $f_y(0, 0)$ exist. Figure 14.63 confirms that near the origin the graph of $z = f(x, y)$ is not well approximated by any plane.

In summary,

- If a function is differentiable at a point, then both partial derivatives exist there.
- Having both partial derivatives at a point does not guarantee that a function is differentiable there.

Continuity and Differentiability

We know that differentiable functions of one variable are continuous. Similarly, it can be shown that if a function of two variables is differentiable at a point, then the function is continuous there.

In Example 3 the function f was continuous at the point where it was not differentiable. Example 4 shows that even if the partial derivatives of a function exist at a point, the function is not necessarily continuous at that point if it is not differentiable there.

Example 4 Suppose that f is the function of two variables defined by

$$f(x,y) = \begin{cases} \dfrac{xy}{x^2 + y^2}, & (x,y) \neq (0,0), \\ 0, & (x,y) = (0,0). \end{cases}$$

Problem 17 on page 642 showed that $f(x,y)$ is not continuous at the origin. Show that the partial derivatives $f_x(0,0)$ and $f_y(0,0)$ exist. Could f be differentiable at $(0,0)$?

Solution From the definition of the partial derivative we see that

$$f_x(0,0) = \lim_{h \to 0} \frac{f(h,0) - f(0,0)}{h} = \lim_{h \to 0} \left(\frac{1}{h} \cdot \frac{0}{h^2 + 0^2} \right) = \lim_{h \to 0} \frac{0}{h} = 0,$$

and similarly

$$f_y(0,0) = 0.$$

So, the partial derivatives $f_x(0,0)$ and $f_y(0,0)$ exist. However, f cannot be differentiable at the origin since it is not continuous there.

In summary,

- If a function is differentiable at a point, then it is continuous there.
- Having both partial derivatives at a point does not guarantee that a function is continuous there.

How Do We Know If a Function Is Differentiable?

Can we use partial derivatives to tell us if a function is differentiable? As we see from Examples 3 and 4, it is not enough that the partial derivatives exist. However, the following theorem gives conditions that *do* guarantee differentiability[3]:

> **Theorem 14.2: Continuity of Partial Derivatives Implies Differentiability**
>
> If the partial derivatives, f_x and f_y, of a function f exist and are continuous on a small disk centered at the point (a,b), then f is differentiable at (a,b).

We will not prove this theorem, although it provides a criterion for differentiability which is often simpler to use than the definition. It turns out that the requirement of continuous partial derivatives is more stringent than that of differentiability, so there exist differentiable functions which do

---
[3] For a proof, see M. Spivak, *Calculus on Manifolds*, p. 31, (New York: Benjamin, 1965).

Example 5 Show that the function $f(x, y) = \ln(x^2 + y^2)$ is differentiable everywhere in its domain.

Solution The domain of f is all of 2-space except for the origin. We shall show that f has continuous partial derivatives everywhere in its domain (that is, the function f is in C^1). The partial derivatives are

$$f_x = \frac{2x}{x^2 + y^2} \quad \text{and} \quad f_y = \frac{2y}{x^2 + y^2}.$$

Since each of f_x and f_y is the quotient of continuous functions, the partial derivatives are continuous everywhere except the origin (where the denominators are zero). Thus, f is differentiable everywhere in its domain.

Most functions built up from elementary functions have continuous partial derivatives, except perhaps at a few obvious points. Thus, in practice, we can often identify functions as being C^1 without explicitly computing the partial derivatives.

Exercises and Problems for Section 14.8

Exercises

In Exercises 1–10, list the points in the xy-plane, if any, at which the function $z = f(x, y)$ is not differentiable.

1. $z = -\sqrt{x^2 + y^2}$
2. $z = \sqrt{(x+1)^2 + y^2}$
3. $z = e^{-(x^2+y^2)}$
4. $z = |x| + |y|$
5. $z = x^{1/3} + y^2$
6. $z = |x + 2| - |y - 3|$
7. $z = (\sin x)(\cos |y|)$
8. $z = |x - 3|^2 + y^3$
9. $z = 4 + \sqrt{(x-1)^2 + (y-2)^2}$
10. $z = 1 + \left((x-1)^2 + (y-2)^2\right)^2$

Problems

For the functions f in Problems 11–14 answer the following questions. Justify your answers.
(a) Use a computer to draw a contour diagram for f.
(b) Is f differentiable at all points $(x, y) \neq (0, 0)$?
(c) Do the partial derivatives f_x and f_y exist and are they continuous at all points $(x, y) \neq (0, 0)$?
(d) Is f differentiable at $(0, 0)$?
(e) Do the partial derivatives f_x and f_y exist and are they continuous at $(0, 0)$?

11. $f(x, y) = \begin{cases} \frac{x}{y} + \frac{y}{x}, & x \neq 0 \text{ and } y \neq 0, \\ 0, & x = 0 \text{ or } y = 0. \end{cases}$

12. $f(x, y) = \begin{cases} \frac{2xy}{(x^2 + y^2)^2}, & (x, y) \neq (0, 0), \\ 0, & (x, y) = (0, 0). \end{cases}$

13. $f(x, y) = \begin{cases} \frac{xy}{\sqrt{x^2 + y^2}}, & (x, y) \neq (0, 0), \\ 0, & (x, y) = (0, 0). \end{cases}$

14. $f(x, y) = \begin{cases} \frac{x^2 y}{x^4 + y^2}, & (x, y) \neq (0, 0), \\ 0, & (x, y) = (0, 0). \end{cases}$

15. Consider the function

$$f(x, y) = \begin{cases} \frac{xy^2}{x^2 + y^2}, & (x, y) \neq (0, 0), \\ 0, & (x, y) = (0, 0). \end{cases}$$

(a) Use a computer to draw the contour diagram for f.
(b) Is f differentiable for $(x, y) \neq (0, 0)$?
(c) Show that $f_x(0, 0)$ and $f_y(0, 0)$ exist.
(d) Is f differentiable at $(0, 0)$?
(e) Suppose $x(t) = at$ and $y(t) = bt$, where a and b are constants, not both zero. If $g(t) = f(x(t), y(t))$, show that

$$g'(0) = \frac{ab^2}{a^2 + b^2}.$$

(f) Show that

$$f_x(0, 0)x'(0) + f_y(0, 0)y'(0) = 0.$$

Does the chain rule hold for the composite function $g(t)$ at $t = 0$? Explain.
(g) Show that the directional derivative $f_{\vec{u}}(0, 0)$ exists for each unit vector $\vec{u}$. Does this imply that f is differentiable at $(0, 0)$?

16. Consider the function

$$f(x,y) = \begin{cases} \dfrac{xy^2}{x^2+y^4}, & (x,y) \neq (0,0), \\ 0, & (x,y) = (0,0). \end{cases}$$

(a) Use a computer to draw the contour diagram for f.
(b) Show that the directional derivative $f_{\vec{u}}(0,0)$ exists for each unit vector $\vec{u}$.
(c) Is f continuous at $(0,0)$? Is f differentiable at $(0,0)$? Explain.

17. Consider the function $f(x,y) = \sqrt{|xy|}$.

(a) Use a computer to draw the contour diagram for f. Does the contour diagram look like that of a plane when we zoom in on the origin?
(b) Use a computer to draw the graph of f. Does the graph look like a plane when we zoom in on the origin?
(c) Is f differentiable for $(x,y) \neq (0,0)$?
(d) Show that $f_x(0,0)$ and $f_y(0,0)$ exist.
(e) Is f differentiable at $(0,0)$? [Hint: Consider the directional derivative $f_{\vec{u}}(0,0)$ for $\vec{u} = (\vec{i}+\vec{j})/\sqrt{2}$.]

18. Suppose a function f is differentiable at the point (a,b). Show that f is continuous at (a,b).

19. Suppose $f(x,y)$ is a function such that $f_x(0,0) = 0$ and $f_y(0,0) = 0$, and $f_{\vec{u}}(0,0) = 3$ for $\vec{u} = (\vec{i}+\vec{j})/\sqrt{2}$.

(a) Is f differentiable at $(0,0)$? Explain.
(b) Give an example of a function f defined on 2-space which satisfies these conditions. [Hint: The function f does not have to be defined by a single formula valid over all of 2-space.]

20. Consider the following function:

$$f(x,y) = \begin{cases} \dfrac{xy(x^2-y^2)}{x^2+y^2}, & (x,y) \neq (0,0), \\ 0, & (x,y) = (0,0). \end{cases}$$

The graph of f is shown in Figure 14.64, and the contour diagram of f is shown in Figure 14.65.

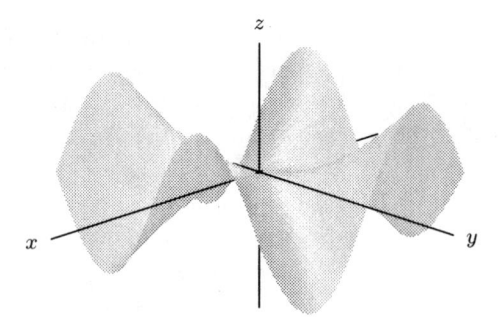

Figure 14.64: Graph of $\dfrac{xy(x^2-y^2)}{x^2+y^2}$

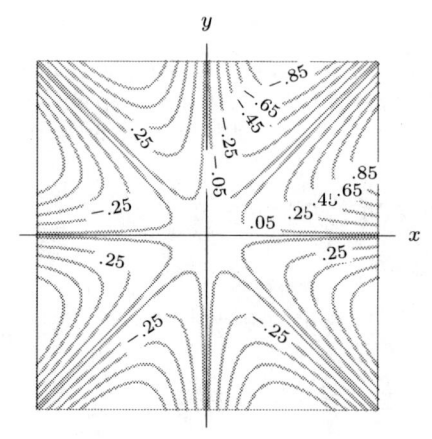

Figure 14.65: Contour diagram of $\dfrac{xy(x^2-y^2)}{x^2+y^2}$

(a) Find $f_x(x,y)$ and $f_y(x,y)$ for $(x,y) \neq (0,0)$.
(b) Show that $f_x(0,0) = 0$ and $f_y(0,0) = 0$.
(c) Are the functions f_x and f_y continuous at $(0,0)$?
(d) Is f differentiable at $(0,0)$?

CHAPTER SUMMARY (see also Ready Reference at the end of the book)

- **Partial Derivatives**
  Definition as a difference quotient, visualizing on a graph, estimating from a contour diagram, computing from a formula, interpreting units, alternative notation.

- **Local Linearity**
  Zooming on a surface, contour diagram or table to see local linearity, the idea of tangent plane, formula for a tangent plane in terms of partials, the differential.

- **Directional Derivatives**
  Definition as a difference quotient, interpretation as a rate of change, computation using partial derivatives.

- **Gradient Vector**
  Definition in terms of partial derivatives, geometric properties of gradient's length and direction, relation to directional derivative, relation to contours and level surfaces.

- **Chain Rule**
  Local linearity and differentials for composition of functions, tree diagrams, chain rule in general, application to physical chemistry.

- **Second and Higher Order Partial Derivatives**
  Interpretations, mixed partials are equal.

- **Taylor Approximations**
  Linear and quadratic polynomial approximations to functions near a point.

REVIEW EXERCISES AND PROBLEMS FOR CHAPTER FOURTEEN

Exercises

For Exercises 1–16, find the partial derivatives. Assume the variables are restricted to a domain on which the function is defined.

1. f_x and f_y if $f(x,y) = x^2 y + x^3 - 7xy^6$
2. $\dfrac{\partial w}{\partial h}$ if $w = 320\pi g h^2 (20 - h)$
3. F_L if $F(L, K) = 3\sqrt{LK}$
4. $\dfrac{\partial B}{\partial t}$ and $\dfrac{\partial B}{\partial r}$ when $B = P(1+r)^t$.
5. f_x and f_y if $f(x,y) = \dfrac{x^2 y}{x^2 + y^2}$
6. $\dfrac{\partial F}{\partial r}$ and $\dfrac{\partial F}{\partial r}$ if $F = \dfrac{G\mu my}{(r^2 + y^2)^{3/2}}$
7. f_p and f_q if $f(p, q) = e^{p/q}$
8. f_x if $f(x, y) = e^{xy}(\ln y)$
9. f_N if $f(N, V) = cN^\alpha V^\beta$
10. f_x and f_y if $f(x, y) = \sqrt{(x-a)^2 + (y-b)^2}$
11. $\dfrac{\partial}{\partial x}(e^x \cos(xy) + ay^2)$, $\dfrac{\partial}{\partial y}(e^x \cos(xy) + ay^2)$, $\dfrac{\partial}{\partial a}(e^x \cos(xy) + ay^2)$
12. $\dfrac{\partial}{\partial \theta}(\sin(\pi\theta\phi) + \ln(\theta^2 + \phi))$
13. f_{xx} and f_{xy} if $f(x, y) = 1/\sqrt{x^2 + y^2}$
14. u_{xx} and u_{yy} if $u = e^x \sin y$
15. V_{rr} and V_{rh} if $V = \pi r^2 h$
16. f_{xxy} and f_{yxx} if $f(x, y) = \sin(x - 2y)$

In Exercises 17–20, find the gradient for the given function.

17. $f(x, y, z) = x^3 + z^3 - xyz$
18. $f(x, y) = \sin(y^2 - xy)$
19. $f(x, y) = \ln(x^2 + y^2)$
20. $f(\rho, \phi, \theta) = \rho \sin\phi \cos\theta$

In Exercises 21–26, find the directional derivative for the function.

21. $f(x, y) = x^3 - y^3$ at $(2, -1)$ in the direction of $\vec{i} - \vec{j}$
22. $f(x, y) = xe^y$ at $(3, 0)$ in the direction of $4\vec{i} - 3\vec{j}$
23. $f(x, y, z) = x^2 + y^2 - z^2$ at $(2, 3, 4)$ in the direction of $2\vec{i} - 2\vec{j} + \vec{k}$
24. $f(x, y, z) = 3x^2 y^2 + 2yz$ at $(-1, 0, 4)$ in the direction of $\vec{i} - \vec{k}$
25. $f(x, y, z) = 3x^2 y^2 + 2yz$ at $(-1, 0, 4)$ in the direction of $-\vec{i} + 3\vec{j} + 3\vec{k}$
26. $f(x, y, z) = e^{x+z} \cos y$ at $(1, 0, -1)$ in the direction of $\vec{i} + \vec{j} + \vec{k}$

In Exercises 27–29, find a vector normal to the curve or surface at the point.

27. $x^2 - y^2 = 3$ at $(2, 1)$
28. $xy + xz + yz = 11$ at $(1, 2, 3)$
29. $z^2 - 2xyz = x^2 + y^2$ at $(1, 2, -1)$

In Exercises 30–32, find the quadratic Taylor polynomial for the function.

30. $f(x, y) = (x + 1)^3 (y + 2)$ about $(0, 0)$
31. $f(x, y) = \cos x \cos 3y$ about $(0, 0)$
32. $f(x, y) = \sqrt{2x - y}$ about $(3, 5)$

Problems

33. (a) Let $f(w, z) = e^{w \ln z}$. Use difference quotients with $h = 0.01$ to approximate $f_w(2, 2)$ and $f_z(2, 2)$.
    (b) Now evaluate $f_w(2, 2)$ and $f_z(2, 2)$ exactly.

For Problems 34–38 use Figure 14.66, showing level curves of $f(x, y)$, to estimate the directional derivatives.

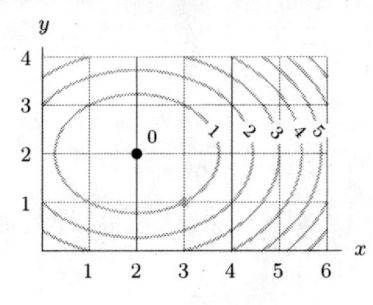

Figure 14.66

34. $f_{\vec{i}}(3, 1)$
35. $f_{\vec{j}}(3, 1)$
36. $f_{\vec{u}}(3, 1)$ where $\vec{u} = (\vec{i} - \vec{j})/\sqrt{2}$
37. $f_{\vec{u}}(3, 1)$ where $\vec{u} = (-\vec{i} + \vec{j})/\sqrt{2}$
38. $f_{\vec{u}}(3, 1)$ with $\vec{u} = (-2\vec{i} + \vec{j})/\sqrt{5}$
39. If $f(x, y) = x^2 y$ and $\vec{v} = 4\vec{i} - 3\vec{j}$, find the directional derivative at the point $(2, 6)$ in the direction of $\vec{v}$.
40. Find the directional derivative of $z = x^2 - y^2$ at the point $(3, -1)$ in the direction making an angle $\theta = \pi/4$ with the x-axis. In which direction is the directional derivative the largest?

The functions $u(x, t)$ in Problems 41–44 satisfy one of the equations (I)–(V), known as *partial differential equations*, for some positive real value of c. Match the function to an equation and give the value of c.

(I) $u_{xx} = c^2 u_{tt}$
(II) $u_{xx} = c^2 (u_t)^2$
(III) $u_{xx} = -c^2 u_{tt}$
(IV) $u_{xx} = -c^2 (u_t)^2$
(V) $(u_{xx})^2 = c^2 u_t$

41. $\sin(x - 2t)$
42. $e^{3t} \sin x$
43. e^{4x-9t}
44. $e^{3x} \sin 2t$

45. Find the point(s) on $x^2 + y^2 + z^2 = 8$ where the tangent plane is parallel to the plane $x - y + 3z = 0$.

46. The vector ∇f at point P and four unit vectors $\vec{u}_1, \vec{u}_2, \vec{u}_3, \vec{u}_4$ are shown in Figure 14.67. Arrange the following quantities in ascending order

$$f_{\vec{u}_1}, \quad f_{\vec{u}_2}, \quad f_{\vec{u}_3}, \quad f_{\vec{u}_4}, \quad \text{the number } 0.$$

The directional derivatives are all evaluated at the point P and the function $f(x, y)$ is differentiable at P.

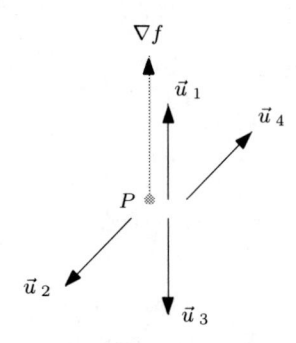

Figure 14.67

47. A differentiable function $f(x, y)$ has $f_x(4, 1) = 2$ and $f_y(4, 1) = -1$. Find the equation of the tangent line to the contour of f through the point $(4, 1)$.

48. The temperature at (x, y) is $T(x, y) = 100 - x^2 - y^2$. In which direction should a heat-seeking bug move from the point (x, y) to increase its temperature fastest?

49. A car is driving northwest at v mph across a sloping plain whose height, in feet above sea level, at a point N miles north and E miles east of a city is given by

$$h(N, E) = 2500 + 100N + 50E.$$

(a) At what rate is the height above sea level changing with respect to distance in the direction the car is driving?
(b) Express the rate of change of the height of the car with respect to time in terms of v.

50. The quantity Q (in pounds) of beef that a certain community buys during a week is a function $Q = f(b, c)$ of the prices of beef, b, and chicken, c, during the week. Do you expect $\partial Q/\partial b$ to be positive or negative? What about $\partial Q/\partial c$?

51. The cost of producing one unit of a product is given by

$$c = a + bx + ky,$$

where x is the amount of labor used (in man hours) and y is the amount of raw material used (by weight) and a and b and k are constants. What does $\partial c/\partial x = b$ mean? What is the practical interpretation of b?

52. People commuting to a city can choose to go either by bus or by train. The number of people who choose either method depends in part upon the price of each. Let $f(P_1, P_2)$ be the number of people who take the bus when P_1 is the price of a bus ride and P_2 is the price of a train ride. What can you say about the signs of $\partial f/\partial P_1$ and $\partial f/\partial P_2$? Explain your answers.

53. Figure 14.68 shows a contour diagram for the temperature T (in °C) along a wall in a heated room as a function of distance x along the wall and time t in minutes. Estimate $\partial T/\partial x$ and $\partial T/\partial t$ at the given points. Give units and interpret your answers.

(a) $x = 15, t = 20$ (b) $x = 5, t = 12$

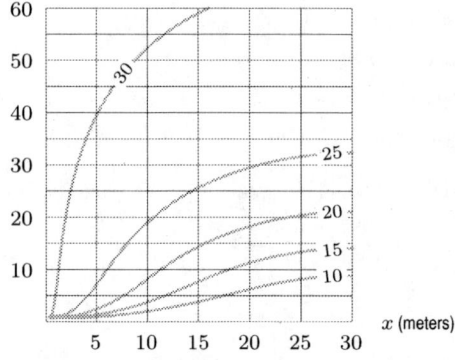

Figure 14.68

54. The acceleration g due to gravity, at a distance r from the center of a planet of mass m, is given by

$$g = \frac{Gm}{r^2},$$

where G is the universal gravitational constant.

(a) Find $\partial g/\partial m$ and $\partial g/\partial r$.
(b) Interpret each of the partial derivatives you found in part (a) as the slope of a graph in the plane and sketch the graph.

55. In analyzing a factory and deciding whether or not to hire more workers, it is useful to know under what circumstances productivity increases. Suppose $P = f(x_1, x_2, x_3)$ is the total quantity produced as a function of x_1, the number of workers, and any other variables x_2, x_3. We define the average productivity of a worker as P/x_1. Show that the average productivity increases as x_1 increases when marginal production, $\partial P/\partial x_1$, is greater than the average productivity, P/x_1.

56. For the Cobb-Douglas production function $P = 40L^{0.25}K^{0.75}$, find the differential dP when $L = 2$ and $K = 16$.

57. The period, T, of a pendulum is $T = 2\pi\sqrt{l/g}$. If the approximate length is $l = 2$ meters, find the approximate error in T if the true length is $l = 1.99$ and we take $g = 9.8$ as an approximation for $g = 9.81$ m/s^2.

58. Figure 14.69 shows the monthly payment, m, on a 5-year car loan if you borrow P dollars at r percent interest. Find a formula for a linear function which approximates m. What is the practical significance of the constants in your formula?

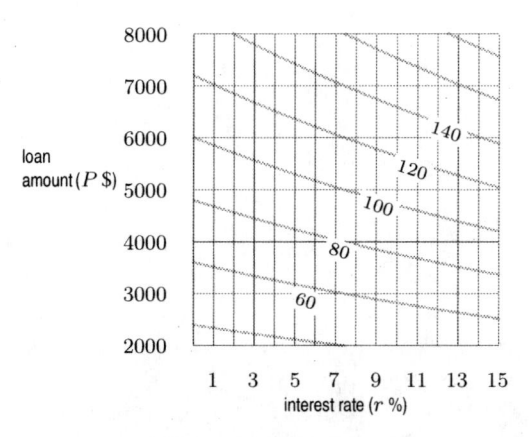

Figure 14.69

In Problems 59–66, the function f is differentiable and $f_x(2, 1) = -3$, $f_y(2, 1) = 4$, and $f(2, 1) = 7$.

59. (a) Give an equation for the tangent plane to the graph of f at $x = 2, y = 1$.
    (b) Give an equation for the tangent line to the contour for f at $x = 2, y = 1$.

60. (a) Find a vector perpendicular to the tangent plane to the graph of f at $x = 2, y = 1$.
    (b) Find a vector perpendicular to the tangent line to the contour for f at $x = 2, y = 1$.

61. Near $x = 2$ and $y = 1$, how far apart are the contours $f(x, y) = 7$ and $f(x, y) = 7.3$?

62. Give an approximate table of values of f for $x = 1.8, 2.0, 2.2$ and $y = 0.9, 1.0, 1, 1$.

63. Give an approximate contour diagram for f for $1 \leq x \leq 3$, $0 \leq y \leq 2$, using contour values $\ldots 5, 6, 7, 8, 9 \ldots$.

64. The function f gives temperature in °C and x and y are in centimeters. A bug leaves $(2, 1)$ at 3 cm/min so that it cools off as fast as possible. In which direction does the bug head? At what rate does it cool off, in °C/min?

65. Find $f_r(2, 1)$ and $f_\theta(2, 1)$, where r and θ are polar coordinates, $x = r\cos\theta$ and $y = r\sin\theta$. If $\vec{u}$ is the unit vector in the direction $2\vec{i} + \vec{j}$, show that $f_{\vec{u}}(2, 1) = f_r(2, 1)$ and explain why this should be the case.

66. Find approximately the largest value of f on or inside the circle of radius 0.1 about the point $(2, 1)$. At what point does f achieve this value?

67. Values of the function $f(x, y)$ near the point $x = 2$, $y = 3$ are given in Table 14.8. Estimate the following.

   (a) $\left.\dfrac{\partial f}{\partial x}\right|_{(2,3)}$ and $\left.\dfrac{\partial f}{\partial y}\right|_{(2,3)}$.
   (b) The rate of change of f at $(2, 3)$ in the direction of the vector $\vec{i} + 3\vec{j}$.
   (c) The maximum possible rate of change of f as you move away from the point $(2, 3)$. In which direction should you move to obtain this rate of change?
   (d) Write an equation for the level curve through the point $(2, 3)$.
   (e) Find a vector tangent to the level curve of f through the point $(2, 3)$.
   (f) Find the differential of f at the point $(2, 3)$. If $dx = 0.03$ and $dy = 0.04$, find df. What does df represent in this case?

Table 14.8

		x	
		2.00	2.01
y	3.00	7.56	7.42
	3.02	7.61	7.47

68. Find the quadratic Taylor polynomial about $(0, 0)$ for $f(x, y) = \cos(x + 2y)\sin(x - y)$.

69. Suppose $f(x, y) = e^{(x-1)^2 + (y-3)^2}$.

   (a) Find the first-order Taylor polynomial about $(0, 0)$.
   (b) Find the second-order (quadratic) Taylor polynomial about the point $(1, 3)$.
   (c) Find a 2-vector perpendicular to the level curve through $(0, 0)$.
   (d) Find a 3-vector perpendicular to the surface $z = f(x, y)$ at the point $(0, 0)$.

70. Figure 14.70 shows a contour diagram for a vibrating string function, $f(x,t)$.

(a) Is $f_t(\pi/2, \pi/2)$ positive or negative? How about $f_t(\pi/2, \pi)$? What does the sign of $f_t(\pi/2, b)$ tell you about the motion of the point on the string at $x = \pi/2$ when $t = b$?

(b) Find all t for which f_t is positive, for $0 \leq t \leq 5\pi/2$.

(c) Find all x and t such that f_x is positive.

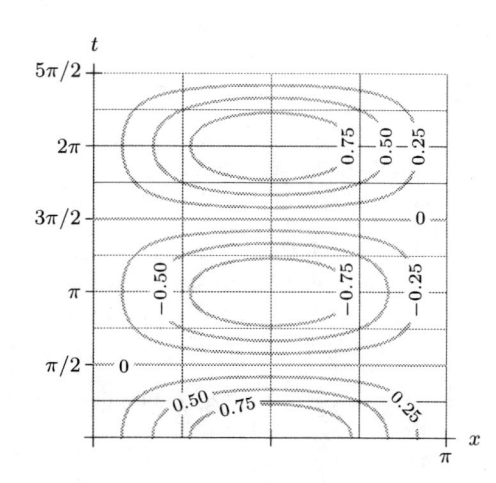

Figure 14.70

71. Each diagram (I) – (IV) in Figure 14.71 represents the level curves of a function $f(x,y)$. For each function f, consider the point above P on the surface $z = f(x,y)$ and choose from the lists which follow:

(a) A vector which could be the normal to the surface at that point;

(b) An equation which could be the equation of the tangent plane to the surface at that point.

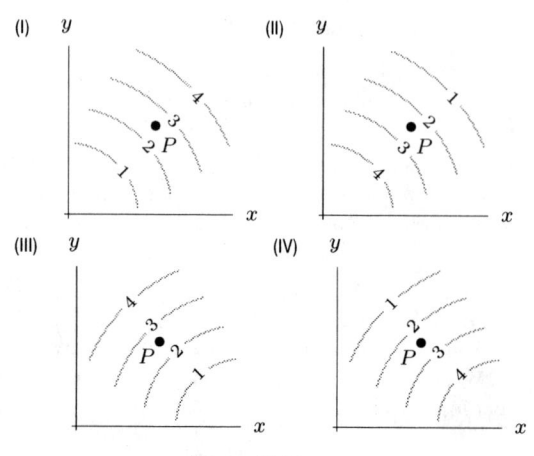

Figure 14.71

Vectors
(E) $2\vec{i} + 2\vec{j} - 2\vec{k}$
(F) $2\vec{i} + 2\vec{j} + 2\vec{k}$
(G) $2\vec{i} - 2\vec{j} + 2\vec{k}$
(H) $-2\vec{i} + 2\vec{j} + 2\vec{k}$

Equations
(J) $x + y + z = 4$
(K) $2x - 2y - 2z = 2$
(L) $-3x - 3y + 3z = 6$
(M) $-\dfrac{x}{2} + \dfrac{y}{2} - \dfrac{z}{2} = -7$

CAS Challenge Problems

72. (a) Find the quadratic Taylor polynomial about $(0,0)$ of
$$f(x,y) = \dfrac{e^x(1 + \sin(3y))^2}{5 + e^{2x}}.$$

(b) Find the quadratic Taylor polynomial about 0 of the one-variable functions $g(x) = e^x/(5 + e^{2x})$ and $h(y) = (1 + \sin(3y))^2$. Multiply these polynomials together and compare with your answer to part (a).

(c) Show that if $f(x,y) = g(x)h(y)$, then the quadratic Taylor polynomial of f about $(0,0)$ is the product of the quadratic Taylor polynomials of $g(x)$ and $h(y)$ about 0. [If you use a computer algebra system, make sure that f, g and h do not have any previously assigned formula.]

73. Let
$$f(x,y) = A_0 + A_1 x + A_2 y + A_3 x^2 + A_4 xy + A_5 y^2,$$
$$g(t) = 1 + B_1 t + B_2 t^2,$$
$$h(t) = 2 + C_1 t + C_2 t^2.$$

(a) Find $L(x,y)$, the linear approximation to $f(x,y)$ at the point $(1,2)$. Also find $m(t)$ and $n(t)$, the linear approximations to $g(t)$ and $h(t)$ at $t = 0$.

(b) Calculate $(d/dt)f(g(t), h(t))|_{t=0}$ and $(d/dt)L(m(t), n(t))|_{t=0}$. Describe what you notice and explain it in terms of the chain rule.

74. Let $f(x,y) = A_0 + A_1 x + A_2 y + A_3 x^2 + A_4 xy + A_5 y^2$.

(a) Find the quadratic Taylor approximation for $f(x,y)$ near the point $(1,2)$, and expand the result in powers of x and y.

(b) Explain what you notice in part (a) and formulate a generalization to points other than $(1,2)$.

(c) Repeat part (a) for the linear approximation. How does it differ from the quadratic?

75. Suppose that $w = f(x,y,z)$, that x and y are functions of u and v, and that z, u, and v are functions of t. Use a computer algebra system to find the derivative
$$\dfrac{d}{dt} f(x(u(t), v(t)), y(u(t), v(t)), z(t))$$
and explain the answer using a tree diagram.

CHECK YOUR UNDERSTANDING

Are the statements in Problems 1–40 true or false? Give reasons for your answer.

1. If $f(x,y)$ is a function of two variables and $f_x(10, 20)$ is defined, then $f_x(10, 20)$ is a scalar.

2. If $f(x,y) = x^2 + y^2$, then $f_y(1,1) < 0$.

3. If the graph of $f(x,y)$ is a hemisphere centered at the origin, then $f_x(0,0) = f_y(0,0) = 0$.

4. If $P = f(T, V)$ is a function expressing the pressure P (in grams/cm^3) of gas in a piston in terms of the temperature T (in degrees °C) and volume V (in cm^3), then $\partial P/\partial V$ has units of grams.

5. If $f_x(a,b) > 0$, then the values of f decrease as we move in the negative x-direction near (a,b).

6. If $g(r,s) = r^2 + s$, then for fixed s, the partial derivative g_r increases as r increases.

7. If $g(u,v) = (u+v)^u$, then $2.3 \leq g_u(1,1) \leq 2.4$.

8. Let $P = f(m,d)$ be the purchase price (in dollars) of a used car that has m miles on its engine and originally cost d dollars when new. Then $\partial P/\partial m$ and $\partial P/\partial d$ have the same sign.

9. If $f(x,y)$ is a function with the property that $f_x(x,y)$ and $f_y(x,y)$ are both constant, then f is linear.

10. If the point (a,b) is on the contour $f(x,y) = k$, then the slope of the line tangent to this contour at (a,b) is $f_y(a,b)/f_x(a,b)$.

11. There is a function $f(x,y)$ with $f_x(x,y) = y$ and $f_y(x,y) = x$.

12. The function $z(u,v) = u\cos v$ satisfies the equation
$$\cos v \frac{\partial z}{\partial u} - \frac{\sin v}{u}\frac{\partial z}{\partial v} = 1.$$

13. If $f(x,y)$ is a function of two variables and $g(x)$ is a function of a single variable, then
$$\frac{\partial}{\partial y}(g(x)f(x,y)) = g(x)f_y(x,y).$$

14. The function $k(r,s) = rse^s$ is increasing in the s-direction at the point $(r,s) = (-1, 2)$.

15. There is a function $f(x,y)$ with $f_x(x,y) = y^2$ and $f_y(x,y) = x^2$.

16. If f is a symmetric two-variable function, that is $f(x,y) = f(y,x)$, then $f_x(x,y) = f_y(x,y)$.

17. If f is a symmetric two-variable function, that is $f(x,y) = f(y,x)$, then $f_x(x,y) = f_y(y,x)$.

18. If $f(x,y)$ has $f_y(x,y) = 0$ then f must be a constant.

19. If $f(x,y) = ye^{g(x)}$ then $f_x = f$.

20. If $f(x,y)$ has $f_x(a,b) = f_y(a,b) = 0$ at the point (a,b), then f is constant everywhere.

21. An equation for the tangent plane to the surface $z = x^2 + y^3$ at $(1,1)$ is $z = 2 + 2x(x-1) + 3y^2(y-1)$.

22. There is a function $f(x,y)$ which has a tangent plane with equation $z = 0$ at a point (a,b).

23. The tangent plane approximation of $f(x,y) = ye^{x^2}$ at the point $(0,1)$ is $f(x,y) \approx y$.

24. If f is a function with differential $df = 2y\,dx + \sin(xy)\,dy$, then f changes by about -0.4 between the points $(1,2)$ and $(0.9, 2.0002)$.

25. The local linearization of $f(x,y) = x^2 + y^2$ at $(1,1)$ gives an overestimate of the value of $f(x,y)$ at the point $(1.04, 0.95)$.

26. If two functions f and g have the same differential at the point $(1,1)$, then $f = g$.

27. If two functions f and g have the same tangent plane at a point $(1,1)$, then $f = g$.

28. If $f(x,y)$ is a constant function, then $df = 0$.

29. If $f(x,y)$ is a linear function, then df is a linear function of dx and dy.

30. If you zoom close enough near a point (a,b) on the contour diagram of any differentiable function, the contours will be *precisely* parallel and *exactly* equally spaced.

31. The gradient vector grad $f(a,b)$ is a vector in 3-space.

32. The gradient vector grad $f(a,b)$ is tangent to the contour of f at (a,b).

33. If you know the gradient vector of f at (a,b) then you can find the directional derivative $f_{\vec{u}}(a,b)$ for any unit vector $\vec{u}$.

34. If you know the directional derivative $f_{\vec{u}}(a,b)$ for all unit vectors $\vec{u}$ then you can find the gradient vector of f at (a,b).

35. The directional derivative $f_{\vec{u}}(a,b)$ is parallel to $\vec{u}$.

36. The gradient grad $f(3,4)$ is perpendicular to the vector $3\vec{i} + 4\vec{j}$.

37. If grad $f(1,2) = \vec{i}$, then f decreases in the $-\vec{i}$ direction at $(1,2)$.

38. If grad $f(1,2) = \vec{i}$, then $f(10,2) > f(1,2)$.

39. At the point $(3,0)$, the function $g(x,y) = x^2 + y^2$ has the same maximal rate of increase as that of the function $h(x,y) = 2xy$.

40. If $f(x,y) = e^{x+y}$, then the directional derivative in any direction $\vec{u}$ (with $\|\vec{u}\| = 1$) at the point $(0,0)$ is always less than or equal to $\sqrt{2}$.

Assume that $f(x,y)$ is a differentiable function. Are the statements in Problems 41–47 true or false? Explain your answer.

41. $f_{\vec{u}}(x_0, y_0)$ is a scalar.

42. $f_{\vec{u}}(a,b) = \|\nabla f(a,b)\|$

43. If $\vec{u}$ is tangent to the level curve of f at some point, then grad $f \cdot \vec{u} = 0$ there.

44. Suppose that f is differentiable at (a,b). Then there is always a direction in which the rate of change of f at (a,b) is 0.

45. There is a function with a point in its domain where $\|\operatorname{grad} f\| = 0$ and where there is a nonzero directional derivative.

46. There is a function with $\|\operatorname{grad} f\| = 4$ and $f_{\vec{i}} = 5$ at some point.

47. There is a function with $\|\operatorname{grad} f\| = 4$ and $f_{\vec{j}} = -3$ at some point.

PROJECTS FOR CHAPTER FOURTEEN

1. Heat Equation

The function $T(x, y, z, t)$ is a solution to the *heat equation*

$$T_t = K(T_{xx} + T_{yy} + T_{zz}),$$

and gives the temperature at the point (x, y, z) in 3-space and time t. The constant K is the *thermal conductivity* of the medium through which the heat is flowing.

(a) Show that the function

$$T(x, y, z, t) = \frac{1}{(4\pi K t)^{3/2}} e^{-(x^2 + y^2 + z^2)/4Kt}$$

is a solution to the heat equation for all (x, y, z) in 3-space and $t > 0$.

(b) For each fixed time t, what are the level surfaces of the function $T(x, y, z, t)$ in 3-space?

(c) Regard t as fixed and compute grad $T(x, y, z, t)$. What does grad $T(x, y, z, t)$ tell us about the direction and magnitude of the heat flow?

2. Matching Birthdays

Consider a class of m students and a year with n days. Let $q(m, n)$ denote the probability, expressed as a number between 0 and 1, that at least two students have the same birthday. Surprisingly, $q(23, 365) \approx 0.5073$. (This means that there is slightly better than an even chance that at least two students in a class of 23 have the same birthday.) A general formula for q is complicated, but it can be shown that

$$\frac{\partial q}{\partial m} \approx +\frac{m}{n}(1-q) \quad \text{and} \quad \frac{\partial q}{\partial n} \approx -\frac{m^2}{2n^2}(1-q).$$

(These approximations hold when n is a good deal larger than m, and m is a good deal larger than 1.)

(a) Explain why the $+$ and $-$ signs in the approximations for $\partial q/\partial m$ and $\partial q/\partial n$ are to be expected.

(b) Suppose there are 21 students in a class. What is the approximate probability that at least two students in the class have the same birthday? (Assume that a year always has 365 days.)

(c) Suppose there is a class of 24 students and you know that no one was born in the first week of the year. (This has the effect of making $n = 358$.) What is the approximate value of q for this class?

(d) If you want to bet that a certain class of 23 students has at least two matching birthdays, would you prefer to have two more students added to the class or to be told that no one in the class was born in December?

(e) (Optional) Find the actual formula for q. [Hint: It's easier to find $1 - q$. There are $n \cdot n \cdot n \cdots n = n^m$ different choices for the students' birthdays. How many such choices have no matching birthdays?]

Chapter Fifteen

OPTIMIZATION: LOCAL AND GLOBAL EXTREMA

In one-variable calculus we saw how to find the maximum and minimum values of a function of one variable. In practice, there are often several variables in an optimization problem. For example, you may have $10,000 to invest in new equipment and advertising for your business. What combination of equipment and advertising will yield the greatest profit? Or, what combination of drugs will lower a patient's temperature the most? In this chapter we consider optimization problems, where the variables are completely free to vary (unconstrained optimization) and where there is a constraint on the variables (for example, a budget constraint).

15.1 LOCAL EXTREMA

Functions of several variables, like functions of one variable, can have *local* and *global* extrema. (That is, local and global maxima and minima.) A function has a local extremum at a point where it takes on the largest or smallest value in a small region around the point. Global extrema are the largest or smallest values anywhere on the domain under consideration. (See Figures 15.1 and 15.2.)

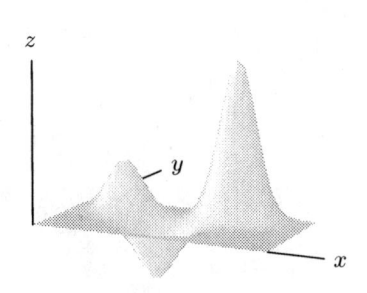

Figure 15.1: Local and global extrema for a function of two variables on $0 \leq x \leq a$, $0 \leq y \leq b$

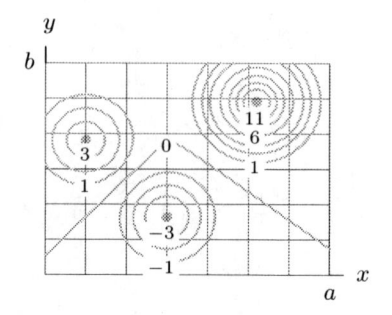

Figure 15.2: Contour map of the function in Figure 15.1

More precisely, considering only points at which f is defined, we say:

- f has a **local maximum** at the point P_0 if $f(P_0) \geq f(P)$ for all points P near P_0.
- f has a **local minimum** at the point P_0 if $f(P_0) \leq f(P)$ for all points P near P_0.

How Do We Detect a Local Maximum or Minimum?

Recall that if the gradient vector of a function is defined and nonzero, then it points in a direction in which the function increases. Suppose that a function f has a local maximum at a point P_0 which is not on the boundary of the domain. If the vector $\operatorname{grad} f(P_0)$ were defined and nonzero, then we could increase f by moving in the direction of $\operatorname{grad} f(P_0)$. Since f has a local maximum at P_0, there is no direction in which f is increasing. Thus, if $\operatorname{grad} f(P_0)$ is defined, we must have

$$\operatorname{grad} f(P_0) = \vec{0}.$$

Similarly, suppose f has a local minimum at the point P_0. If $\operatorname{grad} f(P_0)$ were defined and nonzero, then we could decrease f by moving in the direction opposite to $\operatorname{grad} f(P_0)$, and so we must again have $\operatorname{grad} f(P_0) = \vec{0}$. Therefore, we make the following definition:

Points where the gradient is either $\vec{0}$ or undefined are called **critical points** of the function.

If a function has a local maximum or minimum at a point P_0, not on the boundary of its domain, then P_0 is a critical point. For a function of two variables, we can also see that the gradient vector must be zero or undefined at a local maximum by looking at its contour diagram and a plot of its gradient vectors. (See Figures 15.3 and 15.4.) Around the maximum the vectors are all pointing inward, perpendicularly to the contours. At the maximum the gradient vector must be zero or undefined. A similar argument shows that the gradient must be zero at a local minimum.

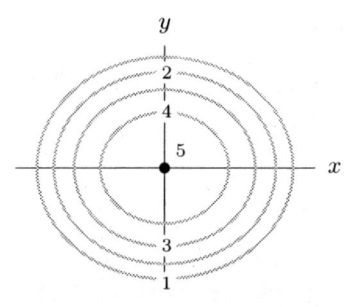

Figure 15.3: Contour diagram around a local maximum of a function

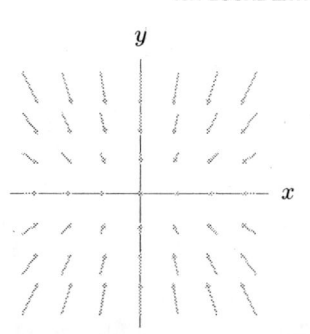

Figure 15.4: Gradients pointing toward the local maximum of the function in Figure 15.3

Finding and Analyzing Critical Points

To find critical points we set grad $f = f_x\vec{i} + f_y\vec{j} + f_z\vec{k} = \vec{0}$, which means setting all the partial derivatives of f equal to zero. We must also look for the points where one or more of the partial derivatives is undefined.

Example 1 Find and analyze the critical points of $f(x, y) = x^2 - 2x + y^2 - 4y + 5$.

Solution To find the critical points, we set both partial derivatives equal to zero:

$$f_x(x, y) = 2x - 2 = 0$$
$$f_y(x, y) = 2y - 4 = 0.$$

Solving these equations gives $x = 1$, $y = 2$. Hence, f has only one critical point, namely $(1, 2)$. To see the behavior of f near $(1, 2)$, look at the values of the function in Table 15.1.

Table 15.1 *Values of $f(x, y)$ near the point $(1, 2)$*

		\multicolumn{5}{c}{x}
		0.8	0.9	1.0	1.1	1.2
	1.8	0.08	0.05	0.04	0.05	0.08
	1.9	0.05	0.02	0.01	0.02	0.05
y	2.0	0.04	0.01	0.00	0.01	0.04
	2.1	0.05	0.02	0.01	0.02	0.05
	2.2	0.08	0.05	0.04	0.05	0.08

The table suggests that the function has a local minimum value of 0 at $(1, 2)$. We can confirm this by completing the square:

$$f(x, y) = x^2 - 2x + y^2 - 4y + 5 = (x - 1)^2 + (y - 2)^2.$$

Figure 15.5 shows that the graph of f is a paraboloid with vertex at the point $(1, 2, 0)$. It is the same shape as the graph of $z = x^2 + y^2$ (see Figure 12.12 on page 612), except that the vertex has been shifted to $(1, 2)$. So the point $(1, 2)$ is a local minimum of f (as well as a global minimum).

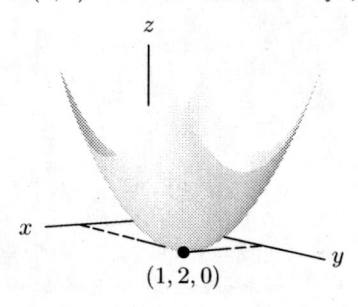

Figure 15.5: The graph of $f(x, y) = x^2 - 2x + y^2 - 4y + 5$ with a local minimum at the point $(1, 2)$

Example 2 Find and analyze any critical points of $f(x, y) = -\sqrt{x^2 + y^2}$.

Solution We look for points where grad $f = \vec{0}$ or is undefined. The partial derivatives are given by

$$f_x(x, y) = -\frac{x}{\sqrt{x^2 + y^2}},$$

$$f_y(x, y) = -\frac{y}{\sqrt{x^2 + y^2}}.$$

These partial derivatives are never simultaneously zero; but they are undefined at $x = 0$, $y = 0$. Thus, $(0, 0)$ is a critical point and a possible extreme point. The graph of f (see Figure 15.6) is a cone, with vertex at $(0, 0)$. So f has a local and global maximum at $(0, 0)$.

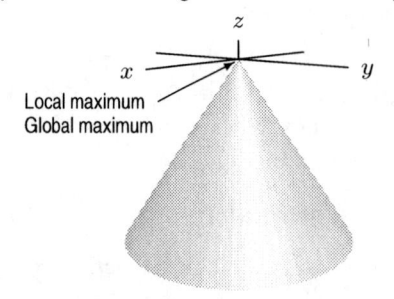

Figure 15.6: Graph of $f(x, y) = -\sqrt{x^2 + y^2}$

Example 3 Find and analyze any critical points of $g(x, y) = x^2 - y^2$.

Solution To find the critical points, we look for points where both partial derivatives are zero:

$$g_x(x, y) = 2x = 0$$

$$g_y(x, y) = -2y = 0.$$

Solving gives $x = 0$, $y = 0$, so the origin is the only critical point.

Figure 15.7 shows that near the origin g takes on both positive and negative values. Since $g(0, 0) = 0$, the origin is a critical point which is neither a local maximum nor a local minimum. The graph of g looks like a saddle.

The previous examples show that critical points can occur at local maxima or minima, or at points which are neither: The functions g and h in Figures 15.7 and 15.8 both have critical points at the origin. Figure 15.9 shows level curves of g. They are hyperbolas showing both positive and negative values of g near $(0, 0)$. Contrast this with the level curves of h near the local minimum in Figure 15.10.

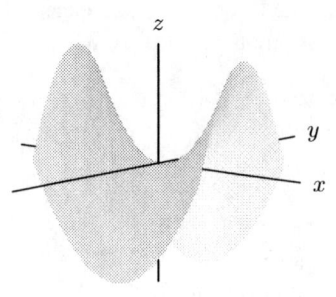

Figure 15.7: Graph of $g(x, y) = x^2 - y^2$, showing saddle shape at the origin

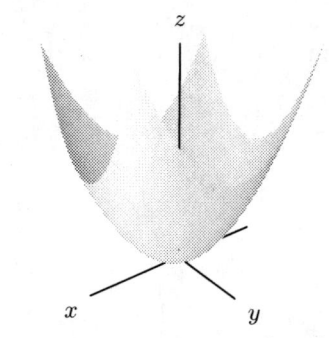

Figure 15.8: Graph of $h(x, y) = x^2 + y^2$, showing minimum at the origin

15.1 LOCAL EXTREMA

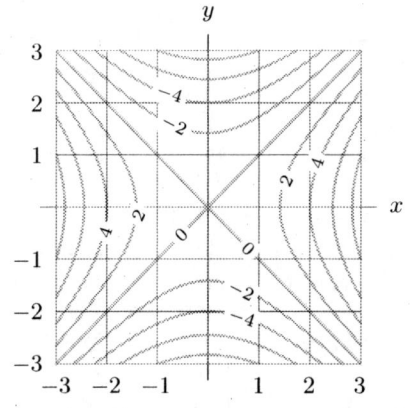

Figure 15.9: Contours of $g(x,y) = x^2 - y^2$, showing a saddle shape at the origin

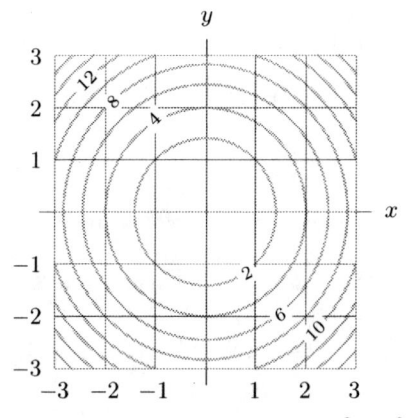

Figure 15.10: Contours of $h(x,y) = x^2 + y^2$, showing a local minimum at the origin

Example 4 Find the local extrema of the function $f(x,y) = 8y^3 + 12x^2 - 24xy$.

Solution We begin by looking for critical points:

$$f_x(x,y) = 24x - 24y,$$
$$f_y(x,y) = 24y^2 - 24x.$$

Setting these expressions equal to zero gives the system of equations

$$x = y, \quad x = y^2,$$

which has two solutions, $(0,0)$ and $(1,1)$. Are these local maxima, local minima or neither? Figure 15.11 shows contours of f near the points. Notice that $f(1,1) = -4$ and that there is no other -4 contour. The contours near $(1,1)$ suggests that f has a local minimum at the point $(1,1)$.

We have $f(0,0) = 0$ and the contours near $(0,0)$ show that f takes both positive and negative values nearby. This suggests that $(0,0)$ is a critical point which is neither a local maximum nor a local minimum.

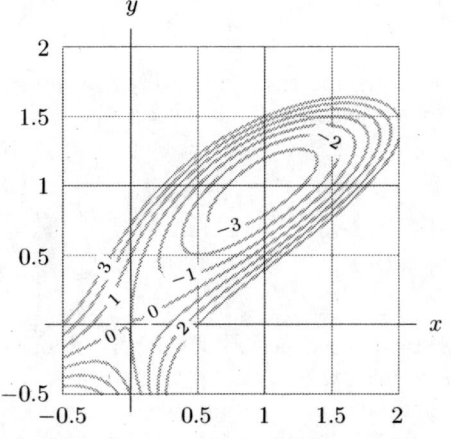

Figure 15.11: Contour diagram of $f(x,y) = 8y^3 + 12x^2 - 24xy$ showing critical points at $(0,0)$ and $(1,1)$

Classifying Critical Points

We can see whether a critical point of a function, f, is a local maximum, local minimum, or neither by looking at the contour diagram. There is also an analytic method for making this distinction.

Quadratic Functions of the form $f(x,y) = ax^2 + bxy + cy^2$

Near most critical points, a function has the same behavior as its quadratic Taylor approximation about that point. Thus, we start by investigating critical points of quadratic functions of the form $f(x,y) = ax^2 + bxy + cy^2$, where a, b and c are constants.

Example 5 Find and analyze the local extrema of the function $f(x, y) = x^2 + xy + y^2$.

Solution To find critical points, we set

$$f_x(x, y) = 2x + y = 0,$$
$$f_y(x, y) = x + 2y = 0.$$

The only critical point is $(0,0)$, and the value of the function there is $f(0,0) = 0$. If f is always positive or zero near $(0,0)$, then $(0,0)$ is a local minimum; if f is always negative or zero near $(0,0)$, it is a local maximum; if f takes both positive and negative values it is neither. The graph in Figure 15.12 suggests that $(0, 0)$ is a local minimum.

How can we be sure that $(0, 0)$ is a local minimum? We complete the square. Writing

$$f(x, y) = x^2 + xy + y^2 = \left(x + \frac{1}{2}y\right)^2 + \frac{3}{4}y^2,$$

shows that $f(x, y)$ is a sum of two squares, so it is always greater than or equal to zero. Thus, the critical point is both a local and a global minimum.

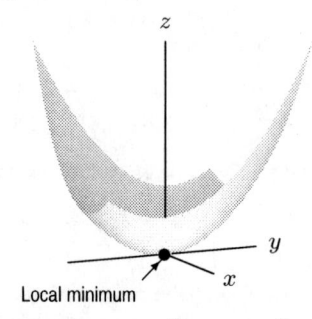

Figure 15.12: Graph of $f(x, y) = x^2 + xy + y^2 = (x + \frac{1}{2}y)^2 + \frac{3}{4}y^2$ showing local minimum at the origin

The Shape of the Graph of $f(x, y) = ax^2 + bxy + cy^2$

In general, a function of the form $f(x, y) = ax^2 + bxy + cy^2$ has one critical point at $(0, 0)$. Assuming $a \neq 0$, we complete the square and write

$$ax^2 + bxy + cy^2 = a\left[x^2 + \frac{b}{a}xy + \frac{c}{a}y^2\right] = a\left[\left(x + \frac{b}{2a}y\right)^2 + \left(\frac{c}{a} - \frac{b^2}{4a^2}\right)y^2\right]$$

$$= a\left[\left(x + \frac{b}{2a}y\right)^2 + \left(\frac{4ac - b^2}{4a^2}\right)y^2\right].$$

The shape of the graph of f depends on whether the coefficient of y^2 is positive, negative, or zero. The sign of the *discriminant*, $D = 4ac - b^2$, determines the sign of the coefficient of y^2.

- If $D > 0$, then the expression inside the square brackets is positive or zero, so the function has a local maximum or a local minimum.
  - If $a > 0$, the function has a local minimum, since the graph is a right-side-up paraboloid, like $z = x^2 + y^2$. (See Figure 15.13.)
  - If $a < 0$, the function has a local maximum, since the graph is an upside-down paraboloid, like $z = -x^2 - y^2$. (See Figure 15.14.)
- If $D < 0$, then the function goes up in some directions and goes down in others, like $z = x^2 - y^2$. We call this a *saddle point*. (See Figure 15.15.)
- If $D = 0$, then the quadratic function is $a(x + by/2a)^2$, whose graph is a parabolic cylinder. (See Figure 15.16.)

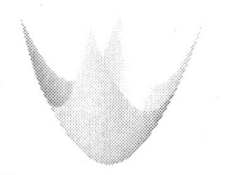

Figure 15.13: Concave up: $D > 0$ and $a > 0$

Figure 15.14: Concave down: $D > 0$ and $a < 0$

Figure 15.15: Saddle point: $D < 0$

Figure 15.16: Parabolic cylinder: $D = 0$

More generally, the graph of $g(x, y) = a(x - x_0)^2 + b(x - x_0)(y - y_0) + c(y - y_0)^2$ has the same shape as the graph of $f(x, y) = ax^2 + bxy + cy^2$, except that the critical point is at (x_0, y_0) rather than $(0, 0)$.[1]

Classifying the Critical Points of a Function

Suppose that f is any function with grad $f(0, 0) = \vec{0}$. Its quadratic Taylor polynomial near $(0, 0)$,

$$f(x, y) \approx f(0, 0) + f_x(0, 0)x + f_y(0, 0)y$$
$$+ \frac{1}{2}f_{xx}(0,0)x^2 + f_{xy}(0,0)xy + \frac{1}{2}f_{yy}(0,0)y^2,$$

can be simplified using $f_x(0, 0) = f_y(0, 0) = 0$, which gives

$$f(x, y) - f(0, 0) \approx \frac{1}{2}f_{xx}(0,0)x^2 + f_{xy}(0,0)xy + \frac{1}{2}f_{yy}(0,0)y^2.$$

The discriminant of this quadratic polynomial is

$$D = 4ac - b^2 = 4\left(\frac{1}{2}f_{xx}(0,0)\right)\left(\frac{1}{2}f_{yy}(0,0)\right) - (f_{xy}(0,0))^2,$$

which simplifies to

$$D = f_{xx}(0,0)f_{yy}(0,0) - (f_{xy}(0,0))^2.$$

There is a similar formula for D if the critical point is at (x_0, y_0). An analogy with quadratic functions suggests the following test for classifying a critical point of a function of two variables:

Second Derivative Test for Functions of Two Variables

Suppose (x_0, y_0) is a point where grad $f(x_0, y_0) = \vec{0}$. Let

$$D = f_{xx}(x_0, y_0)f_{yy}(x_0, y_0) - (f_{xy}(x_0, y_0))^2.$$

- If $D > 0$ and $f_{xx}(x_0, y_0) > 0$, then f has a local minimum at (x_0, y_0).
- If $D > 0$ and $f_{xx}(x_0, y_0) < 0$, then f has a local maximum at (x_0, y_0).
- If $D < 0$, then f has a saddle point at (x_0, y_0).
- If $D = 0$, anything can happen: f can have a local maximum, or a local minimum, or a saddle point, or none of these, at (x_0, y_0).

Example 6 Find the local maxima, minima, and saddle points of $f(x, y) = \frac{1}{2}x^2 + 3y^3 + 9y^2 - 3xy + 9y - 9x$.

Solution The partial derivatives of f give

$$f_x(x, y) = x - 3y - 9 = 0,$$
$$f_y(x, y) = 9y^2 + 18y + 9 - 3x = 0.$$

---

[1] We assumed that $a \neq 0$. If $a = 0$ and $c \neq 0$, the same argument works. If both $a = 0$ and $c = 0$, then $f(x, y) = bxy$, which has a saddle point.

Eliminating x gives $9y^2 + 9y - 18 = 0$, with solutions $y = -2$ and $y = 1$. The corresponding values of x are $x = 3$ and $x = 12$, so the critical points of f are $(3, -2)$ and $(12, 1)$. The discriminant is

$$D(x, y) = f_{xx}f_{yy} - f_{xy}^2 = (1)(18y + 18) - (-3)^2 = 18y + 9.$$

Since $D(3, -2) = -36 + 9 < 0$, we know that $(3, -2)$ is a saddle point of f. Since $D(12, 1) = 18 + 9 > 0$ and $f_{xx}(12, 1) = 1 > 0$, we know that $(12, 1)$ is a local minimum of f.

The second derivative test does not give any information if $D = 0$. However, as the following example illustrates, we may still be able to classify the critical points.

Example 7 Classify the critical points of $f(x, y) = x^4 + y^4$, and $g(x, y) = -x^4 - y^4$, and $h(x, y) = x^4 - y^4$.

Solution Each of these functions has a critical point at $(0, 0)$. Since all the second partial derivatives are 0 there, each function has $D = 0$. Near the origin, the graphs of f, g and h look like the surfaces in Figures 15.13–15.15, respectively, so f has a local minimum at $(0, 0)$, and g has a local maximum at $(0, 0)$, and h is saddle-shaped at $(0, 0)$.

We can get the same results algebraically. Since $f(0, 0) = 0$ and $f(x, y) > 0$ elsewhere, f has a local minimum at the origin. Since $g(0, 0) = 0$ and $g(x, y) < 0$ elsewhere, g has a local maximum at the origin. Lastly, h is saddle-shaped at the origin since $h(0, 0) = 0$ and, away from the origin, $h(x, y) > 0$ on the x-axis and $h(x, y) < 0$ on the y-axis.

Exercises and Problems for Section 15.1

Exercises

1. Which of the points A, B, C in Figure 15.17 appear to be critical points? Classify those that are critical points.

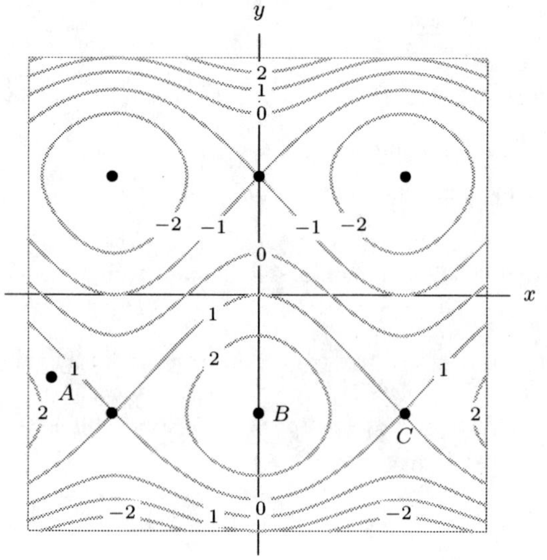

Figure 15.17

For Exercises 2–11, find the critical points and classify them as local maxima, local minima, saddle points, or none of these.

2. $f(x, y) = x^2 - 2xy + 3y^2 - 8y$

3. $f(x, y) = 400 - 3x^2 - 4x + 2xy - 5y^2 + 48y$

4. $f(x, y) = x^3 + y^2 - 3x^2 + 10y + 6$

5. $f(x, y) = x^3 - 3x + y^3 - 3y$

6. $f(x, y) = x^3 + y^3 - 3x^2 - 3y + 10$

7. $f(x, y) = x^3 + y^3 - 6y^2 - 3x + 9$

8. $f(x, y) = (x + y)(xy + 1)$

9. $f(x, y) = 8xy - \frac{1}{4}(x + y)^4$

10. $f(x, y) = 5 + 6x - x^2 + xy - y^2$

11. $f(x, y) = e^{2x^2 + y^2}$

Each function in Exercises 12–15 has a critical point at $(0, 0)$. What sort of critical point is it?

12. $f(x, y) = x^6 + y^6$

13. $g(x, y) = x^4 + y^3$

14. $h(x, y) = \cos x \cos y$

15. $k(x, y) = \sin x \sin y$

Problems

For Problems 16–18, use the contours of f in Figure 15.18.

16. Decide whether you think each point is a local maximum, local minimum, saddle point, or none of these.

(a) P (b) Q (c) R (d) S

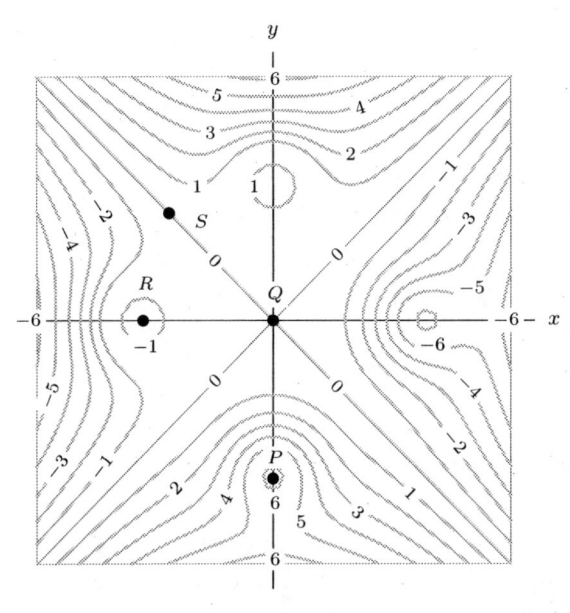

Figure 15.18

17. Sketch the direction of ∇f at several points around each of P, Q, and R.

18. Put arrows showing the direction of ∇f at the points where $\|\nabla f\|$ is largest.

For Problems 19–22, find critical points and classify them as local maxima, local minima, saddle points, or none of these.

19. $f(x,y) = x^3 + e^{-y^2}$
20. $E(x,y) = 1 - \cos x + y^2/2$
21. $f(x,y) = e^x(1 - \cos y)$
22. $f(x,y) = \sin x \sin y$
23. For $f(x,y) = A - (x^2 + Bx + y^2 + Cy)$, what values of A, B, and C give f a local maximum value of 15 at the point $(-2, 1)$?
24. At the point $(1,3)$, suppose that $f_x = f_y = 0$ and $f_{xx} > 0$, $f_{yy} > 0$, $f_{xy} = 0$.

(a) What can you conclude about the behavior of the function near the point $(1,3)$?
(b) Sketch a possible contour diagram.

25. At the point (a,b), suppose that $f_x = f_y = 0$, $f_{xx} > 0$, $f_{yy} = 0$, $f_{xy} > 0$.

(a) What can you conclude about the shape of the graph of f near the point (a,b)?
(b) Sketch a possible contour diagram.

26. Draw a possible contour diagram of f such that $f_x(-1,0) = 0$, $f_y(-1,0) < 0$, $f_x(3,3) > 0$, $f_y(3,3) > 0$, and f has a local maximum at $(3, -3)$.

27. Draw a possible contour diagram of a function with a saddle point at $(2,1)$, a local minimum at $(2,4)$, and no other critical points. Label the contours.

28. (a) Find the critical point of $f(x,y) = (x^2 - y)(x^2 + y)$.
(b) Show that at the critical point, the discriminant $D = 0$, so the second derivative test gives no information about the nature of the critical point.
(c) Sketch contours near the critical point to determine whether it is a local maximum, a local minimum, a saddle point, or none of these.

29. Let $f(x,y) = kx^2 + y^2 - 4xy$. Determine the values of k (if any) for which the critical point at $(0,0)$ is:

(a) A saddle point
(b) A local maximum
(c) A local minimum

30. On a computer, draw contour diagrams for functions

$$f(x,y) = k(x^2 + y^2) - 2xy$$

for $k = -2, -1, 0, 1, 2$. Use these figures to classify the critical point at $(0,0)$ for each value of k. Explain your observations using the discriminant, D.

31. The behavior of a function can be complicated near a critical point where $D = 0$. Suppose that

$$f(x,y) = x^3 - 3xy^2.$$

Show that there is one critical point at $(0,0)$ and that $D = 0$ there. Show that the contour for $f(x,y) = 0$ consists of three lines intersecting at the origin and that these lines divide the plane into six regions around the origin where f alternates from positive to negative. Sketch a contour diagram for f near $(0,0)$. The graph of this function is called a *monkey saddle*.

15.2 OPTIMIZATION

Suppose we want to find the highest and the lowest points in Colorado. A contour map is shown in Figure 15.19. The highest point is the top of a mountain peak (point A on the map, Mt. Elbert,

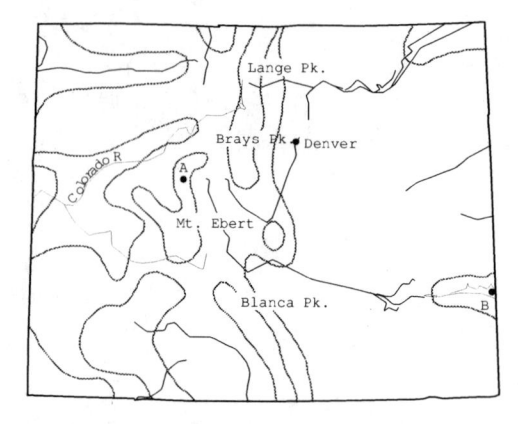

Figure 15.19: The highest and lowest points in the state of Colorado

14,431 feet high). What about the lowest point? Colorado does not have large pits without drainage, like Death Valley in California. A drop of rain falling at any point in Colorado will eventually flow out of the state. If there is no local minimum inside the state, where is the lowest point? It must be on the state boundary at a point where a river is flowing out of the state (point B where the Arkansas River leaves the state, 3,400 feet high). The highest point in Colorado is a global maximum for the elevation function in Colorado and the lowest point is the global minimum.

In general, if we are given a function f defined on a region R, we say:

- f has a **global maximum on** R at the point P_0 if $f(P_0) \geq f(P)$ for all points P in R.
- f has a **global minimum on** R at the point P_0 if $f(P_0) \leq f(P)$ for all points P in R.

The process of finding a global maximum or minimum for a function f on a region R is called *optimization*. If the region R is the entire xy-plane, we have *unconstrained optimization*; if the region R is not the entire xy-plane, that is, if x or y is restricted in some way, then we have *constrained optimization*. If the region R is not stated explicitly, it is understood to be the whole xy-plane.

How Do We Find Global Maxima and Minima?

As the Colorado example illustrates, a global extremum can occur either at a critical point inside the region or at a point on the boundary of the region. This is analogous to single-variable calculus, where a function achieves its global extrema on an interval either at a critical point inside the interval or at an endpoint of the interval. Optimization for functions of more than one variable, however, is more difficult because regions in 2-space can have very complicated boundaries.

For an Unconstrained Optimization Problem

- Find the critical points.
- Investigate whether the critical points give global maxima or minima.

Not all functions have a global maximum or minimum: it depends on the function and the region. First, we consider applications in which global extrema are expected from practical considerations. At the end of this section, we examine the conditions that lead to global extrema. In general, the fact that a function has a single local maximum or minimum does not guarantee that the point is the global maximum or minimum. (See Problem 32.) An exception is if the function is quadratic, in which case the local maximum or minimum is the global maximum or minimum. (See Example 1 on page 749 and Example 5 on page 752.)

Maximizing Profit and Minimizing Cost

In planning production of an item, a company often chooses the combination of price and quantity that maximizes its profit. We use

$$\text{Profit} = \text{Revenue} - \text{Cost},$$

and, provided the price is constant,

$$\text{Revenue} = \text{Price} \cdot \text{Quantity} = pq.$$

In addition, we need to know how the cost and price depend on quantity.

Example 1 A company manufactures two items which are sold in two separate markets. The quantities, q_1 and q_2, demanded by consumers, and the prices, p_1 and p_2 (in dollars), of each item are related by

$$p_1 = 600 - 0.3q_1 \quad \text{and} \quad p_2 = 500 - 0.2q_2.$$

Thus, if the price for either item increases, the demand for it decreases. The company's total production cost is given by

$$C = 16 + 1.2q_1 + 1.5q_2 + 0.2q_1q_2.$$

To maximize its total profit, how much of each product should be produced? What is the maximum profit? [2]

Solution The total revenue, R, is the sum of the revenues, p_1q_1 and p_2q_2, from each market. Substituting for p_1 and p_2, we get

$$R = p_1q_1 + p_2q_2 = (600 - 0.3q_1)q_1 + (500 - 0.2q_2)q_2$$
$$= 600q_1 - 0.3q_1^2 + 500q_2 - 0.2q_2^2.$$

Thus, the total profit P is given by

$$P = R - C = 600q_1 - 0.3q_1^2 + 500q_2 - 0.2q_2^2 - (16 + 1.2q_1 + 1.5q_2 + 0.2q_1q_2)$$
$$= -16 + 598.8q_1 - 0.3q_1^2 + 498.5q_2 - 0.2q_2^2 - 0.2q_1q_2.$$

To maximize P, we compute partial derivatives and set them equal to 0:

$$\frac{\partial P}{\partial q_1} = 598.8 - 0.6q_1 - 0.2q_2 = 0,$$

$$\frac{\partial P}{\partial q_2} = 498.5 - 0.4q_2 - 0.2q_1 = 0.$$

Since grad P is defined everywhere, the only critical points of P are those where grad $P = \vec{0}$. Thus, solving for q_1, q_2, we find that

$$q_1 = 699.1 \quad \text{and} \quad q_2 = 896.7.$$

The corresponding prices are

$$p_1 = 390.27 \quad \text{and} \quad p_2 = 320.66.$$

To see whether or not we have found a maximum, we compute second partial derivatives:

$$\frac{\partial^2 P}{\partial q_1^2} = -0.6, \quad \frac{\partial^2 P}{\partial q_2^2} = -0.4, \quad \frac{\partial^2 P}{\partial q_1 \partial q_2} = -0.2,$$

---
[2] Adapted from M. Rosser, *Basic Mathematics for Economists*, p. 316 (New York: Routledge, 1993).

so,

$$D = \frac{\partial^2 P}{\partial q_1^2}\frac{\partial^2 P}{\partial q_2^2} - \left(\frac{\partial^2 P}{\partial q_1 \partial q_2}\right)^2 = (-0.6)(-0.4) - (-0.2)^2 = 0.2.$$

Therefore we have found a local maximum. The graph of P is an upside-down paraboloid, so $(699.1, 896.7)$ is a global maximum. The company should produce 699.1 units of the first item priced at \$390.27 per unit, and 896.7 units of the second item priced at \$320.66 per unit. The maximum profit $P(699.1, 896.7) \approx \$433,000$.

Example 2 Twenty cubic meters of gravel are to be delivered to a landfill. The trucker plans to purchase an open-top box in which to transport the gravel in numerous trips. The total cost to the trucker is the cost of the box plus \$2 per trip. The box must have height 0.5 m, but the trucker can choose the length and width. The cost of the box is \$20/m² for the ends and \$10/m² for the bottom and sides. Notice the tradeoff: A smaller box is cheaper to buy but requires more trips. What size box should the trucker buy to minimize the total cost?[3]

Solution We first get an algebraic expression for the trucker's cost. Let the length of the box be x meters and the width be y meters and the height be 0.5 m (See Figure 15.20.)

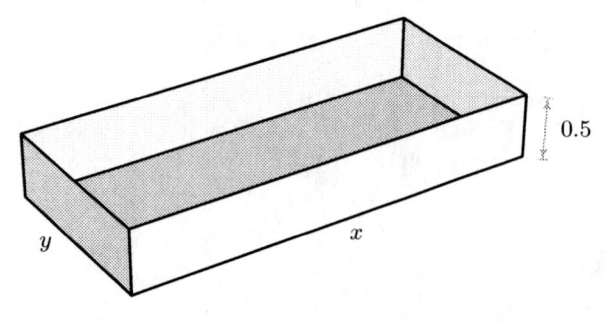

Figure 15.20: The box for transporting gravel

Table 15.2 *Trucker's itemized cost*

$20/(0.5xy)$ at \$2/trip	$80/(xy)$
2 ends at \$20/m² · $0.5y$ m²	$20y$
2 sides at \$10/m² · $0.5x$ m²	$10x$
1 bottom at \$10/m² · xy m²	$10xy$
Total cost	$f(x,y)$

The volume of the box is $0.5xy$ m³, so delivery of 20 m³ of gravel requires $20/(0.5xy)$ trips. The trucker's cost is itemized in Table 15.2. The problem is to minimize

$$\text{Total cost} = f(x,y) = \frac{80}{xy} + 20y + 10x + 10xy.$$

The critical points of this function occur where

$$f_x(x,y) = -\frac{80}{x^2 y} + 10 + 10y = 0$$

$$f_y(x,y) = -\frac{80}{xy^2} + 10x + 20 = 0$$

We put the $80/(x^2y)$ and $80/(xy^2)$ terms on the other side of the the equation, divide, and simplify:

$$\frac{80/(x^2y)}{80/(xy^2)} = \frac{10+10y}{10x+20} \quad \text{so} \quad \frac{y}{x} = \frac{1+y}{x+2} \quad \text{giving} \quad 2y = x.$$

---
[3] Adapted from Claude McMillan, Jr., *Mathematical Programming*, 2nd ed., p. 156-157 (New York: Wiley, 1978).

Substituting $x = 2y$ in the equation $f_y(x, y) = 0$ gives

$$-\frac{80}{2y \cdot y^2} + 10(2y) + 20 = 0$$

$$y^4 + y^3 - 2 = 0.$$

The only positive solution to this equation is $y = 1$, so the only critical point is $(2, 1)$.

To check that the critical point is a minimum, we use the second derivative test. Since

$$D(2, 1) = f_{xx}f_{yy} - (f_{xy})^2 = \frac{160}{2^3 \cdot 1} \cdot \frac{160}{2 \cdot 1^3} - \left(\frac{80}{2^2 \cdot 1^2} + 10\right)^2 = 700 > 0$$

and $f_{xx}(2, 1) > 0$, the point $(2, 1)$ is a local minimum. Since the value of f increases without bound as x or y increases without bound and as $x \to 0^+$ and $y \to 0^+$, it can be shown that $(2, 1)$ is a global minimum. (See Problem 33.) Thus, the optimal box is 2 meters long and 1 meter wide.

Fitting a Line to Data

Suppose we want to fit the "best" line to some data in the plane. We measure the distance from a line to the data points by adding the squares of the vertical distances from each point to the line. The smaller this sum of squares is, the better the line fits the data. The line with the minimum sum of square distances is called the *least squares line*, or the *regression line*. If the data is nearly linear, the least squares line is a good fit; otherwise it may not be. (See Figure 15.21.)

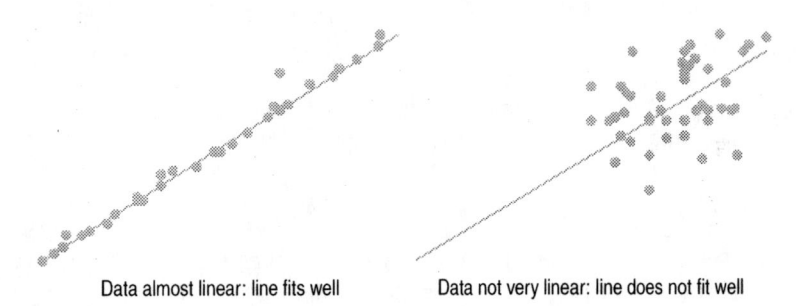

Data almost linear: line fits well   Data not very linear: line does not fit well

Figure 15.21: Fitting lines to data points

Example 3   Find a least squares line for the following data points: $(1, 1)$, $(2, 1)$, and $(3, 3)$.

Solution   Suppose the line has equation $y = b + mx$. If we find b and m then we have found the line. So, for this problem, b and m are the two variables. We want to minimize the function $f(b, m)$ that gives the sum of the three squared vertical distances from the points to the line in Figure 15.22.

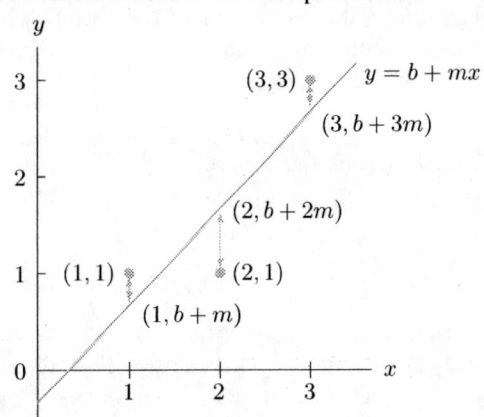

Figure 15.22: The least squares line minimizes the sum of the squares of these vertical distances

The vertical distance from the point $(1, 1)$ to the line is the difference in the y-coordinates $1 - (b + m)$; similarly for the other points. Thus, the sum of squares is

$$f(b, m) = (1 - (b + m))^2 + (1 - (b + 2m))^2 + (3 - (b + 3m))^2.$$

To minimize f we look for critical points. First we differentiate f with respect to b:

$$\frac{\partial f}{\partial b} = -2(1 - (b + m)) - 2(1 - (b + 2m)) - 2(3 - (b + 3m))$$
$$= -2 + 2b + 2m - 2 + 2b + 4m - 6 + 2b + 6m$$
$$= -10 + 6b + 12m.$$

Now we differentiate with respect to m:

$$\frac{\partial f}{\partial m} = 2(1 - (b + m))(-1) + 2(1 - (b + 2m))(-2) + 2(3 - (b + 3m))(-3)$$
$$= -2 + 2b + 2m - 4 + 4b + 8m - 18 + 6b + 18m$$
$$= -24 + 12b + 28m.$$

The equations $\dfrac{\partial f}{\partial b} = 0$ and $\dfrac{\partial f}{\partial m} = 0$ give a system of two linear equations in two unknowns:

$$-10 + 6b + 12m = 0,$$
$$-24 + 12b + 28m = 0.$$

The solution to this pair of equations is the critical point $b = -1/3$ and $m = 1$. Since

$$D = f_{bb}f_{mm} - (f_{mb})^2 = (6)(28) - 12^2 = 24 \quad \text{and} \quad f_{bb} = 6 > 0,$$

we have found a local minimum. The graph of $f(b, m)$ is a parabolic bowl, so the local minimum is the global minimum of f. Thus, the least squares line is

$$y = x - \frac{1}{3}.$$

As a check, notice that the line $y = x$ passes through the points $(1, 1)$ and $(3, 3)$. It is reasonable that introducing the point $(2, 1)$ moves the y-intercept down from 0 to $-1/3$.

The general formulas for the slope and y-intercept of a least squares line are in Project 2 on page 780. Many calculators have built in formulas for b and m, as well as for the *correlation coefficient*, which measures how well the data points fit the least squares line.

How Do We Know Whether a Function Has a Global Maximum or Minimum?

Under what circumstances does a function of two variables have a global maximum or minimum? The next example shows that a function may have both a global maximum and a global minimum on a region, or just one, or neither.

Example 4 Investigate the global maxima and minima of the following functions:
(a) $h(x, y) = 1 + x^2 + y^2$ on the disk $x^2 + y^2 \leq 1$.
(b) $f(x, y) = x^2 - 2x + y^2 - 4y + 5$ on the xy-plane.
(c) $g(x, y) = x^2 - y^2$ on the xy-plane.

Solution (a) The graph of $h(x, y) = 1 + x^2 + y^2$ is a bowl shaped paraboloid with a global minimum of 1 at $(0, 0)$, and a global maximum of 2 on the edge of the region, $x^2 + y^2 = 1$.
(b) The graph of f in Figure 15.5 on page 749 shows that f has a global minimum at the point $(1, 2)$ and no global maximum (because the value of f increases without bound as $x \to \infty, y \to \infty$).

(c) The graph of g in Figure 15.7 on page 750 shows that g has no global maximum because $g(x, y) \to \infty$ as $x \to \infty$ if y is constant. Similarly, g has no global minimum because $g(x, y) \to -\infty$ as $y \to \infty$ if x is constant.

There are, however, conditions that guarantee that a function has a global maximum and minimum. For $h(x)$, a function of one variable, the function must be continuous on a closed interval $a \leq x \leq b$. If h is continuous on a non-closed interval, such as $a \leq x < b$ or $a < x < b$, or on an interval which is not bounded, such as $a < x < \infty$, then h need not have a maximum or minimum value. What is the situation for functions of two variables? As it turns out, a similar result is true for continuous functions defined on regions which are closed and bounded, analogous to the closed and bounded interval $a \leq x \leq b$. In everyday language we say

- A **closed** region is one which contains its boundary;
- A **bounded** region is one which does not stretch to infinity in any direction.

More precise definitions follow. Suppose R is a region in 2-space. A point (x_0, y_0) is a *boundary point* of R if, for every $r > 0$, the disk $(x - x_0)^2 + (y - y_0)^2 < r^2$ with center (x_0, y_0) and radius r contains both points which are in R and points which are not in R. See Figure 15.23. A point (x_0, y_0) can be a boundary point of the region R without belonging to R. A point (x_0, y_0) in R is an *interior point* if it is not a boundary point; thus, for small enough $r > 0$, the disk of radius r centered at (x_0, y_0) lies entirely in the region R. See Figure 15.24. The collection of all the boundary points is the *boundary of R* and the collection of all the interior points is the *interior* of R. The region R is *closed* if it contains its boundary; it is *open* if every point in R is an interior point.

A region R in 2-space is *bounded* if the distance between every point (x, y) in R and the origin is less than some constant K. Closed and bounded regions in 3-space are defined in the same way.

Figure 15.23: Boundary point (x_0, y_0) of R

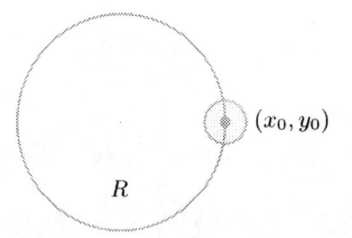

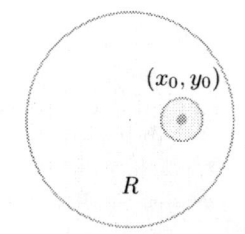

Figure 15.24: Interior point (x_0, y_0) of R

Example 5
(a) The square $-1 \leq x \leq 1$, $-1 \leq y \leq 1$ is closed and bounded.
(b) The first quadrant $x \geq 0$, $y \geq 0$ is closed but is not bounded.
(c) The disk $x^2 + y^2 < 1$ is open and bounded, but is not closed.
(d) The half-plane $y > 0$ is open, but is neither closed nor bounded.

The reason that closed and bounded regions are useful is the following theorem, which is also true for functions of three or more variables:[4]

Theorem 15.1: Extreme Value Theorem for Multivariable Functions

If f is a continuous function on a closed and bounded region R, then f has a global maximum at some point (x_0, y_0) in R and a global minimum at some point (x_1, y_1) in R.

[4] For a proof, see W. Rudin, *Principles of Mathematical Analysis*, 2nd ed., p. 89, (New York: McGraw-Hill, 1976).

Chapter Fifteen OPTIMIZATION: LOCAL AND GLOBAL EXTREMA

If f is not continuous or the region R is not closed and bounded, there is no guarantee that f achieves a global maximum or global minimum on R. In Example 4, the function g is continuous but does not achieve a global maximum or minimum in 2-space, a region which is closed but not bounded. Example 6 illustrates what can go wrong when the region is bounded but not closed.

Example 6 Does the function f have a global maximum or minimum on the region R given by $0 < x^2 + y^2 \le 1$?

$$f(x,y) = \frac{1}{x^2+y^2}$$

Solution The region R is bounded, but it is not closed since it does not contain the boundary point $(0,0)$. We see from the graph of $z = f(x,y)$ in Figure 15.25 that f has a global minimum on the circle $x^2 + y^2 = 1$. However, $f(x,y) \to \infty$ as $(x,y) \to (0,0)$, so f has no global maximum.

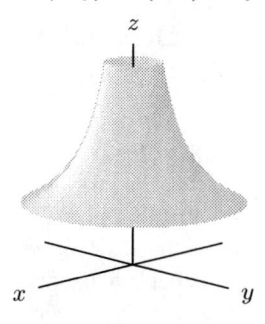

Figure 15.25: Graph showing $f(x,y) = \frac{1}{x^2+y^2}$ has no global maximum on $0 < x^2 + y^2 \le 1$

Exercises and Problems for Section 15.2

Exercises

1. By looking at the weather map in Figure 12.1 on page 604, find the maximum and minimum daily high temperatures in the states of Mississippi, Alabama, Pennsylvania, New York, California, Arizona, and Massachusetts.

2. Compute the regression line for the points $(-1, 2)$, $(0, -1)$, $(1, 1)$ using least squares.

Do the functions in Exercises 3–7 have global maxima and minima?

3. $f(x,y) = x^2 - 2y^2$
4. $g(x,y) = x^2 y^2$
5. $h(x,y) = x^3 + y^3$
6. $f(x,y) = -2x^2 - 7y^2$
7. $f(x,y) = x^2/2 + 3y^3 + 9y^2 - 3x$

In Exercises 8–10, find the global maximum and minimum of the function on $-1 \le x \le 1$, $-1 \le y \le 1$, and say whether it occurs on the boundary of the square. [Hint: Use graphs.]

8. $z = x^2 + y^2$
9. $z = -x^2 - y^2$
10. $z = x^2 - y^2$

In Exercises 11–13, estimate the position and approximate value of the global maxima and minima on the region shown.

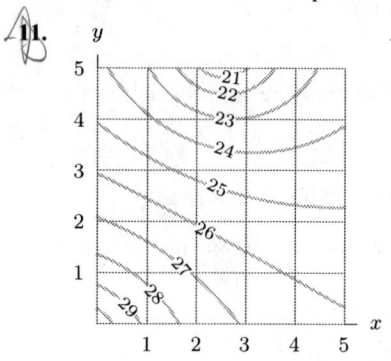

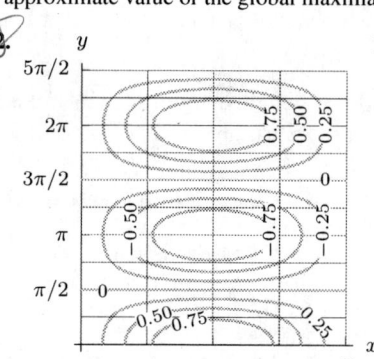

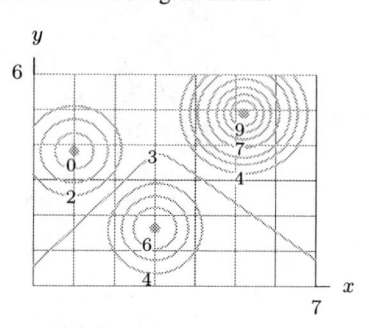

11.

12.

13.

Problems

14. If $a > 0$ then the solution of the linear equation $ax = b$ minimizes the quadratic function $f(x) = (1/2)ax^2 - bx$. This problem shows the two variable analogue.

Suppose that $A > 0$ and $AC - B^2 > 0$. Show that the solution (x, y) of the simultaneous linear equations

$$Ax + By = D$$
$$Bx + Cy = E$$

minimizes the quadratic function $g(x, y) = (1/2)(Ax^2 + 2Bxy + Cy^2) - Dx - Ey$.

15. Find the parabola of the form $y = ax^2 + b$ which best fits the points $(1, 0), (2, 2), (3, 4)$ by minimizing the sum of squares, S, given by

$$S = (a + b)^2 + (4a + b - 2)^2 + (9a + b - 4)^2.$$

A *linear least squares approximation* of a function $f(x)$ on an interval $p \leq x \leq q$ is the linear function $L(x) = b + mx$ for which the point (b, m) minimizes the function $g(b, m) = \int_p^q (f(x) - (b + mx))^2 dx$. Find the linear least squares approximation on the interval $[0, 1]$ for the functions in Problems 16–17.

16. $f(x) = x^2$ **17.** $f(x) = x^3$

18. A missile has a guidance device which is sensitive to both temperature, $t °C$, and humidity, h. The range in km over which the missile can be controlled is given by

$$\text{Range} = 27{,}800 - 5t^2 - 6ht - 3h^2 + 400t + 300h.$$

What are the optimal atmospheric conditions for controlling the missile?

19. The quantity of a product demanded by consumers is a function of its price. The quantity of one product demanded may also depend on the price of other products. For example, the demand for tea is affected by the price of coffee; the demand for cars is affected by the price of gas. The quantities demanded, q_1 and q_2, of two products depend on their prices, p_1 and p_2, as follows

$$q_1 = 150 - 2p_1 - p_2$$
$$q_2 = 200 - p_1 - 3p_2.$$

(a) What does the fact that the coefficients of p_1 and p_2 are negative tell you? Give an example of two products that might be related this way.

(b) If one manufacturer sells both products, how should the prices be set to generate the maximum possible revenue? What is that maximum possible revenue?

20. A company operates two plants which manufacture the same item and whose total cost functions are

$$C_1 = 8.5 + 0.03q_1^2 \quad \text{and} \quad C_2 = 5.2 + 0.04q_2^2,$$

where q_1 and q_2 are the quantities produced by each plant. The total quantity demanded, $q = q_1 + q_2$, is related to the price, p, by

$$p = 60 - 0.04q.$$

How much should each plant produce in order to maximize the company's profit?[5]

21. Two products are manufactured in quantities q_1 and q_2 and sold at prices of p_1 and p_2, respectively. The cost of producing them is given by

$$C = 2q_1^2 + 2q_2^2 + 10.$$

(a) Find the maximum profit that can be made, assuming the prices are fixed.

(b) Find the rate of change of that maximum profit as p_1 increases.

22. Some items are sold at a discount to senior citizens or children. The reason is that these groups are more sensitive to price, so a discount has greater impact on their purchasing decisions. The seller faces an optimization problem: How large a discount to offer in order to maximize profits? Suppose a theater can sell q_c child tickets and q_a adult tickets at prices p_c and p_a, according to the demand functions:

$$q_c = rp_c^{-4} \quad \text{and} \quad q_a = sp_a^{-2},$$

and has operating costs proportional to the total number of tickets sold. What should be the relative price of children's and adults' tickets?

23. A closed rectangular box has volume 32 cm^3. What are the lengths of the edges giving the minimum surface area?

24. An open rectangular box has volume 32 cm^3. What are the lengths of the edges giving the minimum surface area?

25. A closed rectangular box with faces parallel to the coordinate planes has one bottom corner at the origin and the opposite top corner in the first octant on the plane $3x + 2y + z = 1$. What is the maximum volume of such a box?

26. An international airline has a regulation that each passenger can carry a suitcase having the sum of its width, length and height less than or equal to 135 cm. Find the dimensions of the suitcase of maximum volume that a passenger may carry under this regulation.

---

[5] Adapted from M. Rosser, *Basic Mathematics for Economists*, p. 318 (New York: Routledge, 1993).

27. Design a rectangular milk carton box of width w, length l, and height h which holds 512 cm^3 of milk. The sides of the box cost 1 cent/cm^2 and the top and bottom cost 2 cent/cm^2. Find the dimensions of the box that minimize the total cost of materials used.

28. Find the point on the plane $3x+2y+z=1$ that is closest to the origin by minimizing the square of the distance.

29. What is the shortest distance from the surface $xy+3x+z^2=9$ to the origin?

30. The population of the US was about 180 million in 1960, grew to 206 million in 1970, and 226 million in 1980.

    (a) Assuming that the population was growing exponentially, use logarithms and the method of least squares to estimate the population in 1990.
    (b) According to the national census, the 1990 population was 249 million. What does this say about the assumption of exponential growth?
    (c) Predict the population in the year 2010.

31. A company manufactures a product which requires capital and labor to produce. The quantity, Q, of the product manufactured is given by the Cobb-Douglas function
$$Q = AK^a L^b,$$
where K is the quantity of capital and L is the quantity of labor used and A, a, and b are positive constants with $0 < a < 1$ and $0 < b < 1$. One unit of capital costs $\$k$ and one unit of labor costs $\$\ell$. The price of the product is fixed at $\$p$ per unit.

    (a) If $a + b < 1$, how much capital and labor should the company use to maximize its profit?
    (b) Is there a maximum profit in the case $a + b = 1$? What about $a + b \geq 1$? Explain.

32. Let $f(x,y) = x^2(y+1)^3 + y^2$. Show that f has only one critical point, namely $(0,0)$, and that point is a local minimum but not a global minimum. Contrast this with the case of a function with a single local minimum in one-variable calculus.

33. Let $f(x,y) = 80/(xy) + 20y + 10x + 10xy$ in the region R where $x, y > 0$.

    (a) Explain why $f(x,y) > f(2,1)$ at every point in R where
        (i) $x > 20$ (ii) $y > 20$
        (iii) $x < 0.01$ and $y \leq 20$
        (iv) $y < 0.01$ and $x \leq 20$
    (b) Explain why f must have a global minimum at a critical point in R.
    (c) Explain why f must have a global minimum in R at the point $(2, 1)$.

34. Let $f(x,y) = 2/x + 3/y + 4x + 5y$ in the region R where $x, y > 0$.

    (a) Explain why f must have a global minimum at some point in R.
    (b) Find the global minimum.

35. This problem concerns optimization with a parameter. Let $f(x, a)$ give a family of functions of x with parameter a. Suppose that for each a the function $g(x) = f(x, a)$ in the family has a unique maximum $M(a)$ that is reached at a unique critical point $x = h(a)$ of g. Thus $M(a) = f(h(a), a)$. In good situations both M and h are differentiable.

    (a) Show that $M'(a) = f_a(h(a), a)$
    (b) Show that the graph of M is tangent at each point to the graph of one of the cross-sections of f with x fixed. Specifically, show that at the point $(a, M(a))$ the graph of M and the graph of the cross-section of f with $x = h(a)$ are tangent.
    (c) Let $f(x, a) = -(1/2)x^2 + 2ax - a^2$. Find $h(a)$ and $M(a)$. Graph $M(a)$ and the cross-sections $f(x, a)$ with x fixed for $x = -2, 1, 0, 1, 2,$ and 3 on the same axes. Looking at the graphs, why do you think that the equation in part (a) is called the *envelope theorem*?

15.3 CONSTRAINED OPTIMIZATION: LAGRANGE MULTIPLIERS

Many, perhaps most, real optimization problems are constrained by external circumstances. For example, a city wanting to build a public transportation system has only a limited number of tax dollars it can spend on the project. In this section, we see how to find an optimum value under such constraints.

Graphical Approach: Maximizing Production Subject to a Budget Constraint

Suppose we want to maximize the production under a budget constraint. Suppose production, f, is a function of two variables, x and y, which are quantities of two raw materials, and that
$$f(x, y) = x^{2/3} y^{1/3}.$$

If x and y are purchased at prices of p_1 and p_2 thousands of dollars per unit, what is the maximum production f that can be obtained with a budget of c thousand dollars?

To maximize f without regard to the budget, we simply increase x and y. However, the budget constraint prevents us from increasing x and y beyond a certain point. Exactly how does the budget

15.3 CONSTRAINED OPTIMIZATION: LAGRANGE MULTIPLIERS

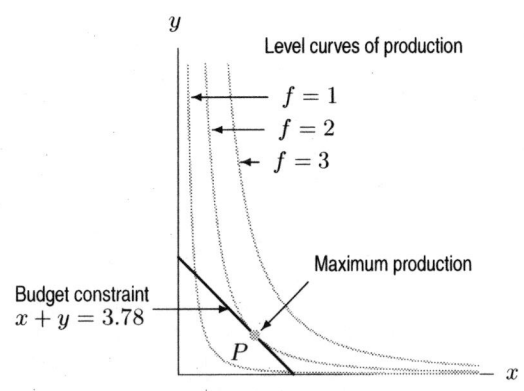

Figure 15.26: Optimal point, P, where budget constraint is tangent to a level of production function

constrain us? With prices of p_1 and p_2, the amount spent on x is $p_1 x$ and the amount spent on y is $p_2 y$, so we must have

$$g(x, y) = p_1 x + p_2 y \leq c,$$

where $g(x, y)$ is the total cost of the raw materials and c is the budget in thousands of dollars.

Let's look at the case when $p_1 = p_2 = 1$ and $c = 3.78$. Then

$$x + y \leq 3.78.$$

Figure 15.26 shows some contours of f and the budget constraint represented by the line $x+y = 3.78$. Any point on or below the line represents a pair of values of x and y that we can afford. A point on the line completely exhausts the budget, while a point below the line represents values of x and y which can be bought without using up the budget. Any point above the line represents a pair of values that we cannot afford.

To maximize f, we find the point which lies on the level curve with the largest possible value of f *and* which lies within the budget. The point must lie on the budget constraint because we should spend all the available money. Unless we are at the point where the budget constraint is tangent to the contour $f = 2$, we can increase f by moving along the line representing the budget constraint in Figure 15.26. For example, if we are on the line to the left of the point of tangency, moving right on the constraint will increase f; if we are on the line to the right of the point of tangency, moving left will increase f. Thus, the maximum value of f on the budget constraint occurs at the point where the budget constraint is tangent to the contour $f = 2$.

Analytical Solution: Lagrange Multipliers

Figure 15.26 suggests that maximum production is achieved at the point where the budget constraint is tangent to a level curve of the production function. The method of Lagrange multipliers uses this fact in algebraic form. Figure 15.27 shows that at the optimum point, P, the gradient of f and the normal to the budget line $g(x, y) = 3.78$ are parallel. Thus, at P, grad f and grad g are parallel, so for some scalar λ, called the *Lagrange multiplier*

$$\text{grad } f = \lambda \text{ grad } g.$$

Since grad $f = \left(\frac{2}{3} x^{-1/3} y^{1/3}\right) \vec{i} + \left(\frac{1}{3} x^{2/3} y^{-2/3}\right) \vec{j}$ and grad $g = \vec{i} + \vec{j}$, equating components gives

$$\frac{2}{3} x^{-1/3} y^{1/3} = \lambda \quad \text{and} \quad \frac{1}{3} x^{2/3} y^{-2/3} = \lambda.$$

Eliminating λ gives

$$\frac{2}{3} x^{-1/3} y^{1/3} = \frac{1}{3} x^{2/3} y^{-2/3}, \quad \text{which leads to} \quad 2y = x.$$

Since the constraint $x + y = 3.78$ must be satisfied, we have $x = 2.52$ and $y = 1.26$. Then
$$f(2.52, 1.26) = (2.52)^{2/3}(1.26)^{1/3} \approx 2.$$

As before, we see that the maximum value of f is approximately 2. Thus, to maximize production on a budget of \$3780, we should use 2.52 units of one raw material and 1.26 units of the other.

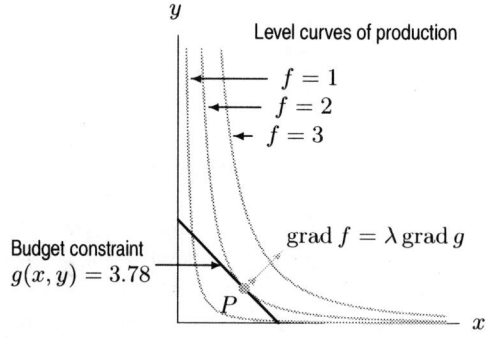

Figure 15.27: At the point, P, of maximum production, the vectors grad f and grad g are parallel

Lagrange Multipliers in General

Suppose we want to optimize an *objective function* $f(x, y)$ subject to a *constraint* $g(x, y) = c$. We look for extrema among the points which satisfy the constraint. We make the following definition.

> Suppose P_0 is a point satisfying the constraint $g(x, y) = c$.
> - f has a **local maximum** at P_0 **subject to the constraint** if $f(P_0) \geq f(P)$ for all points P near P_0 satisfying the constraint.
> - f has a **global maximum** at P_0 **subject to the constraint** if $f(P_0) \geq f(P)$ for all points P satisfying the constraint.
>
> Local and global minima are defined similarly.

As we saw in the production example, constrained extrema occur at points where the contours of f are tangent to the contours of g; they can also occur at endpoints of the constraint. In general, at any point, grad f is perpendicular to the directions in which the directional derivative of f is zero (provided grad $f \neq \vec{0}$ there). Thus if grad f is not perpendicular to the constraint, it is possible to increase and decrease f by moving along the constraint. Therefore, at a constrained extremum, grad f must be perpendicular to the constraint and parallel to grad g. (See Figure 15.28.) Thus, provided grad $g \neq \vec{0}$ at P_0, there is a constant λ such that grad $f = \lambda$ grad g at P_0.

> **Optimizing f Subject to the Constraint $g = c$:**
> If a smooth function, f, has a maximum or minimum subject to a smooth constraint $g = c$ at a point P_0, then either P_0 satisfies the equations
> $$\text{grad } f = \lambda \text{ grad } g \quad \text{and} \quad g = c,$$
> or P_0 is an endpoint of the constraint, or grad $g(P_0) = \vec{0}$. To find P_0, compare values of f at these points. The number λ is called the **Lagrange multiplier**.

If the set of points satisfying the constraint is closed and bounded, such as a circle or line segment, then there must be a global maximum and minimum of f subject to the constraint. If the constraint is not closed and bounded, such as a line or hyperbola, then there may or may not be a global maximum and minimum.

15.3 CONSTRAINED OPTIMIZATION: LAGRANGE MULTIPLIERS

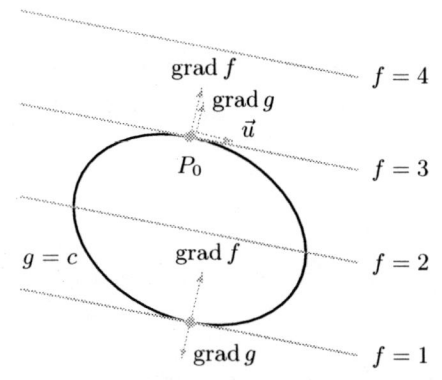

Figure 15.28: Maximum and minimum values of $f(x, y)$ on $g(x, y) = c$ are at points where grad f is parallel to grad g

Example 1 Find the maximum and minimum values of $x + y$ on the circle $x^2 + y^2 = 4$.

Solution The objective function is
$$f(x, y) = x + y,$$
and the constraint is
$$g(x, y) = x^2 + y^2 = 4.$$
Since grad $f = f_x \vec{i} + f_y \vec{j} = \vec{i} + \vec{j}$ and grad $g = g_x \vec{i} + g_y \vec{j} = 2x\vec{i} + 2y\vec{j}$, then grad $f = \lambda$ grad g gives
$$1 = 2\lambda x \quad \text{and} \quad 1 = 2\lambda y,$$
so
$$x = y.$$
We also know that
$$x^2 + y^2 = 4,$$
giving $x = y = \sqrt{2}$ or $x = y = -\sqrt{2}$. The constraint has no endpoints (it's a circle) and grad $g \neq \vec{0}$ on the circle, so we compare values of f at $(\sqrt{2}, \sqrt{2})$ and $(-\sqrt{2}, -\sqrt{2})$. Since $f(x, y) = x + y$, the maximum value of f is $f(\sqrt{2}, \sqrt{2}) = 2\sqrt{2}$; the minimum value is $f(-\sqrt{2}, -\sqrt{2}) = -2\sqrt{2}$. (See Figure 15.29.)

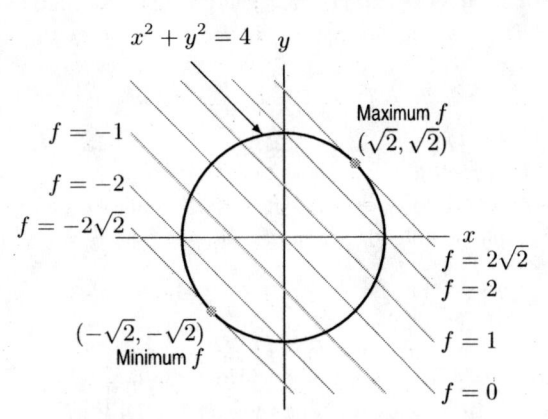

Figure 15.29: Maximum and minimum values of $f(x, y) = x + y$ on the circle $x^2 + y^2 = 4$ are at points where contours of f are tangent to the circle

… Chapter Fifteen OPTIMIZATION: LOCAL AND GLOBAL EXTREMA

How to Distinguish Maxima from Minima

There is a second derivative test[6] for classifying the critical points of constrained optimization problems, but it is more complicated than the test in Section 15.1. However, a graph of the constraint and some contours usually shows which points are maxima, which points are minima, and which are neither.

Optimization with Inequality Constraints

The production problem that we looked at first was to maximize production $f(x, y)$ subject to a budget constraint

$$g(x, y) = p_1 x + p_2 y \leq c.$$

Since the inputs are nonnegative, $x \geq 0$ and $y \geq 0$, we have three inequality constraints, which restrict (x, y) to a region of the plane rather than to a curve in the plane. In principle, we should first check to see whether or not $f(x, y)$ has any critical points in the interior:

$$p_1 x + p_2 y < c, \qquad x > 0 \quad y > 0.$$

However, in the case of a budget constraint, we can see that the maximum of f must occur when the budget is exhausted, so we look for the maximum value of f on the boundary line:

$$p_1 x + p_2 y = c, \qquad x \geq 0 \quad y \geq 0.$$

Strategy for Optimizing $f(x, y)$ Subject to the Constraint $g(x, y) \leq c$

- Find all points in the interior $g(x, y) < c$ where grad f is zero or undefined.
- Use Lagrange multipliers to find the local extrema of f on the boundary $g(x, y) = c$.
- Evaluate f at the points found in the previous two steps and compare the values.

From Section 15.2 we know that if f is continuous on a closed and bounded region, R, then f is guaranteed to attain its global maximum and minimum values on R.

Example 2 Find the maximum and minimum values of $f(x, y) = (x-1)^2 + (y-2)^2$ subject to the constraint $x^2 + y^2 \leq 45$.

Solution First, we look for all critical points for f in the interior of the region. Setting

$$f_x(x, y) = 2(x - 1) = 0$$
$$f_y(x, y) = 2(y - 2) = 0$$

we find f has exactly one critical point at $x = 1, y = 2$. Since $1^2 + 2^2 < 45$, that critical point is in the interior of the region.

Next, we find the local extrema of f on the boundary curve $x^2 + y^2 = 45$. To do this, we use Lagrange multipliers with constraint $g(x, y) = x^2 + y^2 = 45$. Setting grad $f = \lambda$ grad g, we get

$$2(x - 1) = \lambda \cdot 2x,$$
$$2(y - 2) = \lambda \cdot 2y.$$

We can't have $x = 0$ since the first equation would become $-2 = 0$. Similarly, $y \neq 0$. So we can solve each equation for λ by dividing by x and y. Setting the expressions for λ equal gives

$$\frac{x-1}{x} = \frac{y-2}{y},$$

---

[6]See J. E. Marsden and A. J. Tromba, *Vector Calculus*, 4th ed., p.218 (New York: W.H. Freeman, 1996).

so
$$y = 2x.$$
Combining this with the constraint $x^2 + y^2 = 45$, we get
$$5x^2 = 45$$
so
$$x = \pm 3.$$
Since $y = 2x$, we have possible local extrema at $x = 3, y = 6$ and $x = -3, y = -6$.

We conclude that the only candidates for the maximum and minimum values of f in the region occur at $(1, 2)$, $(3, 6)$, and $(-3, -6)$. Evaluating f at these three points we find
$$f(1, 2) = 0, \qquad f(3, 6) = 20, \qquad f(-3, -6) = 80.$$
Therefore, the minimum value of f is 0 at $(1, 2)$ and the maximum value is 80 at $(-3, -6)$.

The Meaning of λ

In the uses of Lagrange multipliers so far, we never found (or needed) the value of λ. However, λ does have a practical interpretation. In the production example, we wanted to maximize
$$f(x, y) = x^{2/3} y^{1/3}$$
subject to the constraint
$$g(x, y) = x + y = 3.78.$$
We solved the equations
$$\frac{2}{3} x^{-1/3} y^{1/3} = \lambda,$$
$$\frac{1}{3} x^{2/3} y^{-2/3} = \lambda,$$
$$x + y = 3.78,$$
to get $x = 2.52, y = 1.26$ and $f(2.52, 1.26) \approx 2$. Continuing to find λ gives us
$$\lambda \approx 0.53.$$
Now we do another, apparently unrelated, calculation. Suppose our budget is increased slightly, from 3.78 to 4.78, giving a new budget constraint of $x + y = 4.78$. Then the corresponding solution is at $x = 3.19$ and $y = 1.59$ and the new maximum value (instead of $f = 2$) is
$$f = (3.19)^{2/3} (1.59)^{1/3} \approx 2.53.$$
Notice that the amount by which f has increased is 0.53, the value of λ. Thus, in this example, the value of λ represents the extra production achieved by increasing the budget by one—in other words, the extra "bang" you get for an extra "buck" of budget. In fact, this is true in general:
- The value of λ is approximately the increase in the optimum value of f when the budget is increased by 1 unit.

More precisely:
- The value of λ represents the rate of change of the optimum value of f as the budget increases.

An Expression for λ

To interpret λ, we look at how the optimum value of the objective function f changes as the value c of the constraint function g is varied. In general, the optimum point (x_0, y_0) depends on the constraint value c. So, provided x_0 and y_0 are differentiable functions of c, we can use the chain rule to differentiate the optimum value $f(x_0(c), y_0(c))$ with respect to c:
$$\frac{df}{dc} = \frac{\partial f}{\partial x} \frac{dx_0}{dc} + \frac{\partial f}{\partial y} \frac{dy_0}{dc}.$$

At the optimum point (x_0, y_0), we have $f_x = \lambda g_x$ and $f_y = \lambda g_y$, and therefore

$$\frac{df}{dc} = \lambda \left(\frac{\partial g}{\partial x} \frac{dx_0}{dc} + \frac{\partial g}{\partial y} \frac{dy_0}{dc} \right) = \lambda \frac{dg}{dc}.$$

But, as $g(x_0(c), y_0(c)) = c$, we see that $dg/dc = 1$, so $df/dc = \lambda$. Thus, we have the following interpretation of the Lagrange multiplier λ:

> The value of λ is the rate of change of the optimum value of f as c increases (where $g(x,y) = c$). If the optimum value of f is written as $f(x_0(c), y_0(c))$, then
>
> $$\frac{d}{dc} f(x_0(c), y_0(c)) = \lambda.$$

Example 3 The quantity of goods produced according to the function $f(x,y) = x^{2/3} y^{1/3}$ is maximized subject to the budget constraint $x + y \leq 3.78$. The budget is increased to allow for a small increase in production. What is the price of the product if the sale of the additional goods covers the budget increase?

Solution We know that $\lambda = 0.53$, which tells us that $df/dc = 0.53$. The constraint corresponds to a budget of \$3.78 thousand. Therefore increasing the budget by \$1000 increases production by about 0.53 units. In order to make the increase in budget profitable, the extra goods produced must sell for more than \$1000. Thus, if p is the price of each unit of the good, then $0.53p$ is the revenue from the extra 0.53 units sold. Thus, we need $0.53p \geq 1000$ so $p \geq 1000/0.53 = \$1890$.

The Lagrangian Function

Constrained optimization problems are frequently solved using a *Lagrangian function*, $\mathcal{L}$. For example, to optimize $f(x, y)$ subject to the constraint $g(x, y) = c$, we use the Lagrangian function

$$\mathcal{L}(x, y, \lambda) = f(x, y) - \lambda(g(x, y) - c).$$

To see how the function $\mathcal{L}$ is used, compute the partial derivatives of $\mathcal{L}$:

$$\frac{\partial \mathcal{L}}{\partial x} = \frac{\partial f}{\partial x} - \lambda \frac{\partial g}{\partial x},$$

$$\frac{\partial \mathcal{L}}{\partial y} = \frac{\partial f}{\partial y} - \lambda \frac{\partial g}{\partial y},$$

$$\frac{\partial \mathcal{L}}{\partial \lambda} = -(g(x, y) - c).$$

Notice that if (x_0, y_0) is an extreme point of $f(x, y)$ subject to the constraint $g(x, y) = c$ and λ_0 is the corresponding Lagrange multiplier, then at the point (x_0, y_0, λ_0) we have

$$\frac{\partial \mathcal{L}}{\partial x} = 0 \quad \text{and} \quad \frac{\partial \mathcal{L}}{\partial y} = 0 \quad \text{and} \quad \frac{\partial \mathcal{L}}{\partial \lambda} = 0.$$

In other words, (x_0, y_0, λ_0) is a critical point for the unconstrained Lagrangian function, $\mathcal{L}(x, y, \lambda)$. Thus, the Lagrangian converts a constrained optimization problem to an unconstrained problem.

Example 4 A company has a production function with three inputs x, y, and z given by

$$f(x, y, z) = 50 x^{2/5} y^{1/5} z^{1/5}.$$

The total budget is \$24,000 and the company can buy x, y, and z at \$80, \$12, and \$10 per unit, respectively. What combination of inputs will maximize production?[7]

---

[7] Adapted from M. Rosser, *Basic Mathematics for Economists*, p. 363 (New York: Routledge, 1993).

Solution

We need to maximize the objective function

$$f(x, y, z) = 50x^{2/5}y^{1/5}z^{1/5},$$

subject to the constraint

$$g(x, y, z) = 80x + 12y + 10z = 24{,}000.$$

Therefore, the Lagrangian function is

$$\mathcal{L}(x, y, z, \lambda) = 50x^{2/5}y^{1/5}z^{1/5} - \lambda(80x + 12y + 10z - 24{,}000),$$

so we solve the system of equations we get from $\operatorname{grad}\mathcal{L} = \vec{0}$:

$$\frac{\partial \mathcal{L}}{\partial x} = 20x^{-3/5}y^{1/5}z^{1/5} - 80\lambda = 0,$$

$$\frac{\partial \mathcal{L}}{\partial y} = 10x^{2/5}y^{-4/5}z^{1/5} - 12\lambda = 0,$$

$$\frac{\partial \mathcal{L}}{\partial z} = 10x^{2/5}y^{1/5}z^{-4/5} - 10\lambda = 0,$$

$$\frac{\partial \mathcal{L}}{\partial \lambda} = -(80x + 12y + 10z - 24{,}000) = 0.$$

We simplify this system to give

$$\lambda = \frac{1}{4}x^{-3/5}y^{1/5}z^{1/5},$$

$$\lambda = \frac{5}{6}x^{2/5}y^{-4/5}z^{1/5},$$

$$\lambda = x^{2/5}y^{1/5}z^{-4/5},$$

$$80x + 12y + 10z = 24{,}000.$$

Eliminating z from the first two equations gives $x = 0.3y$. Eliminating x from the second and third equations gives $z = 1.2y$. Substituting for x and z into $80x + 12y + 10z = 24{,}000$ gives

$$80(0.3y) + 12y + 10(1.2y) = 24{,}000,$$

so $y = 500$. Then $x = 150$ and $z = 600$, and $f(150, 500, 600) = 4{,}622$ units.

The graph of the constraint, $80x + 12y + 10z = 24{,}000$, is a plane. Since the inputs x, y, z must be nonnegative, the graph is a triangle in the first quadrant, with edges on the coordinate planes. On the boundary of the triangle, one (or more) of the variables x, y, z is zero, so the function f is zero. Thus production is maximized within the budget using $x = 150$, $y = 500$, and $z = 600$.

Exercises and Problems for Section 15.3

Exercises

In Exercises 1–15, use Lagrange multipliers to find the maximum and minimum values of f subject to the given constraint, if such values exist.

1. $f(x, y) = x + y$, $\quad x^2 + y^2 = 1$
2. $f(x, y) = 3x - 2y$, $\quad x^2 + 2y^2 = 44$
3. $f(x, y) = xy$, $\quad 4x^2 + y^2 = 8$
4. $f(x, y) = x^2 + y^2$, $\quad x^4 + y^4 = 2$
5. $f(x, y) = x^2 + y^2$, $\quad 4x - 2y = 15$
6. $f(x, y) = x^2 + y$, $\quad x^2 - y^2 = 1$
7. $f(x, y) = x^2 - xy + y^2$, $\quad x^2 - y^2 = 1$
8. $f(x, y, z) = x + 3y + 5z$, $\quad x^2 + y^2 + z^2 = 1$
9. $f(x, y, z) = x^2 - 2y + 2z^2$, $\quad x^2 + y^2 + z^2 = 1$
10. $f(x, y, z) = 2x + y + 4z$, $\quad x^2 + y + z^2 = 16$
11. $f(x, y) = x^2 + 2y^2$, $\quad x^2 + y^2 \leq 4$
12. $f(x, y) = x + 3y$, $\quad x^2 + y^2 \leq 2$
13. $f(x, y) = xy$, $\quad x^2 + 2y^2 \leq 1$
14. $f(x, y) = x^3 - y^2$, $\quad x^2 + y^2 \leq 1$
15. $f(x, y) = x^3 + y$, $\quad x + y \geq 1$

Problems

16. Figure 15.30 shows contours of f. Does f have a maximum value subject to the constraint $g(x, y) = c$ for $x \geq 0, y \geq 0$? If so, approximately where is it and what is its value? Does f have a minimum value subject to the constraint? If so, approximately where and what?

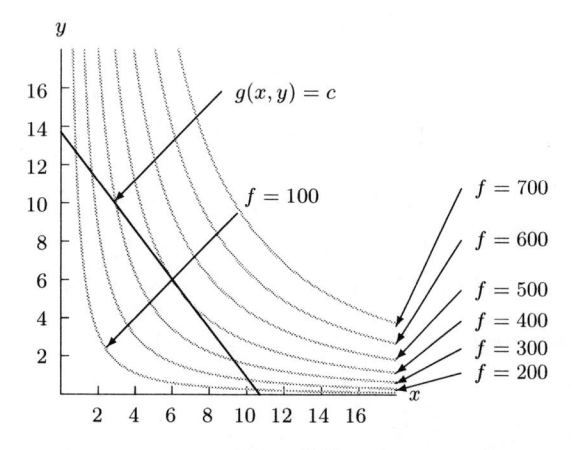

Figure 15.30

17. (a) Draw contours of $f(x, y) = 2x + y$ for $z = -7, -5, -3, -1, 1, 3, 5, 7$.
(b) On the same axes, graph the constraint $x^2 + y^2 = 5$.
(c) Use the graph to approximate the points at which f has a maximum or a minimum value subject to the constraint $x^2 + y^2 = 5$.
(d) Use Lagrange multipliers to find the maximum and minimum values of $f(x, y) = 2x + y$ subject to $x^2 + y^2 = 5$.

18. A firm manufactures a commodity at two different factories. The total cost of manufacturing depends on the quantities, q_1 and q_2, supplied by each factory, and is expressed by the *joint cost function*,

$$C = f(q_1, q_2) = 2q_1^2 + q_1 q_2 + q_2^2 + 500.$$

The company's objective is to produce 200 units, while minimizing production costs. How many units should be supplied by each factory?

19. Design a closed cylindrical container which holds 100 cm^3 and has the minimal possible surface area. What should its dimensions be?

20. An industry manufactures a product from two raw materials. The quantity produced, Q, can be given by the Cobb-Douglas function:

$$Q = cx^a y^b,$$

where x and y are quantities of each of the two raw materials used and a, b, and c are positive constants. The first raw material costs $\$P_1$ per unit and the second costs $\$P_2$ per unit. Find the maximum production possible if no more than $\$K$ can be spent on raw materials.

21. The function $P(x, y)$ gives the number of units produced and $C(x, y)$ gives the cost of production.
(a) A company wishes to maximize production at a fixed cost of $\$50{,}000$. What is the objective function f? What is the constraint equation? What is the meaning of λ in this situation?
(b) A company wishes to minimize costs at a fixed production level of 2000 units. What is the objective function f? What is the constraint equation? What is the meaning of λ in this situation?

22. A chemical manufacturing plant can produce z units of chemical Z given x units of chemical X and y units of chemical Y, where $z = 500x^{0.6}y^{0.3}$. Chemical X costs $\$10$ a unit and chemical Y costs $\$25$ a unit. The company wishes to produce as many units of chemical Z as possible given a fixed budget of $\$2000$.
(a) How many units each of chemical X and chemical Y should the company purchase? How many units of chemical Z can be produced given these purchases?
(b) What is the value of λ? Interpret it in this situation.

23. The quantity, q, of a product manufactured depends on the number of workers, W, and the amount of capital invested, K, and is given by

$$q = 6W^{3/4}K^{1/4}.$$

Labor costs are $\$10$ per worker and capital costs are $\$20$ per unit, and the budget is $\$3000$.
(a) What are the optimum number of workers and the optimum number of units of capital?
(b) Show that at the optimum values of W and K, the ratio of the marginal productivity of labor $(\partial q/\partial W)$ to the marginal productivity of capital $(\partial q/\partial K)$ is the same as the ratio of the cost of a unit of labor to the cost of a unit of capital.
(c) Recompute the optimum values of W and K when the budget is increased by one dollar. Check that increasing the budget by $\$1$ allows the production of λ extra units of the good, where λ is the Lagrange multiplier.

24. The director of a neighborhood health clinic has an annual budget of $\$600{,}000$. He wants to allocate his budget so as to maximize the number of patient visits, V, which is given as a function of the number of doctors, D, and the number of nurses, N, by

$$V = 1000D^{0.6}N^{0.3}.$$

A doctor's salary is $\$40{,}000$; nurses get $\$10{,}000$.
(a) Set up the director's constrained optimization problem.

(b) Describe, in words, the conditions which must be satisfied by $\partial V/\partial D$ and $\partial V/\partial N$ for V to have an optimum value.

(c) Solve the problem formulated in part (a).

(d) Find the value of the Lagrange multiplier and interpret its meaning in this problem.

(e) At the optimum point, what is the marginal cost of a patient visit (that is, the cost of an additional visit)? Will that marginal cost rise or fall with the number of visits? Why?

25. (a) In Problem 23, does the value of λ change if the budget changes from \$3000 to \$4000?

(b) In Problem 24, does the value of λ change if the budget changes from \$600,000 to \$700,000?

(c) What condition must a Cobb-Douglas production function, $Q = cK^a L^b$, satisfy to ensure that the marginal increase of production (that is, the rate of increase of production with budget) is not affected by the size of the budget?

26. Each person tries to balance his or her time between leisure and work. The tradeoff is that as you work less your income falls. Therefore each person has *indifference curves* which connect the number of hours of leisure, l, and income, s. If, for example, you are indifferent between 0 hours of leisure and an income of \$1125 a week on the one hand, and 10 hours of leisure and an income of \$750 a week on the other hand, then the points $l = 0$, $s = 1125$, and $l = 10$, $s = 750$ both lie on the same indifference curve. Table 15.3 gives information on three indifference curves, I, II, and III.

Table 15.3

Weekly income			Weekly leisure hours		
I	II	III	I	II	III
1125	1250	1375	0	20	40
750	875	1000	10	30	50
500	625	750	20	40	60
375	500	625	30	50	70
250	375	500	50	70	90

(a) Graph the three indifference curves.

(b) You have 100 hours a week available for work and leisure combined, and you earn \$10/hour. Write an equation in terms of l and s which represents this constraint.

(c) On the same axes, graph this constraint.

(d) Estimate from the graph what combination of leisure hours and income you would choose under these circumstances. Give the corresponding number of hours per week you would work.

27. Let $f(x, y) = ax^2 + bxy + cy^2$. Show that the maximum value of $f(x, y)$ subject to the constraint $x^2 + y^2 = 1$ is equal to λ, the Lagrange multiplier.

28. (a) If $\sum_{i=1}^{3} x_i = 1$, find the values of x_1, x_2, x_3 making $\sum_{i=1}^{3} x_i^2$ minimum.

(b) Generalize the result of part (a) to find the minimum value of $\sum_{i=1}^{n} x_i^2$ subject to $\sum_{i=1}^{n} x_i = 1$.

29. Figure 15.31 shows ∇f for a function $f(x, y)$ and two curves $g(x, y) = 1$ and $g(x, y) = 2$. Mark the following:

(a) The point(s) A where f has a local maximum.
(b) The point(s) B where f has a saddle point.
(c) The point C where f has a maximum on $g = 1$.
(d) The point D where f has a minimum on $g = 1$.
(e) If you used Lagrange multipliers to find C, what would the sign of λ be? Why?

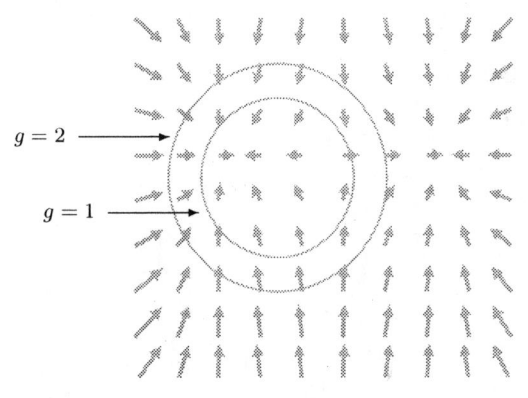

Figure 15.31

30. A mountain climber at the summit of a mountain wants to descend to a lower altitude as fast as possible. The altitude of the mountain is given approximately by

$$h(x, y) = 3000 - \frac{1}{10,000}(5x^2 + 4xy + 2y^2) \text{ meters,}$$

where x, y are horizontal coordinates on the earth (in meters), with the mountain summit located above the origin. In thirty minutes, the climber can reach any point (x, y) on a circle of radius 1000 m. In which direction should she travel in order to descend as far as possible?

31. Let $f(x, y, z) = \sqrt{(x-a)^2 + (y-b)^2 + (z-c)^2}$. Minimize f subject to $Ax + By + Cz + D = 0$. What is the geometric meaning of your solution?

32. An electrical current I flows through a circuit containing the resistors R_1 and R_2 in Figure 15.32. The currents i_1 and i_2 through the individual resistors minimize energy loss $i_1^2 R_1 + i_2^2 R_2$ subject to the constraint $i_1 + i_2 = I$ given by Kirchoff's current law.

(a) Find the currents i_1 and i_2 by the method of Lagrange multipliers.

(b) If you are familiar with Ohm's law, find the meaning of λ.

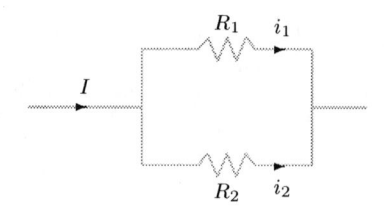

Figure 15.32

33. For constants a, b, c, let $f(x,y) = ax + by + c$ be a linear function, and let R be a region in the xy-plane.

(a) If R is any disk, show that the maximum and minimum values of f on R occur on the boundary of the disk.

(b) If R is any rectangle, show that the maximum and minimum values of f on R occur at the corners of the rectangle. They may occur at other points of the rectangle as well.

(c) Use a graph of the plane $z = f(x,y)$ to explain your answers in parts (a) and (b).

34. The function $f(x,y)$ is defined for all x, y and the origin does not lie on the surface representing $z = f(x,y)$. There is a unique point $P = (a, b, c)$ on the surface which is closest to the origin. Explain why the position vector from the origin to P must be perpendicular to the surface at that point.

35. A company produces one product from two inputs (for example capital and labor). Its production function $g(x,y)$ gives the quantity of the product that can be produced with x units of the first input and y units of the second. The *cost function* (or *expenditure function*) is the three variable function $C(p, q, u)$ where p and q are the unit prices of the two inputs. For fixed p, q, and u, the value $C(p, q, u)$ is the minimum of $f(x,y) = px + qy$ subject to the constraint $g(x,y) = u$.

(a) What is the practical meaning of $C(p, q, u)$?
(b) Find a formula for $C(p, q, u)$ if $g(x,y) = xy$.

36. A *utility function* $U(x,y)$ for two items gives the utility (benefit) to a consumer of x units of item 1 and y units of item 2. The *indirect utility function* is the three variable function $V(p, q, I)$ where p and q are the unit prices of the two items. For fixed p, q, and I, the value $V(p, q, I)$ is the maximum of $U(x,y)$ subject to the constraint $px + qy = I$.

(a) What is the practical meaning of $V(p, q, I)$?
(b) The Lagrange multiplier λ that arises in the maximization defining V is called the *marginal utility of money*. What is its practical meaning?
(c) Find formulas for $V(p, q, I)$ and λ if $U(x,y) = xy$.

37. For each value of λ the function $h(x,y) = x^2 + y^2 - \lambda(2x + 4y - 15)$ has a minimum value $m(\lambda)$.

(a) Find $m(\lambda)$.
(b) For which value of λ is $m(\lambda)$ the largest and what is that maximum value?
(c) Find the minimum value of $f(x,y) = x^2 + y^2$ subject to the constraint $2x + 4y = 15$ using the method of Lagrange multipliers and evaluate λ.
(d) Compare your answers to parts (b) and (c).

38. In a curious game you and your opponent will choose three real numbers. The rules say that you must first choose a value λ, then your opponent is free to choose any values he likes for x and y. Your goal is to make the value of $\mathcal{L}(x, y, \lambda) = 10 - x^2 - y^2 - 2x - \lambda(2x + 2y)$ as small as you can, and your opponent's goal is to make it as large as possible. What value of λ should you choose (assuming you have a brilliant opponent who never makes mistakes)?

CHAPTER SUMMARY (see also Ready Reference at the end of the book)

- **Critical Points**
  Definitions of local extrema, critical points, saddle points, finding critical points algebraically, behavior of contours near critical points, discriminant, second derivative test to classify critical points.

- **Unconstrained Optimization**
  Definitions of global extrema, method of least squares, closed and bounded regions, existence of global extrema.

- **Constrained Optimization**
  Objective function and constraint, definitions of extrema subject to a constraint, geometric interpretation of Lagrange multiplier method, solving Lagrange multiplier problems algebraically, inequality constraints, meaning of the Lagrange multiplier λ, the Lagrangian function.

REVIEW EXERCISES AND PROBLEMS FOR CHAPTER FIFTEEN

Exercises

For Exercises 1–4, find the critical points of the given function and classify them as local maxima, local minima, saddle points, or none of these.

1. $f(x, y) = 2xy^2 - x^2 - 2y^2 + 1$
2. $f(x, y) = x^2y + 2y^2 - 2xy + 6$
3. $f(x, y) = 2x^3 - 3x^2y + 6x^2 - 6y^2$
4. $f(x, y) = \sin x + \sin y + \sin(x + y)$, $0 < x < \pi$, $0 < y < \pi$.

For Exercises 5–8, find the local maxima, local minima, and saddle points of the function. Decide if the local maxima or minima are global maxima or minima. Explain.

5. $f(x, y) = 10 + 12x + 6y - 3x^2 - y^2$
6. $f(x, y) = x^2 + y^3 - 3xy$
7. $f(x, y) = xy + \ln x + y^2 - 10$, $x > 0$
8. $f(x, y) = x + y + \dfrac{1}{x} + \dfrac{4}{y}$

For Exercises 9–15, use Lagrange multipliers to find the maximum and minimum values of f subject to the constraint.

9. $f(x, y) = 3x - 4y$, $x^2 + y^2 = 5$
10. $f(x, y) = x^2 + 2y^2$, $3x + 5y = 200$
11. $f(x, y) = 2xy$, $5x + 4y = 100$
12. $f(x, y) = -3x^2 - 2y^2 + 20xy$, $x + y = 100$?
13. $f(x, y, z) = x^2 - y^2 - 2z$, $x^2 + y^2 = z$
14. $z = 4x^2 - xy + 4y^2$, $x^2 + y^2 \leq 2$
15. $f(x, y) = x^2 - y^2$, $x^2 \geq y$

16. Find the least squares line for the data points $(0, 4)$, $(1, 3)$, $(2, 1)$.

Problems

17. At the point $(1, 3)$, suppose $f_x = f_y = 0$ and $f_{xx} < 0$, $f_{yy} < 0$, $f_{xy} = 0$. Draw a possible contour diagram.

18. A company sells two products which are partial substitutes for each other, such as coffee and tea. If the price of one product rises, then the demand for the other product rises. The quantities demanded, q_1 and q_2, are given as a function of the prices, p_1 and p_2, by

$$q_1 = 517 - 3.5p_1 + 0.8p_2, \quad q_2 = 770 - 4.4p_2 + 1.4p_1.$$

What prices should the company charge in order to maximize the total sales revenue?[8]

19. An international organization must decide how to spend the $2000 they have been allotted for famine relief in a remote area. They expect to divide the money between buying rice at $5/sack and beans at $10/sack. The number, P, of people who would be fed if they buy x sacks of rice and y sacks of beans is given by

$$P = x + 2y + \dfrac{x^2 y^2}{2 \cdot 10^8}.$$

What is the maximum number of people that can be fed, and how should the organization allocate its money?

20. The quantity, Q, of a certain product manufactured depends on the quantity of labor, L, and of capital, K, used according to the function

$$Q = 900 L^{1/2} K^{2/3}.$$

Labor costs $100 per unit and capital costs $200 per unit. What combination of labor and capital should be used to produce 36,000 units of the goods at minimum cost? What is that minimum cost?

21. A company manufactures a product using inputs x, y, and z according to the production function

$$Q(x, y, z) = 20x^{1/2}y^{1/4}z^{2/5}.$$

The prices per unit are $20 for x, and $10 for y, and $5 for z. What quantity of each input should be used in order to manufacture 1,200 units at minimum cost?[9]

22. The quantity q of goods produced is given by

$$q = 80x^{0.75}y^{0.25},$$

where x is the number of units of labor used and y is the number of units of capital used. Labor costs are $6 per unit, capital costs are $4 per unit, and the budget is $8000. Find the optimal values of x and y in order to maximize the quantity produced.

23. A company manufactures x units of one item and y units of another. The total cost in dollars, C, of producing these two items is approximated by the function

$$C = 5x^2 + 2xy + 3y^2 + 800.$$

(a) If the production quota for the total number of items (both types combined) is 39, find the minimum production cost.

(b) Estimate the additional production cost or savings if the production quota is raised to 40 or lowered to 38.

---

[8] Adapted from M. Rosser, *Basic Mathematics for Economists*, p. 318 (New York: Routledge, 1993).
[9] Adapted from M. Rosser, *Basic Mathematics for Economists*, p. 363 (New York: Routledge, 1993).

24. The quantity, Q, of a product manufactured by a company is given by
$$Q = aK^{0.6}L^{0.4},$$
where a is a positive constant, K is the quantity of capital and L is the quantity of labor used. Capital costs are $20 per unit, labor costs are $10 per unit, and the company wants costs for capital and labor combined to be no higher than $150. Suppose you are asked to consult for the company, and learn that 5 units each of capital and labor are being used.

(a) What do you advise? Should the company use more or less labor? More or less capital? If so, by how much?
(b) Write a one sentence summary that could be used to sell your advice to the board of directors.

25. The Cobb-Douglas function models the quantity, q, of a commodity produced as a function of the number of workers, W, and the amount of capital invested, K:
$$q = cW^{1-a}K^a,$$
where a and c are positive constants. Labor costs are $\$p_1$ per worker, capital costs are $\$p_2$ per unit, and there is a fixed budget of $\$b$. Show that when W and K are at their optimal levels, the ratio of marginal productivity of labor to marginal productivity of capital equals the ratio of the cost of one unit of labor to one unit of capital.

26. Figure 15.33 shows contours labeled with values of $f(x, y)$ and a constraint $g(x, y) = c$. Mark the approximate points at which:

(a) $\text{grad } f = \lambda \text{ grad } g$
(b) f has a maximum
(c) f has a maximum on the constraint $g = c$.

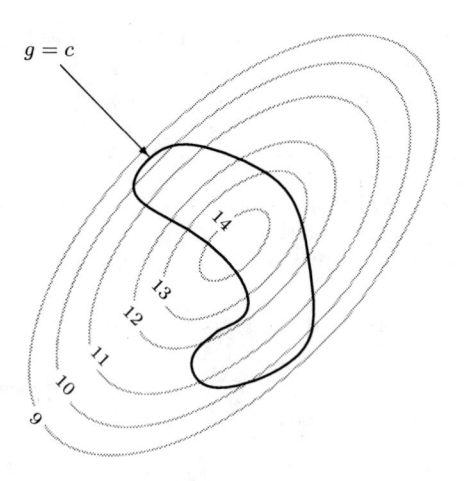

Figure 15.33

27. Let f be differentiable and grad $f(2, 1) = -3\vec{i} + 4\vec{j}$. You want to see if $(2, 1)$ is a candidate for the maximum and minimum values of f subject to a constraint satisfied by the point $(2, 1)$.

(a) Show $(2, 1)$ is not a candidate if the constraint is $x^2 + y^2 = 5$.
(b) Show $(2, 1)$ is a candidate if the constraint is $(x - 5)^2 + (y + 3)^2 = 25$. From a sketch of the contours for f near $(2, 1)$ and the constraint, decide whether $(2, 1)$ is a candidate for a maximum or minimum.
(c) Do the same as part (b), but using the constraint $(x + 1)^2 + (y - 5)^2 = 25$.

28. A biological rule of thumb states that as the area A of an island increases tenfold, the number of animal species, N, living on it doubles. The table contains data for islands in the West Indies. Assume that N is a power function of A.

(a) Use the biological rule of thumb to find
   (i) N as a function of A
   (ii) $\ln N$ as a function of $\ln A$

(b) Using the data given, tabulate $\ln N$ against $\ln A$ and find the line of best fit. Does your answer agree with the biological rule of thumb?

Island	Area (sq km)	Number of species
Redonda	3	5
Saba	20	9
Montserrat	192	15
Puerto Rico	8858	75
Jamaica	10854	70
Hispaniola	75571	130
Cuba	113715	125

29. Figure 15.34 shows two weightless springs with spring constants k_1 and k_2 attached between a ceiling and floor without tension or compression. A mass m is placed between the springs which settle into equilibrium as in Figure 15.35. The magnitudes f_1 and f_2 of the forces of the springs on the mass minimize the complementary energy
$$\frac{f_1^2}{2k_1} + \frac{f_2^2}{2k_2}$$
subject to the force balance constraint $f_1 + f_2 = mg$.

(a) Determine f_1 and f_2 by the method of Lagrange multipliers.
(b) If you are familiar with Hooke's law, find the meaning of λ.

Figure 15.34

Figure 15.35

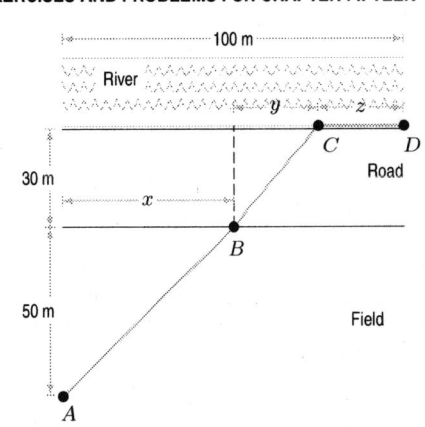

Figure 15.36

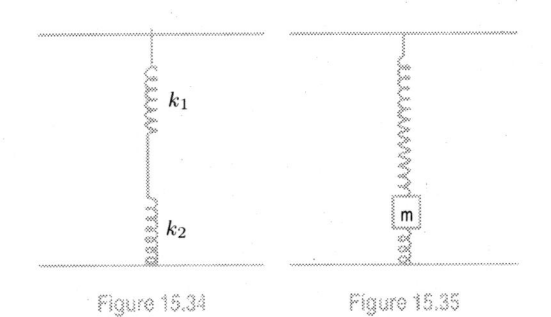

30. Find the point on the surface $z = xy + 1$ which is closest to the origin. Justify your answer.

31. A doctor wants to schedule visits for two patients who have been operated on for tumors so as to minimize the expected delay in detecting a new tumor. Visits for patients 1 and 2 are scheduled at intervals of x_1 and x_2 weeks, respectively. A total of m visits per week is available for both patients combined.

The recurrence rates for tumors for patients 1 and 2 are judged to be v_1 and v_2 tumors per week, respectively. Thus, $v_1/(v_1 + v_2)$ and $v_2/(v_1 + v_2)$ are the probabilities that patient 1 and patient 2, respectively, will have the next tumor. It is known that the expected delay in detecting a tumor for a patient checked every x weeks is $x/2$. Hence, the expected detection delay for both patients combined is given by[10]

$$f(x_1, x_2) = \frac{v_1}{v_1 + v_2} \cdot \frac{x_1}{2} + \frac{v_2}{v_1 + v_2} \cdot \frac{x_2}{2}.$$

Find the values of x_1 and x_2 in terms of v_1 and v_2 that minimize $f(x_1, x_2)$ subject to the fact that m, the number of visits per week, is fixed.

32. What is the value of the Lagrange multiplier in Problem 31? What are the units of λ? What is its practical significance to the doctor?

33. An underground cable is to be laid from point A to point D in Figure 15.36. The cable will go under a field between points A and B, where the cost to lay it is \$500/meter. From points B to C, the cable goes under a road, where the cost to lay it is \$2000/meter. Between points C and D, it goes alongside a stream where the cost is \$300/meter. Find the values of x, y, z in Figure 15.36 minimizing the total cost of laying the cable.

34. An irrigation canal has a trapezoidal cross-section of area 50 m^2, as in Figure 15.37. The average flow rate in the canal is inversely proportional to the wetted perimeter, p, of the canal, that is, to the perimeter of the trapezoid in Figure 15.37, excluding the top. Thus, to maximize the flow rate we must minimize p. Find the depth d, base width w, and angle θ that give the maximum flow rate.[11]

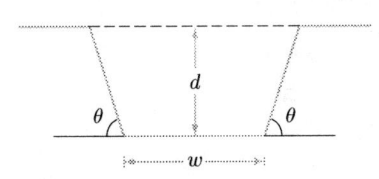

Figure 15.37

35. A light ray crossing the boundary between two different media (for example, vacuum and glass, or air and water) undergoes a change in direction or is *refracted* by an amount which depends on the properties of the media. Suppose that a light ray travels from point A to point B in Figure 15.38, with velocity v_1 in medium 1 and velocity v_2 in medium 2.

(a) Find the time $T(\theta_1, \theta_2)$ taken for the ray to travel from A to B in terms of the angles θ_1, θ_2 and the constants a, b, v_1, and v_2.

(b) Show that the angles θ_1, θ_2 satisfy the constraint

$$a \tan \theta_1 + b \tan \theta_2 = d.$$

(c) What is the effect on the time, T, of letting $\theta_1 \to -\pi/2$ (that is, moving R far to the left of A') or of letting $\theta_1 \to \pi/2$ (that is, moving R far to the right of B')?

---

[10] Adapted from Daniel Kent, Ross Shachter, *et al.*, Efficient Scheduling of Cystoscopies in Monitoring for Recurrent Bladder Cancer in *Medical Decision Making* (Philadelphia: Hanley and Belfus, 1989).

[11] Adapted from Robert M. Stark and Robert L. Nichols, *Mathematical Foundations of Design: Civil Engineering Systems*, (New York: McGraw-Hill, 1972).

(d) *Fermat's Principle* states that the light ray follows a path such that the time taken T is minimized. Use Lagrange multipliers to show that $T(\theta_1, \theta_2)$ is minimized when *Snell's Law of Refraction* holds:

$$\frac{\sin\theta_1}{\sin\theta_2} = \frac{v_1}{v_2}.$$

(The constant v_1/v_2 is called the *index of refraction* of medium 2 with respect to medium 1. For example, the indices of air, water, and glass with respect to a vacuum are approximately 1.0003, 1.33, and 1.52, respectively. The lenses of modern reading "glasses" are made of plastics with a high index of refraction in order to reduce weight and thickness when the prescription is strong.)

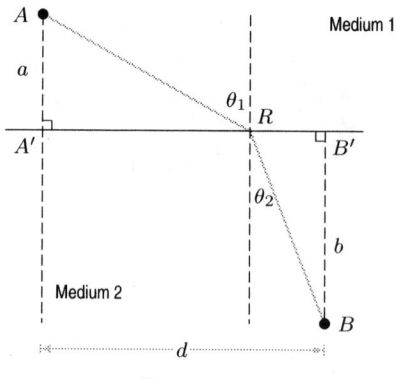

Figure 15.38

CAS Challenge Problems

36. Let $f(x, y) = \dfrac{\sqrt{a + x + y}}{1 + y + \sqrt{a + x}}$, for $x, y > 0$, where a is a positive constant.

    (a) Find the critical points of f and classify them as local maximum, local minimum, saddle point, or none of these.
    (b) Describe how the position and type of the critical points changes with respect to a, and explain this in terms of the graph of f.

37. Students are asked to find the global maximum of $f(x, y) = x^2 + y$ subject to the constraint $g(x, y) = x^2 + 2xy + y^2 - 9 = 0$. Student A uses the method of Lagrange multipliers with the help of a computer algebra system, and says that the global maximum is 11/4. Student B looks at a contour diagram of f and a graph of $g = 0$ and says there is no global maximum. Which student is correct and what mistake is the other one making?

38. Let $f(x, y) = 3x + 2y + 5$, $g(x, y) = 2x^2 - 4xy + 5y^2$.

    (a) Find the maximum of f subject to the constraint $g = 20$.
    (b) Using the value of λ in part (a), estimate the maximum of f subject to each of the constraints $g = 20.5$ and $g = 20.2$.
    (c) Use Lagrange multipliers to find the two maxima in part (b) exactly. Compare them with the estimates.

CHECK YOUR UNDERSTANDING

Are the statements in Problems 1–25 true or false? Give reasons for your answer.

1. If $f_x(P_0) = f_y(P_0) = 0$, then P_0 is a critical point of f.

2. If $f_x(P_0) = f_y(P_0) = 0$, then P_0 is a local maximum or local minimum of f.

3. If P_0 is a critical point of f, then P_0 is either a local maximum or local minimum of f.

4. If P_0 is a local maximum or local minimum of f, and not on the boundary of the domain of f, then P_0 is a critical point of f.

5. The function $f(x, y) = \sqrt{x^2 + y^2}$ has a local minimum at the origin.

6. The function $f(x, y) = x^2 - y^2$ has a local minimum at the origin.

7. If f has a local minimum at P_0 then so does the function $g(x, y) = f(x, y) + 5$.

8. If f has a local minimum at P_0 then the function $g(x, y) = -f(x, y)$ has a local maximum at P_0.

9. Every function has at least one local maximum.

10. If P_0 is a local maximum for f then $f(a, b) \leq f(P_0)$ for all points (a, b) in 2-space.

11. If P_0 is a local maximum of f, then P_0 is also a global maximum of f.

12. If P_0 is a global maximum of f, where f is defined on all of 2-space, then P_0 is also a local maximum of f.

13. Every function has a global maximum.

14. The region consisting of all points (x, y) satisfying $x^2 + y^2 < 1$ is bounded.

15. The region consisting of all points (x, y) satisfying $x^2 + y^2 < 1$ is closed.

16. The function $f(x, y) = x^2 + y^2$ has a global minimum on the region $x^2 + y^2 < 1$.

17. The function $f(x, y) = x^2 + y^2$ has a global maximum on the region $x^2 + y^2 < 1$.

18. If P and Q are two distinct points in 2-space, and f has a global maximum at P, then f cannot have a global maximum at Q.

19. The function $f(x, y) = \sin(1 + e^{xy})$ must have a global minimum in the square region $0 \leq x \leq 1, 0 \leq y \leq 1$.

20. If P_0 is a global minimum of f on a closed and bounded region, then P_0 need not be a critical point of f.

21. If $f(x, y)$ has a local maximum at (a, b) subject to the constraint $g(x, y) = c$, then $g(a, b) = c$.

22. If $f(x, y)$ has a local maximum at (a, b) subject to the constraint $g(x, y) = c$, then $\text{grad} f(a, b) = \vec{0}$.

23. The function $f(x, y) = x + y$ has no global maximum subject to the constraint $x - y = 0$.

24. The point $(2, -1)$ is a local minimum of $f(x, y) = x^2 + y^2$ subject to the constraint $x + 2y = 0$.

25. If $\text{grad} f(a, b)$ and $\text{grad} g(a, b)$ point in opposite directions, then (a, b) is a local minimum of $f(x, y)$ constrained by $g(x, y) = c$.

In Problems 26–33, suppose that M and m are the maximum and minimum values of $f(x, y)$ subject to the constraint $g(x, y) = c$ and that (a, b) satisfies $g(a, b) = c$. Decide whether the statements are true or false. Give an explanation for your answer.

26. If $f(a, b) = M$, then $f_x(a, b) = f_y(a, b) = 0$.

27. If $f(a, b) = M$, then $f(a, b) = \lambda g(a, b)$ for some value of λ.

28. If $\text{grad} f(a, b) = \lambda \text{grad} g(a, b)$, then $f(a, b) = M$ or $f(a, b) = m$.

29. If $f(a, b) = M$ and $f_x(a, b)/f_y(a, b) = 5$, then $g_x(a, b)/g_y(a, b) = 5$.

30. If $f(a, b) = m$ and $g_x(a, b) = 0$, then $f_x(a, b) = 0$.

31. Increasing the value of c will increase the value of M.

32. Suppose that $f(a, b) = M$ and that $\text{grad} f(a, b) = 3 \text{grad} g(a, b)$. Then increasing the value of c by 0.02 increases the value of M by about 0.06.

33. Suppose that $f(a, b) = m$ and that $\text{grad} f(a, b) = 3 \text{grad} g(a, b)$. Then increasing the value of c by 0.02 decreases the value of m by about 0.06.

PROJECTS FOR CHAPTER FIFTEEN

1. **Hockey and Entropy**

   Thirty teams compete for the Stanley Cup in the National Hockey League (after expansion in 2000). At the beginning of the season an experienced fan estimates that the probability that team i will win is some number p_i, where $0 \leq p_i \leq 1$ and

   $$\sum_{i=1}^{30} p_i = 1.$$

   Exactly one team will actually win, so the probabilities have to add to 1. If one of the teams, say team i, is certain to win then p_i is equal to 1 and all the other p_j are zero. Another extreme case occurs if all the teams are equally likely to win, so all the p_i are equal to 1/30, and the outcome of the hockey season is completely unpredictable. Thus, the *uncertainty* in the outcome of the hockey season depends on the probabilities $p_1, \ldots, p_{30}$. In this problem we measure this uncertainty quantitatively using the following function:

   $$S(p_1, \ldots, p_{30}) = -\sum_{i=1}^{30} p_i \frac{\ln p_i}{\ln 2}.$$

   Note that as $p_i \leq 1$, we have $-\ln p_i \geq 0$ and hence $S \geq 0$.

   (a) Show that $\lim_{p \to 0} p \ln p = 0$. (This means that S is a continuous function of the p_i, where $0 \leq p_i \leq 1$ and $1 \leq i \leq 30$, if we set $p \ln p|_{p=0}$ equal to zero. Since S is then a continuous function on a closed and bounded region, it attains a maximum and a minimum value on this region.)

   (b) Find the maximum value of $S(p_1, \ldots, p_{30})$ subject to the constraint $p_1 + \cdots + p_{30} = 1$. What are the values of p_i in this case? What does your answer mean in terms of the uncertainty in the outcome of the hockey season?

   (c) Find the minimum value of $S(p_1, \ldots, p_{30})$, subject to the constraint $p_1 + \cdots + p_{30} = 1$. What are the values of p_i in this case? What does your answer mean in terms of the uncertainty in the outcome of the hockey season?

   [Note: The function S is an example of an *entropy* function; the concept of entropy is used in information theory, statistical mechanics, and thermodynamics when measuring the uncertainty in an experiment (the hockey season in this problem) or physical system.]

2. Fitting a Line to Data Using Least Squares

In this problem you will derive the general formulas for the slope and y-intercept of a least squares line. Assume that you have n data points $(x_1, y_1), (x_2, y_2), \cdots, (x_n, y_n)$. Let the equation of the least squares line be $y = b + mx$.

(a) For each data point (x_i, y_i), show that the corresponding point directly above or below it on the least squares line has y-coordinate $b + mx_i$.

(b) For each data point (x_i, y_i), show that the square of the vertical distance from it to the point found in part (a) is $(y_i - (b + mx_i))^2$.

(c) Form the function $f(b, m)$ which is the sum of all of the n squared distances found in part (b). That is,

$$f(b, m) = \sum_{i=1}^{n} (y_i - (b + mx_i))^2.$$

Show that the partial derivatives $\dfrac{\partial f}{\partial b}$ and $\dfrac{\partial f}{\partial m}$ are given by

$$\frac{\partial f}{\partial b} = -2 \sum_{i=1}^{n} (y_i - (b + mx_i))$$

and

$$\frac{\partial f}{\partial m} = -2 \sum_{i=1}^{n} (y_i - (b + mx_i)) \cdot x_i.$$

(d) Show that the critical point equations $\dfrac{\partial f}{\partial b} = 0$ and $\dfrac{\partial f}{\partial m} = 0$ lead to a pair of simultaneous linear equations in b and m:

$$nb + \left(\sum x_i\right) m = \sum y_i$$
$$\left(\sum x_i\right) b + \left(\sum x_i^2\right) m = \sum x_i y_i$$

(e) Solve the equations in part (d) for b and m, getting

$$b = \left(\sum_{i=1}^{n} x_i^2 \sum_{i=1}^{n} y_i - \sum_{i=1}^{n} x_i \sum_{i=1}^{n} x_i y_i \right) \bigg/ \left(n \sum_{i=1}^{n} x_i^2 - \left(\sum_{i=1}^{n} x_i \right)^2 \right)$$

$$m = \left(n \sum_{i=1}^{n} x_i y_i - \sum_{i=1}^{n} x_i \sum_{i=1}^{n} y_i \right) \bigg/ \left(n \sum_{i=1}^{n} x_i^2 - \left(\sum_{i=1}^{n} x_i \right)^2 \right)$$

(f) Apply the formulas of part (e) to the data points $(1, 1), (2, 1), (3, 3)$ to check that you get the same result as in Example 3 on page 759.

11 True
13 False
15 True
17 True
19 True
21 False
23 False
25 True
27 False
29 False
31 True
33 True
35 $dy/dx = y - x^2$
37 $dy/dx = x/y$

Section 12.1

1 B, B, B
3 $(1, -1, -3)$; Front, left, below
5 Q
7 $x^2 + y^2 + z^2 = 25$
9

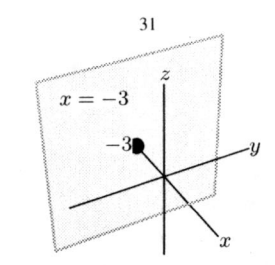

11

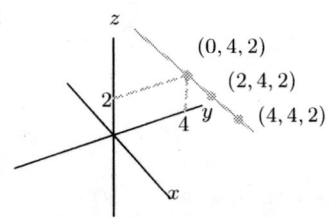

13 predicted high temperature

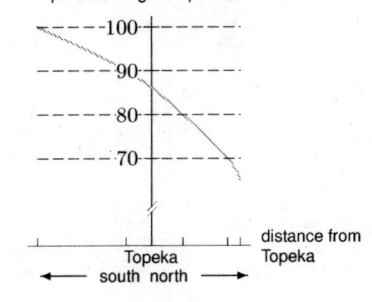

17 Increasing function
19 (a) All increasing
   (b) Deposit of $1250 gives balance $1276 after 25 mos at 1% int
21 (a) Decreasing
   (b) Increasing
23 (a) $-19°F$
   (b) 20 mph
   (c) About 17.5 mph
   (d) About 16.7°F
25 $f(x, y) = y^2$
27 Cylinder radius 2 along x-axis
29 $\sqrt{a^2 + c^2}$
31 $(x - 4)^2 + (y - 7)^2 + (z - 1)^2 = 4$
33 h

Section 12.2

1 (a) (IV)
  (b) (II)
  (c) (I)
  (d) (V)
  (e) (III)
3 Horizontal plane 3 units above the xy-plane

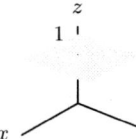

5 Bowl opening up, vertex $(0, 0, 4)$

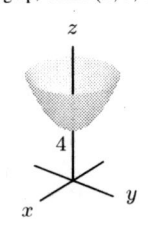

7 Parabolic cylinder extended along x-axis

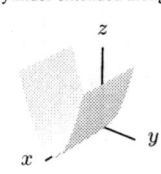

9 Circular cylinder of radius 2 in the z-direction

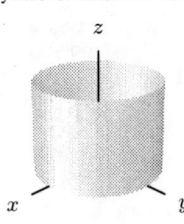

11 (a) Bowl
   (b) Neither
   (c) Plate
   (d) Bowl
   (e) Plate
13 (a)

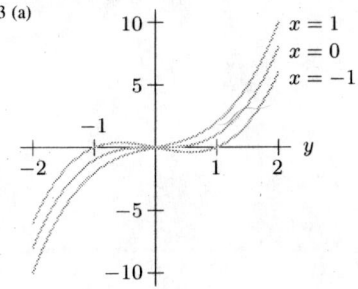

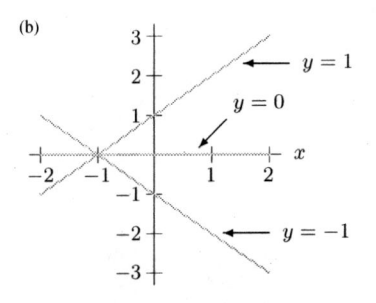

(b)

15 IV
17 (a) C (mg per liter)

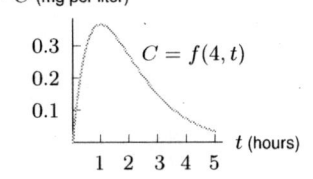

(b) C (mg per liter)

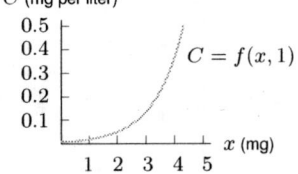

19 (a) IV
   (b) I
   (c) III
21 (a) $y = 0$
   (b) $x = 0$
   (c) $z = 1$
23 (a)

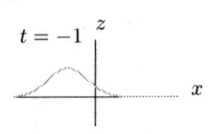

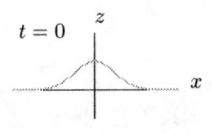

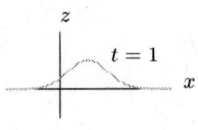

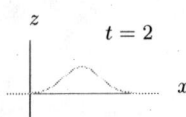

(b) Increasing x
(c)

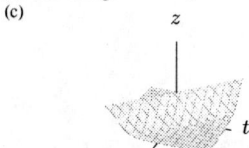

Section 12.3

1 Contours evenly spaced

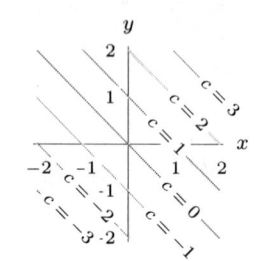

3

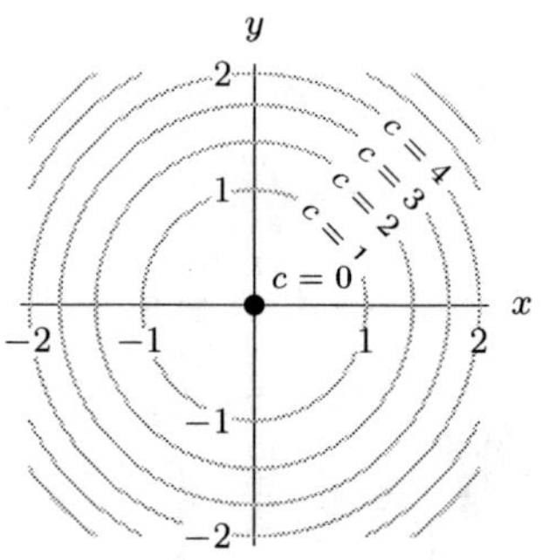

5

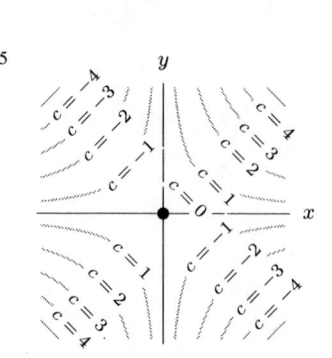

7

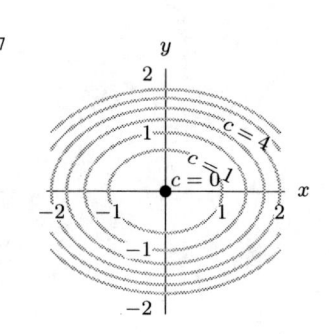

11 Price: x; Advertising: y

13

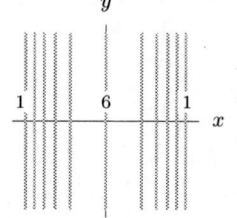

15

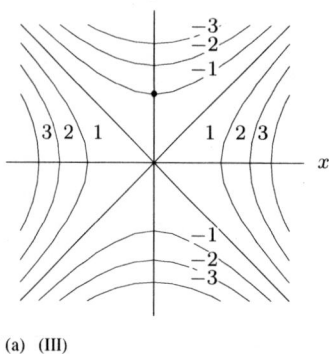

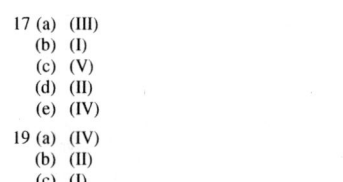

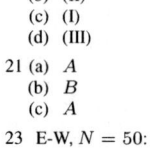

17 (a) (III)
   (b) (I)
   (c) (V)
   (d) (II)
   (e) (IV)

19 (a) (IV)
   (b) (II)
   (c) (I)
   (d) (III)

21 (a) A
   (b) B
   (c) A

23 E-W, $N = 50$:

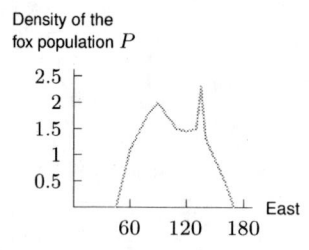

Density of the fox population P

E-W, $N = 100$:

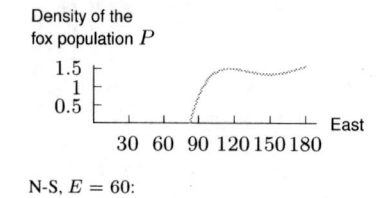

Density of the fox population P

N-S, $E = 60$:

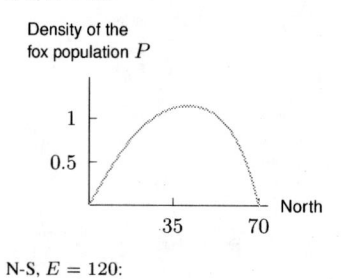

Density of the fox population P

N-S, $E = 120$:

Density of the fox population P

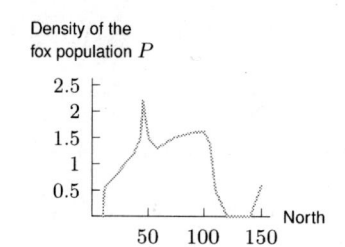

25 Doubles production

27 $\alpha + \beta > 1$: increasing
   $\alpha + \beta = 1$: constant
   $\alpha + \beta < 1$: decreasing

29 (a)

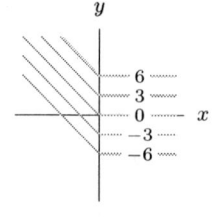

(b)

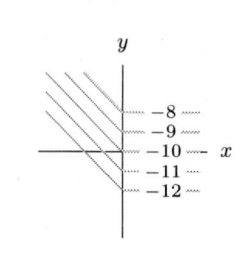

(c)

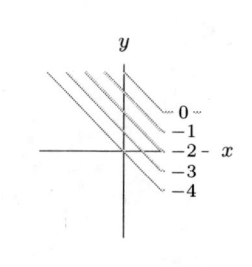

(d)

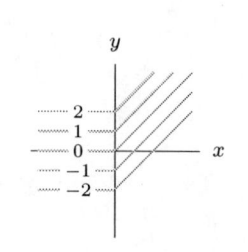

31 (a) (i)

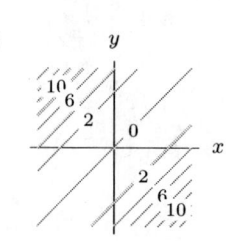

(ii)

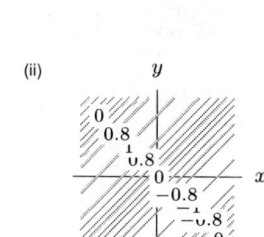

    (b) Lines parallel to $y = x$
33 $z = y(x-y)(x+y)$. Other answers possible

Section 12.4

1 (a) Yes
 (b) Dollars per month
     Dollars per minute
 (c) Charge to connect to service
 (d) $120
3 (a) $z = -13 + 5x - 3y$
 (b)

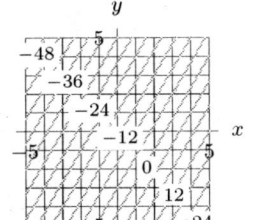

5 Not a linear function
7 Not a linear function
9 $z = 2 - \frac{1}{2}x - \frac{2}{3}y$
11 $z = 4 + 3x + y$
13 Linear
15 Not linear
17 $f(x, y) = -95 + 5x - 5y$
19 $f(x, y) = -0.01x + 0.3y + 1$
21

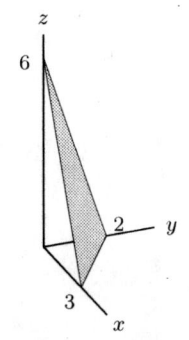

23

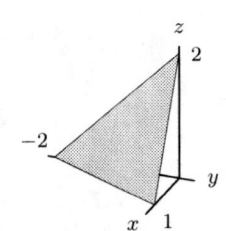

25 180 lb person at 8 mph
   120 lb person at 10 mph
27 (a) Linear
   (b) Linear
   (c) Not linear
29 (a) All $\Delta z = 7$
   (b) All $\Delta z = -5$

Section 12.5

1 (a) I

 (b) II
3 $f(x, y) = 2x - (y/2) - 3$
  $g(x, y, z) = 4x - y - 2z = 6$
5 $f(x, y) = -\sqrt{2(1 - x^2 - y^2)}$
  $g(x, y, z) = x^2 + y^2 + z^2/2 = 1$
7 No
9 No
11 $f(x, y, z) = (x-a)^2 + (y-b)^2 + (z-c)^2$
13 $f(x, y) = (1 - x^2 - y)^2$
15 Hyperboloid of two sheets
17 Elliptic paraboloid
19 $f(x, y, z) = x + 2y - z + 1$
21 Elliptical cylinder along y-axis
23 All planes with normal $\vec{i} + \vec{j} + \vec{k}$
25 Parallel planes
27 (a) Graph of f is graph of
       $x^2 + y^2 + z^2 = 1, z \geq 0$
   (b) $\sqrt{1 - x^2 - y^2} - z = 0$
29 $g(x, y, z) = 1 - \sqrt{x^2 + z^2} - y = 0$
31 Surface of rotation

Section 12.6

1 Not continuous
3 Continuous
5 Not continuous
7 1
9 0
11 1
13
15 $c = 1$
17 (c) No
21 No

Chapter 12 Review

1 Could not be true
3 Might be true
5 True
7 $y = k$

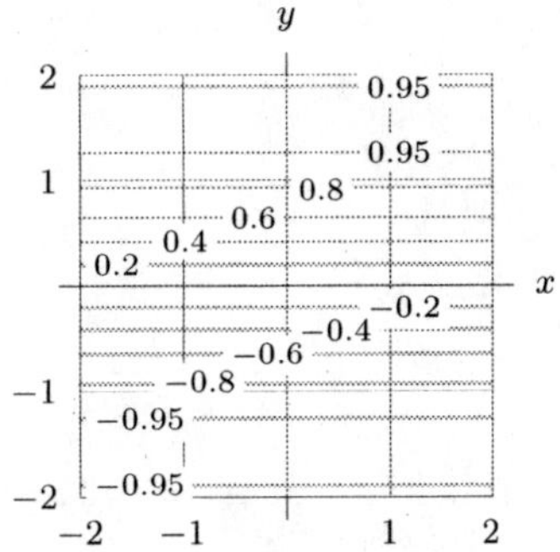

9 $2x^2 + y^2 = k$
11 $h(x, y) = 4 + 2x - y$
13 Vertical line through $(2, 1, 0)$
15 $(-2, 3, -6); 7$
17 $f(x, y) = 2x - y + 2$
19 (a) III

    (b) II
    (c) I
    (d) IV
21 (a)

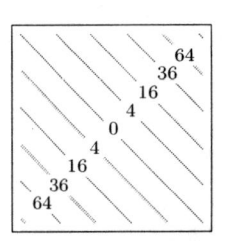

    (b)

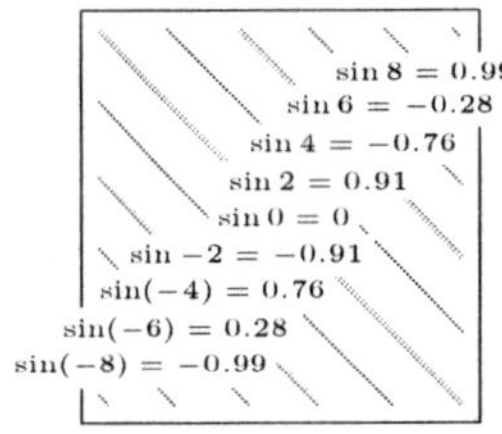

23 $72°$F; $76°$F
25 Infinite families of parallel planes
27 (a)

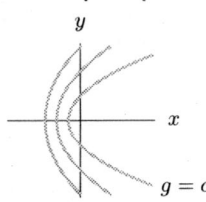

    (b) $c \geq 0$
29 $z = f(x, y) = 4 - 2x - 4y/3$
   $g(x, y, z) = (x/2) + (y/3) + (z/4) = 1$
31 $z = f(x, y) = -\sqrt{4 - (x-3)^2 - y^2}$
   $g(x, y, z) = (x-3)^2 + y^2 + z^2 = 4$
37

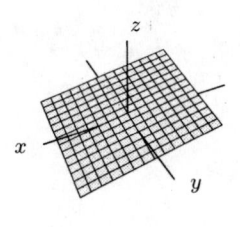

39

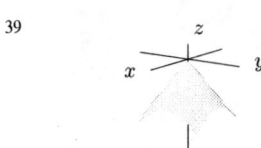

41 (a) $9 + 3x + 4y, 21 + 7x + 8y$
   (b) $189 + 63x + 64y$

Ch. 12 Understanding
1. True
3. False
5. False
7. True
9. True
11. False
13. False
15. True
17. True
19. False
21. True
23. True
25. True
27. False
29. False
31. True
33. False
35. False
37. False
39. True
41. False
43. True
45. True
47. True
49. False
51. True
53. True
55. True
57. False
59. False

Section 13.1
1. $\vec{i} + 3\vec{j}$
3. $-4.5\vec{i} + 8\vec{j} + 0.5\vec{k}$
5. $\sqrt{11}$
7. 5.6
9. $-15\vec{i} + 25\vec{j} + 20\vec{k}$
11. $21\vec{j}$
13. $\sqrt{65}$
15. (a)

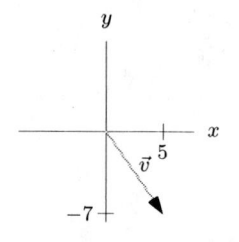

(b) $\|\vec{v}\| = 8.602$
(c) $54.46°$

17. $\vec{a} = -2\vec{j}, \vec{b} = 3\vec{i}, \vec{c} = \vec{i} + \vec{j},$
    $\vec{d} = 2\vec{j}, \vec{e} = \vec{i} - 2\vec{j}, \vec{f} = -3\vec{i} - \vec{j}$
19. $\vec{a} = \vec{b} = \vec{c} = 3\vec{k}$
    $\vec{d} = 2\vec{i} + 3\vec{k}$
    $\vec{e} = \vec{j}$
    $\vec{f} = -2\vec{i}$
21. $\vec{u}$ and $\vec{w}$; $\vec{v}$ and $\vec{q}$.
23. (a) $(3/5)\vec{i} + (4/5)\vec{j}$

(b) $6\vec{i} + 8\vec{j}$

25. (a) $-2\vec{i} + 3\vec{j} + 6\vec{k}$, $2\vec{i} + 2\vec{j} - 10\vec{k}$,
    $-4\vec{i} + \vec{j} + 16\vec{k}$
    (b) Submarine and ship
27. $\vec{k}$
29. $\vec{i} + \vec{k}$
31. $-40\vec{i} - 30\vec{j}$

Section 13.2
1. Scalar
3. Vector
5. Vector
7. $-37.59\vec{i}, -13.68\vec{j}$
9. (a) $50\vec{i}$
   (b) $-50\vec{j}$
   (c) $25\sqrt{2}\vec{i} - 25\sqrt{2}\vec{j}$
   (d) $-25\sqrt{2}\vec{i} + 25\sqrt{2}\vec{j}$
11. $-140.8\vec{i} + 140.8\vec{j} + 18\vec{k}$
13. $4.87°$ north of east
    540.63 km/hr
15. $3.4°$ north of east
17.

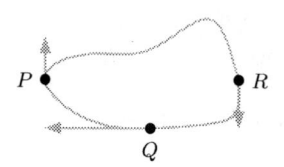

19. (79.00, 79.33, 89.00, 68.33, 89.33)
21. $-11\vec{i} + 4\vec{j}$
23. $-98.76\vec{i} + 18.94\vec{j} + 2998.31\vec{k}$
    2998.31 newtons directly up

Section 13.3
1. -14
3. 14
5. -2
7. $28\vec{j} + 14\vec{k}$
9. 185
11. $3\vec{i} + 4\vec{j} - \vec{k}$ (multiples of)
13. $\vec{i} + 3\vec{j} + 2\vec{k}$ (multiples of)
15. Many answers possible
    (a) $8\vec{i} + 6\vec{j}$
    (b) $-3\vec{i} + 4\vec{j}$
17. Parallel:
    $3\vec{i} + \sqrt{3}\vec{j}$ and $\sqrt{3}\vec{i} + \vec{j}$
    Perpendicular:
    $\sqrt{3}\vec{i} + \vec{j}$ and $\vec{i} - \sqrt{3}\vec{j}$
    $3\vec{i} + \sqrt{3}\vec{j}$ and $\vec{i} - \sqrt{3}\vec{j}$
19. $\vec{v}_1, \vec{v}_4, \vec{v}_8$ all parallel
    $\vec{v}_3, \vec{v}_5, \vec{v}_7$ all parallel
    $\vec{v}_1, \vec{v}_4, \vec{v}_8$ perpendicular to $\vec{v}_3, \vec{v}_5, \vec{v}_7$
    $\vec{v}_2$ and $\vec{v}_9$ perpendicular
21. $\vec{a} = -\frac{8}{21}\vec{d} + (\frac{79}{21}\vec{i} + \frac{10}{21}\vec{j} - \frac{118}{21}\vec{k})$
23. $5x + y - 2z = 3$
25. $2x + 4y - 3z = 5$
27. (a) $\lambda = -2.5$
    (b) $a = -6.5$
29. $38.7°$
31. $50.91°, 53.13°,$ and $75.96°$

33. No
35. Averages; weightings
    85.65%; class average

Section 13.4
1. $-\vec{i}$
3. $-\vec{i} + \vec{j} + \vec{k}$
5. $\vec{i} + 3\vec{j} + 7\vec{k}$
7. $\vec{i}$
9. $2\vec{k}$
11. $\vec{v} \times \vec{w} = -6\vec{i} + 7\vec{j} + 8\vec{k}$
    $\vec{w} \times \vec{v} = 6\vec{i} - 7\vec{j} - 8\vec{k}$
    $\vec{v} \times \vec{w} = -(\vec{w} \times \vec{v})$
13. $x + y + z = 1$
15. (a) 0.6
    (b) 0.540
17. Parallel to the z-axis
19. (a) 1.5
    (b) $y = 1$
21. $4\vec{i} + 26\vec{j} + 14\vec{k}$
23. $4(x - 4) + 26(y - 5) + 14(z - 6) = 0$
27. (b) $\vec{c} \perp \vec{a}$ and $\|\vec{c}\| = \|\vec{a}\|$
    (c) $-a_2 b_1 + a_1 b_2$

Chapter 13 Review
1. $4\vec{i} + \vec{j} + 3\vec{k}$
3. -3
5. $\vec{0}$
7. $-6\vec{i} - 9\vec{j} + 3\vec{k}$
9. 0
11. (a) 4
    (b) $-4\vec{i} - 11\vec{j} - 17\vec{k}$
    (c) $3.64\vec{i} + 2.43\vec{j} - 2.43\vec{k}$
    (d) $79.0°$
    (e) 0.784.
    (f) $2\vec{i} - 2\vec{j} + \vec{k}$ (many answers possible)
    (g) $-4\vec{i} - 11\vec{j} - 17\vec{k}$.
13. $\pm(-\vec{i} + \vec{j} - 2\vec{k})/\sqrt{6}$
15. $-3\vec{i} + 4\vec{j}$
17. (a) 1.625
    (b) 1.019
19. $4x - 3y - z = 0$
21. $\sqrt{6}/2$
23. (a) $\vec{u}$ and $-\vec{u}$ where
    $\vec{u} = \frac{12}{13}\vec{i} - \frac{4}{13}\vec{j} - \frac{3}{13}\vec{k}$
    (b) $\theta \approx 49.76°$
    (c) $13/2$
    (d) $13/\sqrt{29}$

25 $9x - 16y + 12z = 5$
   0.23

27 (a) $53.1°$ east of south
   (b) $37.9°$ west of south; does not move

29 (a) $-500\vec{i}$
   (b) $(300, 60, 4)$
   (c) $-178.7\vec{i} - 89.4\vec{j} - 7.2\vec{k}$

31 $38.7°$ south of east

33 228.43 newtons
   $85.5°$ south of east

39 $0, \vec{0}$

41 (a) $\vec{c} = -\vec{i} - 4\vec{j} + 3\vec{k}$ or $\vec{c} = \vec{i} + 4\vec{j} - 3\vec{k}$
   (b) $\vec{i} + 4\vec{j} - 3\vec{k}$

43 (a) $(tv - sw - ty + wy + sz - vz)\vec{i} +$
       $(-tu + rw + tx - wx - rz + uz)\vec{j} +$
       $(su - rv - sx + vx + ry - uy)\vec{k}$
   (b) $((s(u-x) + vx - uy +$
       $r(-v+y))/c)(a\vec{i} + b\vec{j} + c\vec{k})$

Ch. 13 Understanding

1 False
3 True
5 False
7 False
9 False
11 False
13 True
15 True
17 False
19 True
21 True
23 False
25 False
27 True
29 True

Section 14.1

1 $f_x(3,2) \approx 2, f_y(3,2) \approx -1$

3 (a) Negative
   (b) Positive

5 (a) Concentration/distance
       Rate of change of concentration with distance
       $\partial c/\partial x < 0$
   (b) Concentration/time
       Rate of change of concentration with time
       For small t, $\partial c/\partial t > 0$
       For large t, $\partial c/\partial t < 0$

7 (a) Both negative
   (b) Both negative

9 Positive, Negative, 10, 2, -4

11 $f_T(5, 20) \approx 1.2°F/°F$

13 (A) $0.06, -0.06$
   (B) $0, -0.05$
   (C) $0, 0$

15 (a) $2.5, 0.02$
   (b) $3.33, 0.02$
   (c) $3.33, 0.02$

17 (a) (i) Positive
       (ii) Negative
   (b) $f_x(P)$: from $+$ to $-$
       $f_y(P)$: no change

19 (a) Negative
   (b) Negative
   (c) Negative

21 -2.5

23

	w (gm/m³)		
	0.1	0.2	0.3
10	1300	900	1200
T (°C) 20	800	800	900
30	800	700	800

Section 14.2

1 $f_x = 10xy^3 + 8y^2 - 6x,$
  $f_y = 15x^2y^2 + 16xy$

3 $21x^5y^6 - 96x^4y^2 + 5x$

5 $2xy + 10x^4y$

7 $\dfrac{270x^3y^7 - 168x^2y^6 - 15xy^2 + 16y}{(15xy - 8)^2}$

9 $a/(2\sqrt{x})$

11 g

13 $(a+b)/2$

15 $2B/u_0$

17 $2mv/r$

19 Gm_1/r^2

21 $v_0 + at$

23 $\pi/\sqrt{lg}$

25 $f_a = e^a \sin(a+b) + e^a \cos(a+b)$

27 y

29 $\epsilon_0 E$

31 $c\gamma a_1 b_1 K^{b_1 - 1}(a_1 K^{b_1} + a_2 L^{b_2})^{\gamma - 1}$

33 $m_0 v/(c^2(1 - v^2/c^2)^{3/2})$

35 -2.51

37 (a) $3.3, 2.5$
   (b) $4.1, 2.1$
   (c) $4, 2$

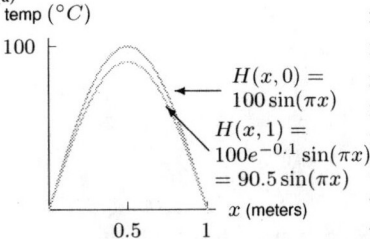

39 (a)
temp (°C)
100

$H(x,0) = 100\sin(\pi x)$
$H(x,1) = 100e^{-0.1}\sin(\pi x) = 90.5\sin(\pi x)$

0.5   1 x (meters)

   (b) $254.2e^{-0.1t}$ °C/m
   (c) $-254.2e^{-0.1t}$ °C/m
   (d) $-10e^{-0.1t}\sin(\pi x)$ °C/sec

41 $F = 684$ newtons,
   $\partial F/\partial m = 9.77$ newtons/kg,
   $\partial F/\partial r = -0.000214$ newtons/meter

43 1.277 m², 0.005 m²/kg, 0.006 m²/cm

45 $f(x,y) = x^4y^2 - 3xy^4 + C$

Section 14.3

1 $z = -4 + 2x + 4y$

3 $z = 6 + 3x + y$

5 $dg = (2u + v)\,du + u\,dv$

7 $dz = -e^{-x}\cos(y)dx - e^{-x}\sin(y)dy$

9 $df = dx - dy$

11 $dg = 4\,dx$

13 $z = 9 + 6(x - 3) + 9(y - 1)$

15 (b) $f(x,y) \approx$
      $0.3345 - 0.3$
   (c) $f(x,y) \approx 0.$
      $0.3345(x -$

17 $P(r, L) \approx$
   $80 + 2.5(r - 8) + 0.02(L - 4000),$
   $P(r, L) \approx$
   $120 + 3.33(r - 8) + 0.02(L - 6000)$
   $P(r, L) \approx$
   $160 + 3.33(r - 13) + 0.02(L - 7000)$

19 $V \approx 0.0305T - 1.27p + 38.61$

21 (a) 20 hours per day
   (b) 18.615 hours per day

23 (a) $d\rho = -\beta \rho\, dT$
   (b) $0.00015, \beta \approx 0.0005$

25 8246.68

Section 14.4

1 Negative

3 Approximately zero

5 Negative

7 $\vec{i}$

9 $-\vec{i}$

11 $\vec{i} + \vec{j}$

13 $-\vec{i} + \vec{j}$

15 $(\frac{15}{2}x^4)\vec{i} - (\frac{24}{7}y^5)\vec{j}$

17 $50\vec{i} + 100\vec{j}$

19 $\nabla z = e^y\vec{i} + e^y(1 + x + y)\vec{j}$

21 $(x\vec{i} + y\vec{j})/\sqrt{x^2 + y^2}$

23 $\sin\theta\vec{i} + r\cos\theta\vec{j}$

25 $\nabla z = \dfrac{1}{y}\cos(\dfrac{x}{y})\vec{i} - \dfrac{x}{y^2}\cos(\dfrac{x}{y})\vec{j}$

27 $\left(\dfrac{-12\beta}{(2\alpha - 3\beta)^2}\right)\vec{i} + \left(\dfrac{12\alpha}{(2\alpha - 3\beta)^2}\right)\vec{j}$

29 $60\vec{i} + 85\vec{j}$

31 $10\pi\vec{i} + 4\pi\vec{j}$

33 $\frac{1}{100}(2\vec{i} - 6\vec{j})$

35 5

37 $-46/5$

39 0.3

41 0.7

43 -0.4

45
y
5
4
3
2
1
0
   1 2 3 4 5 x

47 (a) P
   (b) S
   (c) Q, S
   (d) R

49 Fourth quadrant

51 P

53 (a) -3.268
   (b) -4.919
55 (a) $-16\vec{i} + 12\vec{j}$
   (b) $16\vec{i} - 12\vec{j}$
   (c) $12\vec{i} + 16\vec{j}$; answers may vary
57 $\partial z/\partial x = 1.8$
   $\partial z/\partial y = -0.1$
59 (a) $-5\sqrt{2}/2$
   (b) $4\vec{i} + \vec{j}$
61 $3\vec{i} + 2\vec{j}$; $3(x-2) + 2(y-3) = 0$
63 $-5\vec{i}$; $x = 2$
65 (a) 98.387 ft/mile
   (b) 295.161 ft/hour
67 $5/\sqrt{2}$
69 (a) P, Q
   (b)

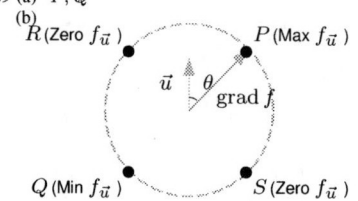

R (Zero $f_{\vec{u}}$) P (Max $f_{\vec{u}}$)

$\vec{u}$ θ grad f

Q (Min $f_{\vec{u}}$) S (Zero $f_{\vec{u}}$)

   (c) $\|\text{grad } f\|$
      $f_{\vec{u}} = \|\text{grad } f\| \cos \theta$

Section 14.5

1 $2x\vec{i} + 3y^2\vec{j} - 4z^3\vec{k}$
3 $-2(x\vec{i} + y\vec{j} + z\vec{k})/(x^2 + y^2 + z^2)^2$
5 $(e^y + 1/x)\vec{i} + xe^y\vec{j} + (1/z)\vec{k}$
7 $(2x_1x_2^3x_3^4)\vec{i} + (3x_1^2x_2^2x_3^4)\vec{j} + (4x_1^2x_2^3x_3^3)\vec{k}$
9 $6\vec{i} + 3\vec{j} + 2\vec{k}$
11 $9/\sqrt{3}$
13 $-1/\sqrt{2}$
15 $-\sqrt{77/2}$
17 $-2\vec{i} - 2\vec{j} + 4\vec{k}$;
   $-2(x+1) - 2(y-1) + 4(z-2) = 0$
19 $2\vec{i} - 2\vec{j} + \vec{k}$;
   $2(x+1) - 2(y-1) + (z-2) = 0$
21 $-2\vec{i} + \vec{k}$; $-2(x+1) + (z-2) = 0$
23 (a) 0
   (b) $24/\sqrt{19}$
25 (a) $4\vec{i} + 6\vec{j}$ (or any positive multiple)
   (b) $-\sqrt{52}$
27 (a) $-\vec{j}$,
      $\frac{2}{3}\vec{i} - \frac{1}{3}\vec{j} + \frac{2}{3}\vec{k}$
   (b) $y = 0$; $2x - y + 2z = 3$
   (c) $x = y = 0$ and $z \neq 0$
29 (a) $6.33\vec{i} + 0.76\vec{j}$
   (b) -34.69
31 $3x + 10y - 5z + 19 = 0$
33 (a) Spheres centered at origin
   (b) $(-2x\vec{i} - 2y\vec{j} - 2z\vec{k})e^{-(x^2+y^2+z^2)}$
   (c) $-3\sqrt{2}e^{-1}$ degrees/sec
35 1.131 atm/sec

Section 14.6

1 $\frac{dz}{dt} = e^{-t}\sin(t)(2\cos t - \sin t)$
3 $2e^{1-t^2}(1 - 2t^2)$

5 $2t\sin(\ln t) + 2t\ln(t)\cos(t^2)$
   $+ t\cos(\ln t) + \frac{\sin t^2}{t}$
7 $\frac{\partial z}{\partial u} = \frac{e^v}{u}$
   $\frac{\partial z}{\partial v} = e^v \ln u$
9 $\frac{\partial z}{\partial u} = 2ue^{(u^2-v^2)}(1 + u^2 + v^2)$
   $\frac{\partial z}{\partial v} = 2ve^{(u^2-v^2)}(1 - u^2 - v^2)$
11 $\frac{\partial z}{\partial u} = \frac{1}{vu}\cos\left(\frac{\ln u}{v}\right)$
   $\frac{\partial z}{\partial v} = -\frac{\ln u}{v^2}\cos\left(\frac{\ln u}{v}\right)$
13 $\frac{\partial z}{\partial u} =$
   $(e^{-v\cos u} - v(\cos u)e^{-u\sin v})\sin v$
   $-(-u(\sin v)e^{-v\cos u} + e^{-u\sin v})v\sin u$
   $\frac{\partial z}{\partial v} =$
   $(e^{-v\cos u} - v(\cos u)e^{-u\sin v})u\cos v$
   $+ (-u(\sin v)e^{-v\cos u} + e^{-u\sin v})\cos u$
15 $\frac{\partial w}{\partial u} = \frac{\partial w}{\partial x}\frac{\partial x}{\partial u} + \frac{\partial w}{\partial y}\frac{\partial y}{\partial u} + \frac{\partial w}{\partial z}\frac{\partial z}{\partial u}$
   $\frac{\partial w}{\partial v} = \frac{\partial w}{\partial x}\frac{\partial x}{\partial v} + \frac{\partial w}{\partial y}\frac{\partial y}{\partial v} + \frac{\partial w}{\partial z}\frac{\partial z}{\partial v}$
17 -0.6
19 $dV/dt = 0.3812$ volts/sec
21 $b \cdot e + d \cdot p$
23 (a) $F_u(x, y, 3)$
   (b) $F_w(3, y, x)$
   (c) $F_u(x, y, x) + F_w(x, y, x)$
   (d) $F_u(x, y, xy) + yF_w(x, y, xy)$
27 (a) $\frac{\partial z}{\partial r} = \cos\theta \frac{\partial z}{\partial x} + \sin\theta \frac{\partial z}{\partial y}$
      $\frac{\partial z}{\partial \theta} = r(\cos\theta \frac{\partial z}{\partial y} - \sin\theta \frac{\partial z}{\partial x})$
   (b) $\frac{\partial z}{\partial y} = \sin\theta \frac{\partial z}{\partial r} + \frac{\cos\theta}{r}\frac{\partial z}{\partial \theta}$
      $\frac{\partial z}{\partial x} = \cos\theta \frac{\partial z}{\partial r} - \frac{\sin\theta}{r}\frac{\partial z}{\partial \theta}$
29 $\left(\frac{\partial U_3}{\partial P}\right)_V$
31 $\left(\frac{\partial U}{\partial T}\right)_V = 7/2$
   $\left(\frac{\partial U}{\partial V}\right)_T = 11/4$
33 $\left(\frac{\partial U}{\partial P}\right)_T = \left(\frac{\partial U}{\partial V}\right)_T \left(\frac{\partial V}{\partial P}\right)_T$
37 $F(b, x)$

Section 14.7

1 $f_{xx} = 6y$
   $f_{xy} = 6x + 15y^2$
   $f_{yx} = 6x + 15y^2$
   $f_{yy} = 30xy$
3 $f_{xx} = 6(x+y)$
   $f_{yy} = 6(x+y)$
   $f_{yx} = 6(x+y)$
   $f_{xy} = 6(x+y)$
5 $f_{xx} = 4y^2e^{2xy}$
   $f_{xy} = 4xye^{2xy} + 2e^{2xy}$
   $f_{yx} = 4xye^{2xy} + 2e^{2xy}$
   $f_{yy} = 4x^2e^{2xy}$
7 $f_{xx} = \frac{y^2}{(x^2+y^2)^{3/2}}$
   $f_{xy} = \frac{-xy}{(x^2+y^2)^{3/2}} = f_{yx}$
   $f_{yy} = \frac{x^2}{(x^2+y^2)^{3/2}}$
9 $f_{xx} = -(\sin(x^2+y^2))4x^2$

   $+ 2\cos(x^2+y^2)$
   $f_{xy} = -(\sin(x^2+y^2))4xy = f_{yx}$
   $f_{yy} = -(\sin(x^2+y^2))4y^2$
   $+ 2\cos(x^2+y^2)$
11 $Q(x, y) = 1 + 2x - 2y + x^2 - 2xy + y^2$
13 $Q(x, y) = 1 - 2x^2 - y^2$
15 $Q(x, y) = 1 + x + x^2/2 - y^2/2$
17 $Q(x, y) = -y + x^2 - y^2/2$
19 (a) Positive
   (b) Zero
   (c) Positive
   (d) Zero
   (e) Zero
21 (a) Negative
   (b) Zero
   (c) Negative
   (d) Zero
   (e) Zero
23 (a) Zero
   (b) Negative
   (c) Zero
   (d) Negative
   (e) Zero
25 (a) Positive
   (b) Positive
   (c) Zero
   (d) Zero
   (e) Zero
27 (a) Positive
   (b) Negative
   (c) Negative
   (d) Negative
   (e) Positive
29 $L(x, y) = y$
   $Q(x, y) = y + 2(x-1)y$
   $L(0.9, 0.2) = 0.2$
   $Q(0.9, 0.2) = 0.16$
   $f(0.9, 0.2) = 0.162$
31 $L(x, y) = 1 + (x-1) - y$
   $Q(x, y) = 1 + (x-1) - y$
   $-(x-1)y + (1/2)y^2$
   $L(0.9, 0.2) = 0.7$
   $Q(0.9, 0.2) = 0.74$
   $f(0.9, 0.2) \approx 0.737$
33 Positive, negative
35 (a) $z_{yx} = 4y$
   (b) $z_{xyx} = 0$
   (c) $z_{xyy} = 4$
37 (a) Increasing, decreasing
   (b) $f_x > 0, f_y < 0$
   (c) $f_{xx} > 0, f_{yy} < 0$
   (d) y

   (e) P
39 (a) xy
      $1 - \frac{1}{2}(x - \frac{\pi}{2})^2 - \frac{1}{2}(y - \frac{\pi}{2})^2$

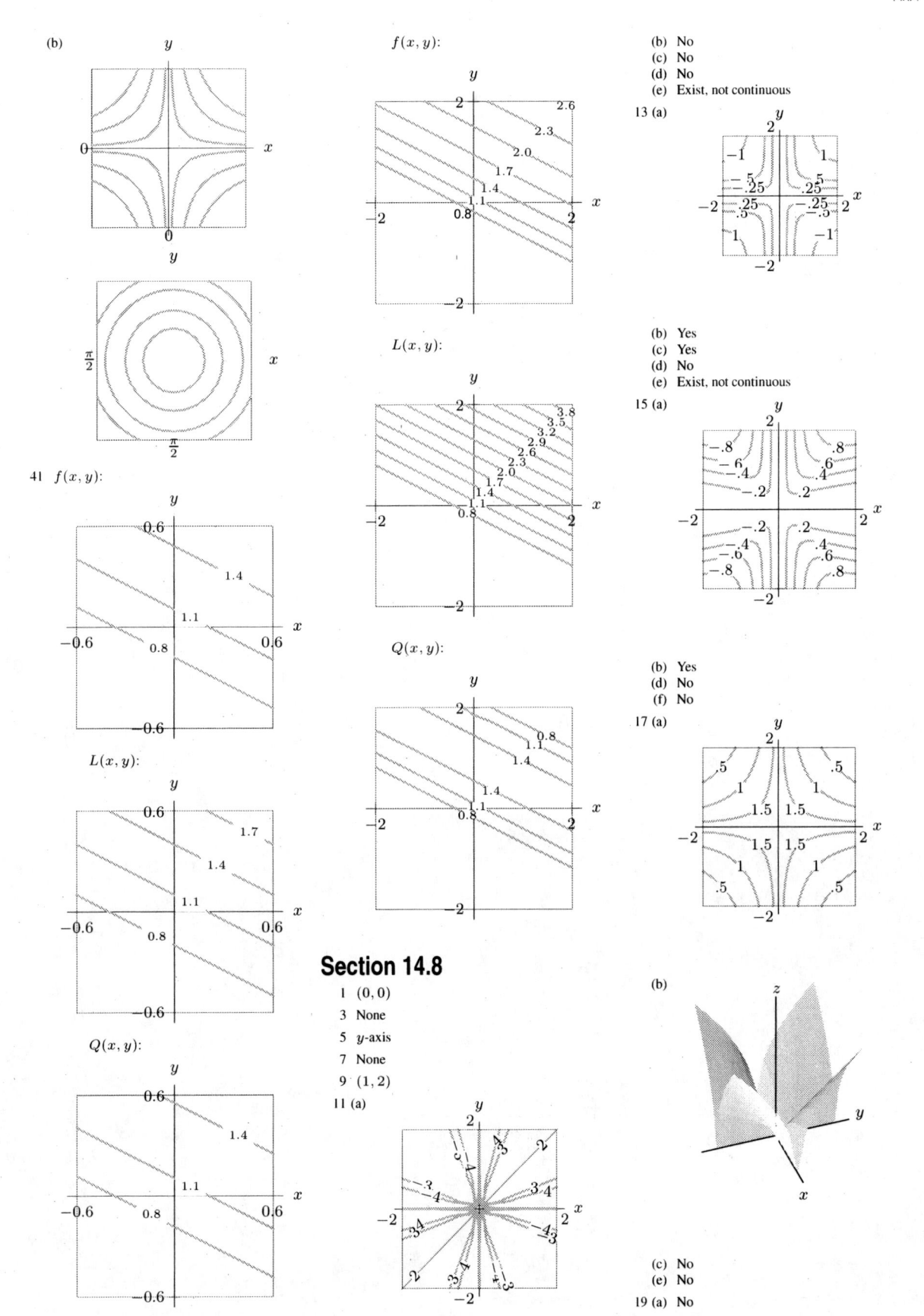

(b)

41 $f(x,y)$:

$L(x,y)$:

$Q(x,y)$:

$f(x,y)$:

$L(x,y)$:

$Q(x,y)$:

Section 14.8

1 $(0,0)$
3 None
5 y-axis
7 None
9 $(1,2)$
11 (a)

(b) No
(c) No
(d) No
(e) Exist, not continuous

13 (a)

(b) Yes
(c) Yes
(d) No
(e) Exist, not continuous

15 (a)

(b) Yes
(d) No
(f) No

17 (a)

(b)

(c) No
(e) No

19 (a) No

Chapter 14 Review

1. $f_x = 2xy + 3x^2 - 7y^6$
   $f_y = x^2 - 42xy^5$
3. $\partial F/\partial L = \frac{3}{2}\sqrt{K/L}$
5. $f_x = \frac{2xy^3}{(x^2+y^2)^2}$, $f_y = \frac{x^4-x^2y^2}{(x^2+y^2)^2}$
7. $\partial f/\partial p = (1/q)e^{p/q}$
   $\partial f/\partial q = -(p/q^2)e^{p/q}$
9. $f_N = c\alpha N^{\alpha-1}V^\beta$
11. $e^x \cos(xy) + e^x \cdot (-\sin(xy)) \cdot y$
    $e^x \cdot (-\sin(xy)) \cdot x + 2ay$
    y^2
13. $f_{xx} = (2x^2 - y^2)(x^2 + y^2)^{-5/2}$
    $f_{xy} = 3xy(x^2 + y^2)^{-5/2}$
15. $V_{rr} = 2\pi h$, $V_{rh} = 2\pi r$
17. $(3x^2 - yz)\vec{i} - xz\vec{j} + (3z^2 - xy)\vec{k}$
19. $2(x\vec{i} + y\vec{j})/(x^2 + y^2)$
21. $15/\sqrt{2}$
23. -4
25. $24/\sqrt{19}$
27. $4\vec{i} - 2\vec{j}$
29. $2\vec{i} - 2\vec{j} - 6\vec{k}$
31. $Q(x, y) = 1 - (1/2)x^2 - (9/2)y^2$
33. (a) $f_w(2,2) \approx 2.78$
       $f_z(2,2) \approx 4.01$
    (b) $f_w(2,2) \approx 2.773$
       $f_z(2,2) = 4$
35. -0.5
37. -0.71
39. $84/5$
41. (I); $c = 1/2$
43. (I); $c = 4/9$
45. $\pm 4\sqrt{\frac{2}{11}}\left(\frac{1}{2}, -\frac{1}{2}, \frac{3}{2}\right)$
47. $y = 2x - 7$
49. (a) 35.355
    (b) $35.355v$
53. (a) $-3/8°$C/m, $2/5°$C/min
    (b) $-1°$C/m, $3/28°$C/min
57. -0.008544
59. (a) $z = 7 - 3(x - 2) + 4(y - 1)$
    (b) $-3(x - 2) + 4(y - 1) = 0$
61. 0.06
63. *y*

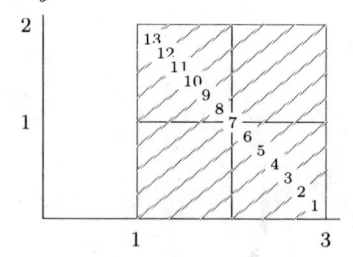

65. $f_r(2, 1) = -2/\sqrt{5}$, $f_\theta(2, 1) = 11$
67. (a) $-14, 2.5$
    (b) -2.055
    (c) 14.221 in direction of $-14\vec{i} + 2.5\vec{j}$
    (d) $f(x, y) = f(2, 3) = 7.56$
    (e) For example, $\vec{v} = 2.5\vec{i} + 14\vec{j}$
    (f) -0.32
69. (a) $e^{10} - 2e^{10}x - 6e^{10}y$
    (b) $1 + (x - 1)^2 + (y - 3)^2$
    (c) $-2\vec{i} - 6\vec{j}$
    (d) $-2e^{10}\vec{i} - 6e^{10}\vec{j} - \vec{k}$
71. (a) (I)-(E), (II)-(F), (III)-(G), (IV)-(H)
    (b) (I)-(L), (II)-(J), (III)-(M), (IV)-(K)
73. (a) $A_0 + A_1 + 2A_2 + A_3 + 2A_4 + 4A_5 + (A_1 + 2A_3 + 2A_4)(x-1) + (A_2 + A_4 + 4A_5)(y - 2), 1 + B_1 t, 2 + C_1 t$
    (b) $A_1 B_1 + 2A_3 B_1 + 2A_4 B_1 + A_2 C_1 + A_4 C_1 + 4A_5 C_1$
75. $\frac{\partial w}{\partial x}\frac{\partial x}{\partial u}\frac{du}{dt} + \frac{\partial w}{\partial y}\frac{\partial y}{\partial u}\frac{du}{dt} + \frac{\partial w}{\partial x}\frac{\partial x}{\partial v}\frac{dv}{dt} + \frac{\partial w}{\partial y}\frac{\partial y}{\partial v}\frac{dv}{dt} + \frac{\partial w}{\partial z}\frac{dz}{dt}$

Ch. 14 Understanding

1. True
3. True
5. True
7. True
9. True
11. True
13. True
15. False
17. True
19. False
21. False
23. True
25. False
27. False
29. True
31. False
33. True
35. False
37. True
39. True
41. True
43. True
45. False
47. True

17.

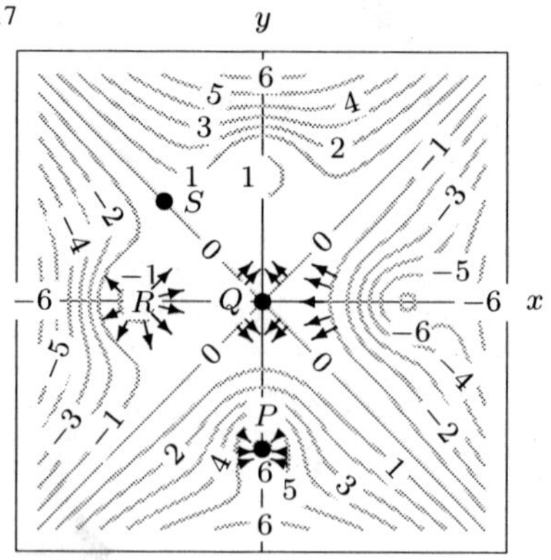

19. Saddle point: $(0, 0)$.
21. $y = 0, \pm 2\pi, \pm 4\pi, \ldots$ Local minima
23. $A = 10$, $B = 4$, $C = -2$
25. (a) (a, b) is a saddle point.
    (b)

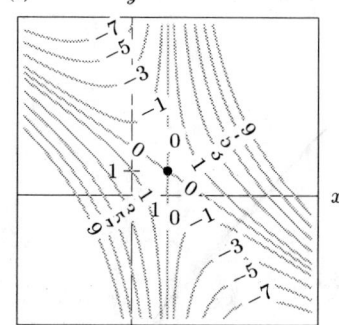

27.

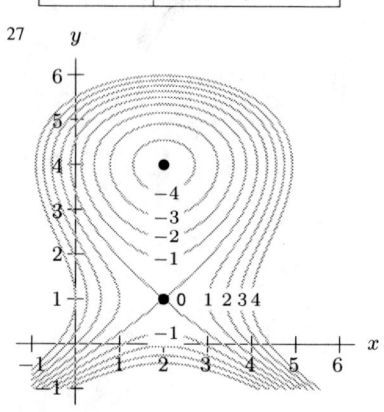

29. (a) $k < 4$
    (b) None
    (c) $k \geq 4$
31.

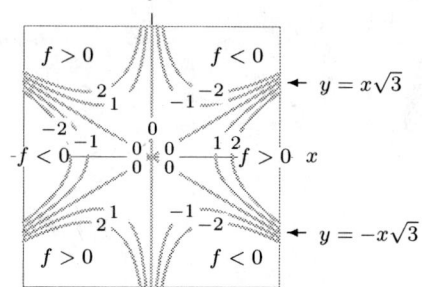

Section 15.1

1. A: no
   B: yes, max
   C: yes, saddle
3. Local max: $(1, 5)$
5. Saddle pts: $(1, -1), (-1, 1)$
   local max $(-1, -1)$
   local min $(1, 1)$.
7. Local maximum at $(-1, 0)$,
   Saddle points at $(1, 0)$ and $(-1, 4)$,
   Local minimum at $(1, 4)$
9. $(0, 0), (1, 1), (-1, -1)$
   saddle point,
   local maximum,
   local maximum
11. Local minimum at $(0, 0)$
13. Saddle point
15. Saddle point